Current Topics in Microbiology and Immunology

Volume 397

More information about this series at http://www.springer.com/series/82

Steffen Backert
Editor

Inflammasome Signaling and Bacterial Infections

Responsible Series Editor: Klaus Aktories

Springer

Editor
Steffen Backert
Division of Microbiology,
Department of Biology
Friedrich Alexander University
Erlangen/Nuremberg
Erlangen
Germany

ISSN 0070-217X ISSN 2196-9965 (electronic)
Current Topics in Microbiology and Immunology
ISBN 978-3-319-41170-5 ISBN 978-3-319-41171-2 (eBook)
DOI 10.1007/978-3-319-41171-2

Library of Congress Control Number: 2016943412

Printed on acid-free paper

This Springer imprint is published by Springer Nature
The registered company is Springer International Publishing AG Switzerland

Preface

One of the most fascinating riddles in immune biology is how vertebrates can sense the presence of a broad variety of highly diverse pathological agents. In response to this challenge, the two branches of the immune system, innate and adaptive immunity, evolved during evolution. The innate immune system represents the more ancient and more rapid defense instrument against infectious microbes or other pathophysiological stimuli. This system relies on a repertoire of germline-encoded receptors that sense pathogen-associated molecules ("non-self") and endogenous stress signals ("danger"). About 15 years ago, the term "inflammasome" was introduced as a new concept in this fascinating research field. Inflammasomes represent a conserved set of macromolecular signaling complexes in immune cells that sense "hazard" and trigger the inflammatory machinery. The inflammasome concept contributed significantly to a renaissance in the fields of innate immunity and cell death. In recent years, the overall importance of inflammasomes was and still is increasingly recognized, as the complex biology of various pathological scenarios is better understood. The concerted actions of inflammasomes and other components of the innate immune system are of great significance for responding adequately to harmful microbial infections. However, accumulating data also support the view that deregulated inflammasome signaling is connected to a variety of inflammatory human pathologies, including inflammatory bowel disease, rheumatoid arthritis as well as hereditary periodic fever syndromes. Unlike this high importance for human and animal health, it appears that a comprehensive volume on inflammasome biology is still not available.

With the breathtaking expansion of research on inflammasome signaling cascades in recent years, this is an opportune time to review present knowledge about inflammasome action. Accordingly, a comprehensive collection of reviews on the multiple facets of inflammasome signal transduction seems both timely and appropriate for a book series. The present volume on "Inflammasome Signaling and Bacterial Infections" summarizes our current scientific understanding of inflammasome biology in 13 chapters by experts in this research area. It is designed to provide an introduction to inflammasomes and bacterial pathogenesis for advanced undergraduates, graduate students, medical students, postdocs, and (bio)medical investigators, who are interested in infectious diseases and immunology. We discuss the most recent insights in the major components of known canonical and non-canonical inflammasome complexes and highlight their mechanism of action, in particular in response to infection with important bacterial model organisms and the corresponding disease pathologies.

The first two chapters are designed to provide the necessary background and a general overview for understanding the topics covered in the following chapters. This introduction includes advances in understanding the inflammasome structure at the molecular level and general strategies of up- and downstream signaling events. In the subsequent chapters, we specifically discuss the composition and activity of distinct inflammasomes during infection with various gut pathogens (*Salmonella*, *Shigella*, *Yersinia*, *Listeria*, and *Helicobacter*), respiratory pathogens (*Mycobacterium*, *Legionella*, *Burkholderia*, and *Streptococcus*) as well as skin and soft tissue pathogens (*Francisella* and *Staphylococcus*). As will become evident from these detailed review articles, there is much more complexity in inflammasome signaling pathways than was originally anticipated, adding greatly to the interest into these signaling factor cascades. Within the individual chapters, readers will find not only consensus and paradigm, but also differing perspectives on the regulation and functions of the multitude of inflammasome factors. Importantly, all of the reviews point out specific areas, where the lack of sufficient knowledge and understanding raises intriguing new questions for further experimentation. These outstanding questions often pertain to the increasingly complex biological functions of inflammasome components and diverse mechanisms of regulation in a variety of systems, ranging from mouse models to humans. Recurring themes are (i) how the regulation and function of inflammasome proteins are highly dependent on the cell type and activating stimulus, (ii) how pyroptotic cell death is regulated in detail, and (iii) how persistent pathogens can dampen the canonical inflammasome machinery in order to establish long-term infection. In the future, better characterization of the cellular and molecular biology of the inflammasomes will pinpoint important new therapeutic targets for the treatment and prevention of multiple infectious diseases and pathological conditions. If this comprehensive collection of reviews on inflammasomes stimulates fresh new thinking and research on the involved signaling pathways, this book will have accomplished its goal.

The above-discussed advances in the field have helped to shape the core of this volume. I am very grateful to the scientific contributors from around the globe, who have participated in the preparation of these outstanding chapters covering our growing knowledge of this important innate immune mechanism. I hope that this volume brings an invaluable resource to readers new to the field and expands the resources for those professionals already working in the inflammasome area. I would like to thank all participants for their support and help in making this book a success.

Erlangen, Germany
May 2016

Steffen Backert

Contents

Abbreviations

AhR	Aryl hydrocarbon receptor
AIM2	Absent in melanoma 2
ALR	AIM2-like receptor
AP1	Activator protein 1
Apaf-1	Apoptotic peptidase activating factor 1
ASC	Apoptosis-associated speck-like protein containing a caspase recruitment domain
BID	BH3 interacting domain
BIR	Baculovirus inhibitor repeat
BMDM	Bone marrow-derived macrophage
CAPS	Cryopyrin-associated periodic syndrome
CARD	Caspase recruitment domain
cGAS	cyclic GMP-AMP synthase
CIITA	MHC class II trans-activator
CLR	C-type lectin receptor
DAMP	Danger-associated molecular pattern
DC	Dendritic cell
FCAS	Familial cold auto-inflammatory syndrome
FIIND	'Function to find' domain
FMF	Familial Mediterranean fever
FPI	*Francisella* pathogenicity island
GBP	Guanylate-binding protein
HD1	Helical domain 1
Hin200	Hematopoietic IFN-inducible nuclear protein with a 200-amino acid repeat
HMGB1	High-mobility group box 1
Icm/Dot	Intracellular multiplication/defective organelle trafficking
IEC	Intestinal epithelial cell
IFI16	Interferon gamma-inducible protein 16
Ifi204	Murine homologue of human IFI16

IFN	Interferon
IKK	IκB kinase
IL	Interleukin
IRAK-1	IL-1 receptor-associated kinase 1
IRF	Interferon regulatory factor
ISG	Interferon-stimulated gene
LAI-1	*Legionella* autoinducer-1
LCV	*Legionella*-containing vacuole
LeTx	Lethal toxin
LF	Lethal factor
LPS	Lipopolysaccharide
Lqs	*Legionella* quorum sensing
LRR	Leucine-rich repeat
LVS	Live vaccine strain
MAPK	Mitogen-activated protein kinase
MLN	Mesenteric lymph nodes
Mtb	*Mycobacterium tuberculosis*
MWS	Muckle-Well syndrome
MyD88	Myeloid differentiation primary response gene 88
NACHT	NAIP-CIITA-HET-E domain
NAIP	NLR family, apoptosis inhibiting protein
NBD	Nucleotide-binding domain
NEK	NIMA-related kinase
NF-κB	Nuclear factor-kappa B
NLR	Nucleotide-binding domain and leucine-rich-repeat-containing proteins
NLRC	NLR family, CARD domain containing
NLRP	NLR family, pyrin domain containing
NOD	Nucleotide-binding oligomerization domain
NOMID	Neonatal-onset multisystem inflammatory disease
PAMP	Pathogen-associated molecular pattern
PI	Phosphoinositide
PRR	Pattern recognition receptor
PtdIns	Phosphatidylinositol
PYD	Pyrin domain
RACK1	Receptor for activated C-kinase
RIPK	Receptor-interacting serine/threonine-protein kinase
RLR	RIG-I-like receptor
ROS	Reactive oxygen species
SCV	*Salmonella*-containing vacuole
SNPs	Single-nucleotide polymorphisms
SPI	*Salmonella* pathogenicity island
STAND	Signal transduction ATPases with numerous domains
T3SS	Type III secretion system
T4SS	Type IV secretion system
T6SS	Type VI secretion system

T7SS	Type VII secretion system
TB	Tuberculosis
TLR	Toll-like receptor
TNF-α	Tumor necrosis factor-alpha
TP1	Telomerase-associated protein
TRIM	Tripartite motif-containing protein
WHD	Winged helical domain

General Strategies in Inflammasome Biology

Hanne Dubois, Andy Wullaert and Mohamed Lamkanfi

Abstract The complementary actions of the innate and adaptive immune systems often provide effective host defense against microbial pathogens and harmful environmental agents. Germline-encoded pattern recognition receptors (PRRs) endow the innate immune system with the ability to detect and mount a rapid response against a given threat. Members of several intracellular PRR families, including the nucleotide-binding domain and leucine-rich repeat containing receptors (NLRs), the AIM2-like receptors (ALRs), and the tripartite motif-containing (TRIM) protein Pyrin/TRIM20, nucleate the formation of inflammasomes. These cytosolic scaffolds serve to recruit and oligomerize the cysteine protease caspase-1 in filaments that promote its proximity-induced autoactivation. This oligomerization occurs either directly or indirectly through intervention of the bipartite adaptor protein ASC, apoptosis-associated speck-like protein containing a caspase recruitment domain (CARD), which is needed for the domain interaction. Caspase-1 cleaves the precursors of the inflammatory cytokines interleukin (IL)-1β and IL-18 and triggers their release into the extracellular space, where they act on effector cells to promote both local and systemic immune responses. Additionally, inflammasome activation gives rise to a lytic mode of cell death, named pyroptosis, which is thought to contribute to initial host defense against infection by eliminating replication niches of intracellular pathogens and exposing them to the immune system. Inflammasome-induced host defense responses are the subject of intense investigation, and understanding their physiological roles during infection and the regulatory circuits that are involved is becoming increasingly detailed. Here, we discuss current understanding of the activation mechanisms and biological outcomes of inflammasome activation.

H. Dubois · A. Wullaert · M. Lamkanfi (✉)
NOD-like Receptor and Inflammasome Laboratory, Inflammation Research Center, VIB, 9052 Zwijnaarde, Belgium
e-mail: mohamed.lamkanfi@irc.vib-ugent.be

H. Dubois · A. Wullaert · M. Lamkanfi
Department of Internal Medicine, Ghent University, 9000 Ghent, Belgium

S. Backert (ed.), *Inflammasome Signaling and Bacterial Infections*, Current Topics in Microbiology and Immunology 397,
DOI 10.1007/978-3-319-41171-2_1

Contents

1 Introduction: PRRs, PAMPs, and DAMPs

The complementary qualities of the innate and adaptive immune systems allow vertebrates to mount a highly tailored and efficacious host defense against intruding pathogens, while avoiding immunopathology. The efficacy of such combined immune response for defense against pathogens and harmful environmental agents is illustrated by the presence of innate and adaptive immune subsystems in early vertebrates such as the lamprey and mammals alike (Boehm 2012). The innate immune system detects and mounts the initial response to the threat. In mammals, the innate immune system encompasses a diversity of physical and chemical barriers. This includes mucosal membranes that interface the extracellular environment, a complement system that tags invading agents for removal, and professional phagocytes that clear the infectious agent. Phagocytes such as macrophages and neutrophils also release inflammatory mediators to recruit additional immune cells into the affected area (Palm and Medzhitov 2009a). Adaptive immunity is directed by dendritic cells and other antigen-presenting cells (APCs) that relay information about the harmful agent to lymphocytes. Pathogen-derived peptide antigens are presented to T-lymphocytes in association with major histocompatibility complex (MHC) proteins, whereas multiple mechanisms may govern how B-lymphocytes encounter antigens (Cyster 2010; Palm and Medzhitov 2009b). The highly specific adaptive immune system produces B- and T-lymphocytes with diverse antigen receptors, and clonal expansion of the cells recognizing foreign material culminates in its targeted removal. In addition, the adaptive immune system is capable of immune memory that provides protection from reinfection with the same pathogen (Koch and Radtke 2011).

While antigen receptor gene rearrangements in lymphocytes enable the adaptive immune system to recognize seemingly any antigen, innate immune cells rely on only a fixed set of germline-encoded 'pattern recognition receptors' (PRRs) to detect pathogens (Takeuchi and Akira 2010). PRRs are expressed on many cell types that may come in contact with microbes, including hematopoietic cells, fibroblasts, endothelial cells, and epithelial cells that line mucosal membranes.

Given the limited repertoire of PRRs that is available to the host, it may be unsurprising that—rather than signaling out a particular microbe—PRRs guard conserved microbial signatures termed 'pathogen-associated molecular patterns' (PAMPs) that may signal infection by a certain class of pathogens. Microbial nucleic acids, bacterial secretion systems, and components of the microbial cell wall that are not produced by eukaryotes are the examples of such conserved microbial factors that are sensed by PRRs. Nevertheless, damaged host cells as well may trigger PRR activation by releasing danger-associated molecular patterns (DAMPs) such as uric acid crystals, ATP, high-mobility group box 1 (HMGB1), and the heat-shock proteins Hsp70 and Hsp90 (Takeuchi and Akira 2010). Detection of DAMPs by PRRs is thought to primarily promote tissue repair, although excessive release might elicit a severe inflammatory response exacerbating tissue damage in several infectious and autoinflammatory and autoimmune diseases (Lotze et al. 2007).

PRR families may be subdivided into genuine transmembrane receptors that survey the extracellular environment and endosomes for DAMPs and PAMPs, and those that reside in cytosolic compartments. Toll-like receptors (TLRs) and C-type lectin receptors (CLRs) are prime examples of the first class, while the RIG-I-like receptor (RLR), the AIM2-like receptor (ALR), and the nucleotide-binding domain and leucine-rich repeat containing (NLR) proteins all respond to pathogens at intracellular compartments (Takeuchi and Akira 2010). PRRs are also frequently classified according to the PAMPs and DAMPs they sense, or the immune signaling pathways they control. For instance, members of several PRR families may detect microbial DNA or RNA molecules and engage signaling cascades that culminate in activation of members of the nuclear factor-kappa B (NF-κB), activator protein 1 (AP1), and interferon regulatory factor (IRF) transcription factor families (Battistini 2009; Kim and Choi 2010; Vallabhapurapu and Karin 2009). The concerted activities of these key inflammatory transcription factors lead to the production of type I interferons (IFN-I), inflammatory cytokines, and other pro-inflammatory or microbicidal proteins (Takeuchi and Akira 2010). In addition to these comprehensive transcriptional reprograming events, innate immune cells are equipped with PRRs that may assemble cytosolic multi-protein complexes called 'inflammasomes.' Inflammasomes are regarded as key elements in the innate immune response of mammalian hosts in providing protection against invading micro-organisms. By definition, inflammasomes are scaffolds for activation of the inflammatory cysteine-dependent aspartate-specific protease caspase-1. This protease is chiefly known for its key role in maturation and secretion of the inflammatory cytokines interleukin IL-1β and IL-18 (Lamkanfi and Dixit 2014). Additionally, caspase-1 activation may result in a programmed, lytic cell death of myeloid cells that has been named 'pyroptosis' (Cookson and Brennan 2001). Based on a wealth of experimental evidence gathered in the past 2 decades, these two inflammasome-dependent biological responses (cytokine production and pyroptosis) contribute importantly to the immune system's ability to resolve the threat and restore homeostasis. Here, we will review and discuss current understanding of inflammasome biology with an emphasis on recent developments in control of microbial infections. A brief introduction of the different inflammasomes characterized to date will be followed by a discussion of the roles of particular inflammasome complexes in microbial infections. Finally, mechanisms regulating inflammasome activation will be discussed along with specific examples illustrating the importance of tight regulation of inflammasome activation.

2 PRRs as Inflammasome Scaffolds

Selected members of the 'Hematopoietic interferon-inducible nuclear protein family' (Hin200) and the 'Tripartite Motif Family' (TRIM) PRR families assemble inflammasomes in their own right (Fig. 1), but the majority of inflammasomes relies on NLR proteins for pathogen sensing and scaffold assembly. NLRs are defined by the combined presence of a nucleotide-binding and oligomerization domain (NACHT/NBD) and leucine-rich repeat (LRR) motifs, typically located in the central and most carboxy-terminal regions of the proteins (Kanneganti et al. 2007). Most NLRs in addition contain an amino-terminal protein interaction domain of the baculovirus inhibitor repeat (BIR), Pyrin (PYD), and caspase recruitment domain (CARD) types, thus allowing their subclassification based on domain architecture (Fig. 1). The NLR family consists of 22 human and 34 murine members that play diverse roles in the mammalian immune and reproductive systems (Kanneganti et al. 2007; Van Gorp et al. 2014). NLRs are considered an evolutionary ancient PRR family as supported by the identification of over 200 NLR genes in the genome of the sea urchin *Strongylocentrotus purpuratus* (Rast et al. 2006), and the presence of NLRs in zebrafish (Laing et al. 2008; Stein et al. 2007) and tetrapods (Hansen et al. 2011). However, NLR genes appear to have been lost during speciation of particular animal species. For instance, NLRs are absent from the genome of the model nematode species *Caenorhabditis elegans* and the insects *Drosophila melanogaster* (fruit fly) and *Apis mellifera* (honey bee). However, they are present in other insects such as *Culex quinquefasciatus* (southern house mosquito) and *Aedes aegypti* (yellow fever mosquito) (Lange et al. 2011). Although genuine NLR genes have so far only been identified in animals, homologs with functionally related domains and motifs have been described in *Hydra*, fungi, and plants as well (Lange et al. 2011). The comparable domain architecture of the pathogen-resistant (R-) proteins of higher plants in which carboxy-terminal LRR motifs combine with a centrally located NB-ARC domain that is structurally related to the NACHT ATPase of NLRs represents an nice example of convergent evolution (Chisholm et al. 2006). Moreover, also R-proteins are central mediators of antimicrobial resistance mechanisms that are collective referred to as the 'hypersensitive response' in plants (Hulbert et al. 2001).

The physiological roles of most mammalian NLRs are largely obscure, but some members have well-defined roles in the regulation of inflammatory gene transcription. Among the best characterized examples are the intracellular peptidoglycan

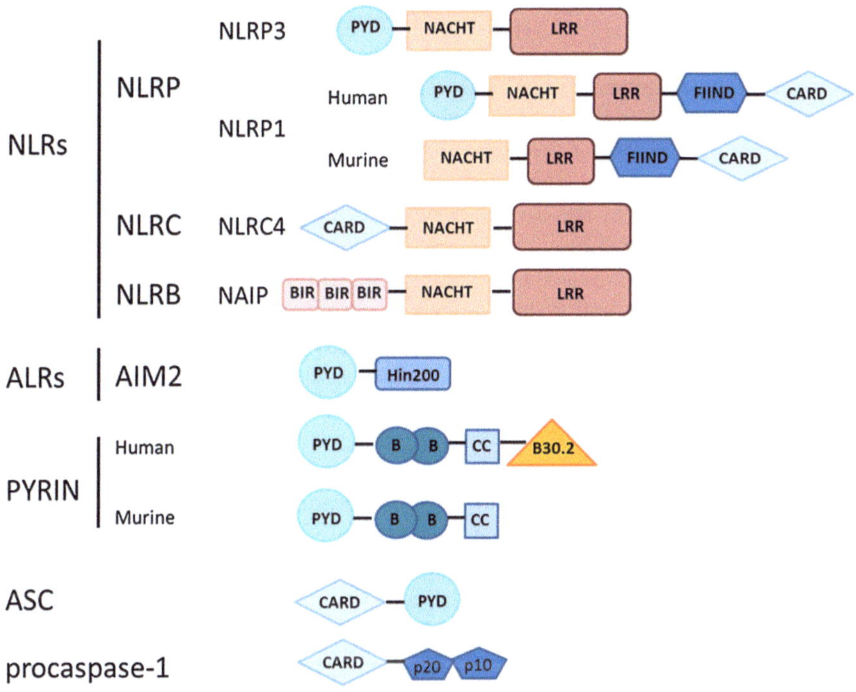

Fig. 1 Inflammasome components and their domain architecture. Members of the nucleotide-binding domain and leucine-rich repeat containing receptors (NLR) family contain a central 'neuronal apoptosis inhibitor protein (NAIP), CIITA (MHC class II transcription activator), HET-E (incompatibility locus protein from *Podospora anserina*) and TP1 (telomerase-associated protein)' (NACHT) ATPase domain, and leucine-rich repeat (LRR) motifs. NLR family members may be further subdivided into NLRP, containing a Pyrin domain (PYD); NLRC, containing a caspase recruitment domain (CARD); and NLRB, containing a Baculovirus inhibitor repeat (BIR) domain. The latter subset is frequently referred to as the NAIP proteins. The AIM2-like receptor (ALR) family is composed of a PYD and a dsDNA-binding 'hematopoietic interferon-inducible nuclear protein with a 200 amino acid repeat' (Hin200) domain. Finally, the inflammasome scaffold protein Pyrin consists of PYD, B-Box-type zinc finger (BB), and coiled-coil (CC) domains that are followed by a carboxy-terminal B30.2 domain in the human, but not its murine ortholog. Inflammasome assembly involves homotypic CARD and PYD interactions between its components

receptors NOD1 and NOD2 (nucleotide-binding oligomerization domain containing proteins 1/2) that promote RIPK2- and CARD9-dependent transcription of NF-κB- and AP1-target genes (Wilmanski et al. 2008). Additionally, transcription of major histocompatibility class II (MHC II) genes in antigen-presenting cells requires the NLR protein class II trans-activator (CIITA) (Wilmanski et al. 2008). Another subset of NLRs—namely NLRP1, NLRP3, NLRC4, and NAIP—promotes immune responses at the post-translational level by initiating inflammasome signaling (Lamkanfi and Dixit 2014).

Signal-induced PRR clustering is thought to recruit and promote the nucleation of procaspase-1 filaments in which the protease zymogens undergo proximity-induced autoactivation. Caspase-1 may be recruited in these inflammasome scaffolds through direct homotypic interactions involving the CARD motifs of procaspase-1 and that of the nucleating PRR (NLRC4, NLRP1b), or indirectly through homotypic interactions with the bipartite PYD/CARD inflammasome adaptor protein ASC in the case of PYD-containing NLRP3, AIM2, and Pyrin receptors. Notably, inflammasome-induced activation of caspase-1 by each of these inflammasomes is associated with ASC-dependent autoproteolytic cleavage of caspase-1, but automaturation does not appear essential for pyroptosis induction in the context of the NLRP1b and NLRC4 inflammasomes (Broz et al. 2010b; Van Opdenbosch et al. 2014). Nevertheless, ASC does contribute to efficient secretion of bioactive IL-1β and IL-18 in response to triggers of these respective inflammasomes.

IL-1β is responsible for generation of fever, lymphocyte activation, and guiding the transmigration of leukocytes into the stress location (Dinarello 2009). As such, IL-1β induces both a systemic and a local response to infection and injury. IL-18 does not have this pyrogenic activity, but it orchestrates IFN gamma production, leading to control over the Th1 population. Depending on the cytokine environment, IL-18 can as well orchestrate the Th2 population. Furthermore, IL-18 regulates ROS production, expression of cell adhesion molecules and expression of other chemokines/cytokines (Dinarello 2009). Considering the broad impact of these proinflammatory cytokines on the hosts' immune responses, a strict control of their activity is needed. One mechanism by which this is accomplished involves transcriptional control of their expression levels. ProIL-1β is virtually absent in naïve myeloid cells, but its mRNA levels are highly responsive to NF-κB-dependent transcriptional upregulation. By contrast, proIL-18 is constitutively present in the cytosol of naïve macrophages and dendritic cells. The differential regulation of proIL-1β and proIL-18 at the transcriptional level is illustrated by the observation that the NLRP1b and NLRC4 inflammasomes can be induced to secrete mature IL-18—but not IL-1β—in the absence of prior TLR engagement (Nystrom et al. 2013; Van Opdenbosch et al. 2014). In this context, the requirement for cytokine maturation may be regarded as another safeguard against accidental release of IL-1β and IL-18. Indeed, unlike most other cytokines that are secreted through the classical secretory pathway, proIL-1β and proIL-18 are produced as biologically inactive cytosolic precursors that await caspase-1-dependent cleavage for release of their bioactive forms (Gu et al. 1997; Lamkanfi 2011). Several mechanisms have been proposed by which the latter cytokines may be released, the most recent being another prominent outcome of inflammasome activation, namely pyroptosis. Pyroptosis is a lytic form of cell death characterized by cytoplasmic swelling and early rupture of the plasma membrane that requires the protease activities of either caspase-1 or caspase-11 (Cookson and Brennan 2001; Kayagaki et al. 2011). Pyroptosis is thought to constitute a defensive innate immune strategy of the host that eliminates the replicative niche of intracellularly replicating pathogens and exposes them to other immune cells (Aachoui et al. 2013; Casson et al. 2013; Miao et al. 2010a). Additionally, pyroptosis-associated cell lysis further releases DAMPs such as IL-1α and HMGB1 into the extracellular environment (de Gassart and Martinon 2015; Lamkanfi et al. 2010).

3 Inflammasome Activation Mechanisms

3.1 The NLRP3 Inflammasome

The NLRP3 inflammasome is by far responding to the largest set of activating agents (Lamkanfi and Dixit 2014). It is also rather unique among inflammasomes in that it requires an NF-κB-mediated signal prior to its activation. This so-called 'signal 1' or 'priming' step involves transcriptional upregulation of NLRP3 together with proIL-1β (Bauernfeind et al. 2009) and is defective in mice lacking the NF-κB regulator A20/TNFAIP3 (Vande Walle et al. 2014). Non-transcriptional mechanisms that involve its de-ubiquitination and/or IL-1 receptor-associated kinase (IRAK-1) kinase activity have also been proposed to additionally control NLRP3 inflammasome activation (Juliana et al. 2012; Lin et al. 2014; Lopez-Castejon et al. 2013; Py et al. 2013). Together, these priming mechanisms establish effective checkpoints that prevent accidental NLRP3 inflammasome activation.

In the presence of such priming signals, NLRP3 inflammasome assembly and activation is induced when the host cell is exposed to a 'signal 2'. NLRP3 inflammasome activation may be induced by the components of bacterial [*Staphylococcus aureus, Streptococcus pneumoniae, Listeria monocytogenes* (Mariathasan et al. 2006; McNeela et al. 2010; Wu et al. 2010)]; viral [Influenza A virus (IAV); Encephalomyocarditis virus (EMCV); Vesicular stomatitis virus (VSV)] (Allen et al. 2009; Kanneganti et al. 2006; Rajan et al. 2011); as well as fungal (*Candida albicans, Aspergillus fumigatus, Cryptococcus neoformans*) (Gross et al. 2009; Guo et al. 2014; Hise et al. 2009; Said-Sadier et al. 2010) origin. In addition, DAMPs such as millimolar concentrations extracellular ATP, calcium pyrophosphate dehydrate and monosodium urate (Mariathasan et al. 2006; Martinon et al. 2006, 2009), environmental crystals (alum, silica, asbestos) (Dostert et al. 2008; Eisenbarth et al. 2008; Hornung et al. 2008; Martinon et al. 2006), β-fibrils and aggregates (β-amyloid, β-glucans) (Halle et al. 2008; Kumar et al. 2009), and ionophores (nigericin, maitotoxin) (Mariathasan et al. 2006; Perregaux and Gabel 1994) all engage the NLRP3 inflammasome. The NLRP3 inflammasome may also be engaged by intracellular LPS. Unlike for the 'canonical' stimuli above, caspase-11 and its human orthologs caspases 4 and 5 are required for cytosolic LPS-induced NLRP3 inflammasome activation (Baker et al. 2015; Hagar et al. 2013; Kayagaki et al. 2011; Kayagaki et al. 2013; Schmid-Burgk et al. 2015; Shi et al. 2014). In this 'non-canonical' signaling pathway, caspase-11 induces pyroptosis independently of the NLRP3 inflammasome, while cleavage of proIL-1β and proIL-18 is relayed through the NLRP3 inflammasome (Kayagaki et al. 2011).

The importance of non-canonical NLRP3 inflammasome signaling in Gram-negative infections is highlighted by the resistance of caspase-11-deficient mice to LPS-induced lethality (Kayagaki et al. 2011; Wang et al. 1998), and their increased susceptibility to enteric pathogens (Broz et al. 2012; Gurung et al. 2012; Knodler et al. 2014). Recent work revealed cleavage of the gasdermin D as a central commonality of pyroptosis induction by caspases 1 and 11 (Kayagaki et al. 2015; Shi et al. 2016, 2015). Gasdermin D cleavage also links LPS-induced caspase-11 activation with engagement of the NLRP3 inflammasome, but how it integrates with mechanisms of canonical inflammasome activation is not fully clarified.

Since the diverse canonical NLRP3 inflammasome-activating agents listed above are structurally and chemically unrelated, a direct ligand sensing model for the NLRP3 inflammasome is highly unlikely. Instead, NLRP3 activation is thought to involve a defined cellular event or secondary messenger that is commonly and selectively triggered by these NLRP3-activating agents. In this respect, K^+ is among the most frequently cited ions in regulation of NLRP3 activation. A drop in intracellular K^+ levels is regarded as a prerequisite for activation of NLRP3 because it accompanies NLRP3 inflammasome activation by a diversity of agents, and because preventing K^+ efflux by cultivating cells in media containing high extracellular K^+ concentrations prevents caspase-1 activation by the NLRP3 inflammasome (Franchi et al. 2007; Munoz-Planillo et al. 2013; Perregaux and Gabel 1994; Petrilli et al. 2007). However, also activation of the Nlrp1b inflammasome was reported to be sensitive to high K^+ concentrations (Fink et al. 2008; Wickliffe et al. 2008). Na^+, Ca^{2+}, and Cl^- are other candidates for ion flux-mediated NLRP3 activation. For instance, exchanging Na^+ for Li^+, choline or K^+ in iso-osmotic media of LPS-primed macrophages was shown to prevent ATP-induced caspase-1 activation (Perregaux and Gabel 1998). The specific role of intracellular Ca^{2+} is a matter of debate, with reports arguing against its requirement but rather correlating its involvement with simultaneous K^+ efflux induction (Katsnelson et al. 2015; Munoz-Planillo et al. 2013); or implicating Ca^{2+} fluxing as a critical step in NLRP3 inflammasome activation (Lee et al. 2012; Yaron et al. 2015).

Mitochondrial dysfunction—sometimes coupled with ionic flux deregulation—has also been proposed to control NLRP3 activation, with potential involvement of cardiolipin release, oxidized mitochondrial DNA, and loss of the mitochondrial membrane potential as secondary messengers for NLRP3 activation (Iyer et al. 2013; Nakahira et al. 2011; Shimada et al. 2012). Despite the ambiguities surrounding the molecular events regulating NLRP3 activation, Nek7 was recently established as a key NLRP3-binding partner that selectively controls activation of the NLRP3 inflammasome (He et al. 2016; Schmid-Burgk et al. 2016; Shi et al. 2016, 2015). This finding may provide a novel grasping point for further dissection of NLRP3 activation mechanisms.

The need for tight regulation of NLRP3 inflammasome activation is best highlighted by the existence of autoinflammatory diseases caused by gain-of-function mutations in NLRP3. These cryopyrin-associated periodic syndromes (CAPS) cover three autoinflammatory diseases. Familial cold autoinflammatory syndrome (FCAS) is associated with cold-induced fevers, rash, and constitutional symptoms.

Muckle–Wells syndrome (MWS) is not triggered by exposure to cold and additionally features the possible occurrence of hearing loss (Hoffman et al. 2001). Fever, chronic meningitis, eye inflammation, hearing loss, skin rash, and a deforming arthropathy all are clinical aspects of neonatal-onset multisystem inflammatory disease (NOMID) (Aksentijevich et al. 2002). Over 80 disease-associated NLRP3 mutations have been reported, most of them being situated within or in the close vicinity of the central NACHT domain. The effectiveness of anti-IL-1 therapies in CAPS patients illustrates the pathogenic role of IL-1β in these diseases (Yu and Leslie 2011). The identification of small-molecule inhibitors that prevent NLRP3 inflammasome activation might one day enable therapeutic strategies that selectively prevent cytokine secretion by this inflammasome only (Coll et al. 2015; Lamkanfi et al. 2009).

3.2 *The NLRP1 Inflammasome*

NLRP1 was one of the first NLRs reported to assemble an inflammasome (Martinon et al. 2002). NLRP1 undergoes autocleavage in a unique 'function to find' domain (FIIND) that lays between its carboxy-terminal CARD and LRR motifs (Chavarria-Smith and Vance 2013; Finger et al. 2012; Frew et al. 2012). Whereas humans encode a single NLRP1 gene, mice may express up to three different NLRP1 isoforms, known as Nlrp1a, Nlrp1b, and Nlrp1c (Boyden and Dietrich 2006). Moreover, human NLRP1 is equipped with both an amino-terminal PYD and a carboxy-terminal CARD, the former being absent in its murine homologs. Nevertheless, the murine paralogs appear to have non-redundant roles in immune signaling. Nlrp1b is critical for inflammasome assembly in macrophages upon exposure to lethal toxin (LeTx) of *Bacillus anthracis*, the causative agent of anthrax (Fig. 2). LeTx is a bipartite toxin consisting of a protective antigen (PA) and a lethal factor (LF) subunit, in which PA functions as a pore-forming unit enabling cytosolic delivery of the zinc metalloprotease LF subunit (Bann 2012). LF protease activity is required for Nlrp1b inflammasome activation because catalytically inactive mutants of LF fail to activate the Nlrp1b inflammasome (Levinsohn et al. 2012). This suggests that instead of directly sensing the presence of LF, Nlrp1b monitors LF protease activity in the cytosol. Notably, the inflammasome adaptor ASC is critical for LeTx-induced caspase-1 autocleavage, but the induction of pyroptosis proceeded unhampered in the absence of ASC specks (Guey et al. 2014; Van Opdenbosch et al. 2014). ASC was as well dispensable for Nlrp1b-dependent IL-1β and IL-18 secretion, although it could enhance this output. These observations suggest that direct recruitment of procaspase-1 to Nlrp1b platforms induces conformational changes in caspase-1 that suffice for its activation, while ASC-dependent caspase-1 autocleavage may serve to lock the protease in its active conformation. Next to its role in sensing LeTx activity, Nlrp1b was recently identified as a type 1 diabetes-susceptibility gene in the NOD mouse model (Motta et al. 2015), and implicated in protection from dextran sodium sulfate-induced

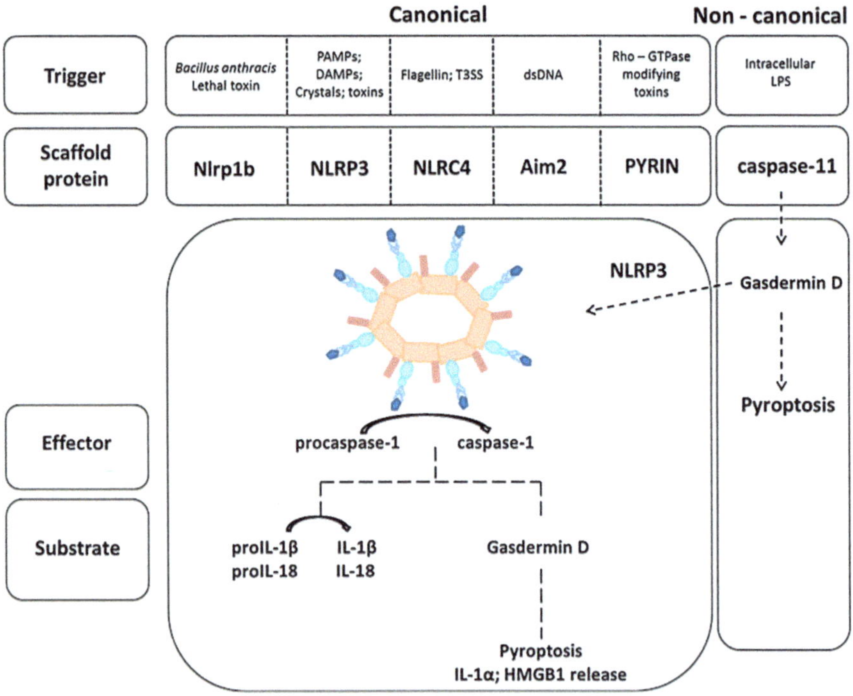

Fig. 2 Simplified scheme of inflammasome triggers and signaling events. AIM2, Pyrin, and inflammasome-assembling NLR family members detect—directly or through a secondary messenger—the presence of a pathogen and/or cellular damage. This results in their oligomerization and inflammasome assembly in which inactive procaspase-1 precursors are recruited directly or through the bipartite adaptor protein ASC. Oligomerization of procaspase-1 leads to its autoactivation, and active caspase-1 catalyzes the maturation and secretion of the inflammatory cytokines interleukin IL-1β and IL-18. Additionally, inflammasome activation may result in a lytic mode of programmed cell death termed pyroptosis. A non-canonical mode of inflammasome signaling is engaged upon detection of bacterial LPS in the cytosolic compartment by murine caspase-11 and its human orthologs caspases 4 and 5. Caspase-11 directly induces pyroptosis through cleavage of gasdermin D and promotes secretion of IL-1β and IL-18 through the NLRP3 inflammasome

colitis in mice (Williams et al. 2015). Also the Nlrp1a isoform was shown to induce inflammasome responses in vivo (Masters et al. 2012). Hematopoietic progenitor cells expressing gain-of-function mutations in Nlrp1a had ASC-independent, but caspase-1-mediated induction of pyroptosis, which prevented their proliferation and differentiation. These mice consequently suffered from leukopenia and anemia during hematopoietic stress (Masters et al. 2012). Identified single-nucleotide polymorphisms (SNPs) in the human NLRP1 locus were shown to confer an increased risk for the development of autoimmune diseases as well, with genetic variants being linked to Addison's disease, generalized vitiligo and type I diabetes (Jin et al. 2007; Spritz 2007; Zurawek et al. 2010). Other studies further linked

NLRP1 variants to systemic lupus erythematosus and rheumatoid arthritis (Levandowski et al. 2013; Magitta et al. 2009).

3.3 *The NLRC4 Inflammasome*

The upstream mechanisms promoting assembly and activation of the NLRC4 inflammasome probably are the best understood among inflammasome scaffolds (Fig. 2). NLRC4 has an amino-terminal CARD motif that allows direct recruitment of procaspase-1 in the absence of ASC (Broz et al. 2010b; Van Opdenbosch et al. 2014). Also the molecular constituents by which bacterial pathogens such as *Salmonella enterica* serovar Typhimurium (Lara-Tejero et al. 2006; Mariathasan et al. 2004; Miao et al. 2006), *Legionella pneumophila* (Amer et al. 2006; Zamboni et al. 2006), *Pseudomonas auruginosa* (Miao et al. 2008; Sutterwala et al. 2007), *Shigella flexneri* (Suzuki et al. 2007, 2014), and *Listeria monocytogenes* (Wu et al. 2010) trigger NLRC4 inflammasome activation have been mapped. The virulence of these bacteria strongly depends on flagellin-mediated motility and the delivery of pathogenic effector molecules in the host cell cytosol through bacterial type III and IV secretion systems (T3SS and T4SS, respectively). Intracellular detection of flagellin and/or T3SS components is what engages the NLRC4 inflammasome. Notably, NLRC4 does not detect the presence of these virulence factors directly, but their unwarranted presence in the cytosol is signaled by NLR family apoptosis inhibiting protein (NAIP) isoforms that act upstream of NLRC4 in the pathway. A full-length human NAIP isoform as well as mouse Naip5 and Naip6 serve as cytosolic flagellin receptors (Kofoed and Vance 2011; Kortmann et al. 2015; Miao et al. 2006). A shorter isoform of human NAIP and its murine Naip1 ortholog recognizes T3SS needle proteins (Miao et al. 2010b; Zhao et al. 2011), whereas Naip2 binds T3SS rod proteins (Kofoed and Vance 2011; Suzuki et al. 2014; Zhao et al. 2011). Notably, biochemical studies with chimeric Naip fusions pointed to the region surrounding the central NACHT as the bacterial ligand-binding domain of Naip proteins (Tenthorey et al. 2014). In addition to NAIP-mediated detection of flagellin and the bacterial T3SS, phosphorylation of NLRC4 at Ser533 was shown to be required for inducing NLRC4 inflammasome activation upon *S.* Typhimurium infection (Qu et al. 2012). Subsequent studies in Naip5-deficient macrophages showed that flagellin-dependent NLRC4 phosphorylation proceeds independently of Naip5, and that flagellin mutants which were unable to induce NLRC4 inflammasome activation retained their ability to induce potent NLRC4 Ser533 phosphorylation (Matusiak et al. 2015). Together, these findings suggest a two-step Nlrc4 activation mechanism in which Ser533 phosphorylation primes NLRC4 for subsequent Naip5-dependent activation of the NLRC4 inflammasome.

Notably, the NLRC4 inflammasome is not only linked to in vivo host defense against bacterial pathogens (Broz et al. 2010a; Miao et al. 2010a), but also may lead to the development of autoinflammatory disease in patients. Recently, three de novo SNPs in the NACHT domain of the NLRC4 gene were described to associate with autoinflammatory disease: Nlrc4-MAS (NLRC4^{T337S}) (Canna et al. 2014), SCAN4 (NLRC4^{V341A}) (Romberg et al. 2014), and FCAS-like syndrome (NLRC4^{H443P}) (Kitamura et al. 2014). Although the three identified missense mutations induce somewhat distinct clinical syndromes, they have spontaneous activation of the NLRC4 inflammasome as the unifying underlying etiology. Nlrc4-MAS and SCAN4 patients further produce extremely high levels of circulating IL-18, and increased inflammasome-induced macrophage cell death was observed in SCAN4 patients (Canna et al. 2014; Romberg et al. 2014).

3.4 The AIM2 Inflammasome

Absent in melanoma 2 (AIM2) is a member of the HIN200 family/AIM-2-like receptor (ALR) family of PRRs that are characterized by an N-terminal PYD and the presence of one or two DNA-binding hematopoietic IFN inducible nuclear protein with 200 amino acid (Hin200) domains (Hornung et al. 2009). AIM2 assembles an inflammasome scaffold when dsDNA of bacterial or viral origin is bound to its Hin200 domain (Choubey 2012; Fernandes-Alnemri et al. 2009; Jin et al. 2012; Rathinam et al. 2010; Sauer et al. 2010). As such, AIM2 endows myeloid cells with the ability to produce IL-1β and IL-18 and induce pyroptosis upon recognition of bacterial and viral nucleic acids in the cytosol of infected macrophages (Rathinam et al. 2010; Sagulenko et al. 2013). The AIM2 inflammasome was shown to be critically involved in controlling infection by *Francisella tularensis*, the causative agent of tularemia (Fernandes-Alnemri et al. 2010; Jones et al. 2010; Rathinam et al. 2010), and further mediates host defense against *Listeria monocytogenes* (Tsuchiya et al. 2010; Wu et al. 2010) and the viral pathogens mouse cytomegalovirus (MCMV) (Alnemri 2010; Kanneganti 2010; Rathinam et al. 2010) and vaccinia virus (VV) (Fernandes-Alnemri et al. 2010; Rathinam et al. 2010).

3.5 The Pyrin Inflammasome

The human *MEFV* gene encodes Pyrin/TRIM20, a member of the TRIM family that is composed of an amino-terminal PYD motif that is followed by a centrally located coiled-coil and B-box region that mediates its oligomerization (Vajjhala et al. 2014; Yu et al. 2007). Unlike in rodents, this architecture is further extended to the carboxy-terminus with a B30.2 domain in human Pyrin (Fig. 1). The expression of Pyrin is largely restricted to myeloid cells with studies showing that monocyte differentiation toward macrophages significantly reduces the expression of Pyrin (Seshadri et al. 2007). Recently, Pyrin was shown to form an inflammasome that activates caspase-1 and mediates secretion of IL-1β in response to RhoA-inactivating bacterial toxins (Xu et al. 2014). As a result, the Pyrin inflammasome is engaged in macrophages and monocytes that have been infected with *Clostridium difficile* or *Burkholderia cenocepacia* (Gavrilin et al. 2012; Xu et al. 2014). The former is a Gram-positive obligate anaerobic pathogen that is the primary cause of nosocomial diarrhea in hospitalized patients and elderly people. *B. cenocepacia* is an opportunistic pathogen that causes progressive respiratory inflammation in immunocompromised patients. The above studies suggest that Pyrin may indirectly respond to the enzymatic activity of RhoA-modifying toxins, but more studies are needed to clarify the molecular chain of events leading to Pyrin inflammasome activation.

Notably, more than 280 mutations have been identified in *MEFV* that cause familial Mediterranean fever (FMF), the most common monogenic autoinflammatory disease worldwide. FMF has a largely autosomal recessive inheritance, although patients with apparent dominant inheritance have also been documented (Balow et al. 1997; Consortium 1997). Notably, the majority of FMF-associated mutations is clustered in the carboxy-terminal B30.2 domain of human Pyrin that is absent in its mouse ortholog (Touitou et al. 2004). As the name suggests, FMF is particularly common in Southern Europe, the Mediterranean Basin, the Middle East, and the Caucasus, frequently affecting Jewish, Turkish, Armenian, Arab, and Italian populations. In these regions, the prevalence of FMF is between 1 in 500 and 1 in 1000, and *MEFV* mutations are very common, with the carrier rate reaching up to 1:5 in these endemic regions. This restricted geographical distribution suggests the existence of an as yet unknown selective advantage for heterozygous carriers in the Mediterranean Basin. Although the disease is less prevalent in Northern Europe (with estimated frequencies of 1:75,000), the disease has spread over the world with migrations of South European, North African, and Middle Eastern populations over the past centuries and in more recent times (Ozen and Bilginer 2014). FMF usually has a childhood onset and is characterized by recurrent attacks of fever associated with serositis. Its main long-term complication is amyloid A (AA) amyloidosis, a severe manifestation with poor prognosis. The microtubule polymerization inhibitor colchicine remains the therapeutic choice to prevent both FMF attacks and complications in most patients, and IL-1-targeting biologicals such as anakinra appear effective in patients with colchicine-refractory disease (Chae et al. 2006; Masters et al. 2009; Ng et al. 2010).

4 Concluding Remarks

Understanding of inflammasome biology is increasing at an impressive pace, with work during the past few years revealing the ligands and detailing the signaling mechanisms of the NLRC4, Pyrin, AIM2, and other inflammasomes. Activation of the different inflammasomes is increasingly recognized to be regulated by a variety of post-translational mechanisms, with each scaffold having its private peculiarities. In addition, disease-associated mutations in Pyrin, NLRP3, NLRC4, and NLRP1 have been shown to cause autoinflammatory diseases with variable clinical presentations. This realization has also led to studies demonstrating the remarkable efficacy of anti-IL-1 therapies in these rare genetic syndromes. Despite the tremendous progress made on these and other fronts, quite some gaps remain in our understanding of inflammasome biology. Further progress is needed in understanding the molecular mechanisms promoting caspase-1 activation by the different inflammasome platforms. The mechanisms by which gasdermin D cleavage promotes pyroptosis warrant further investigation, and the physiological role of pyroptosis in release of cytokines and DAMPs in the context of infections, autoinflammation, and autoimmune diseases is an exciting area of continued study. Newly gained insight in these and other aspects of inflammasome biology likely is to offer novel opportunities for targeted treatment of inflammatory diseases.

Acknowledgments We apologize to colleagues whose work is not cited due to space constraints. A.W. is supported in part by a fellowship from the Fund for Scientific Research-Flanders. Work in ML's laboratory is supported by grants from VIB, Ghent University (BOF 01N02313, BOF 01J11113, BOF14/GOA/013), the Odysseus Foundation (grant Nr. G0C4913N to A.W. and M.L.), the Fund for Scientific Research-Flanders (grant G011315N), and the European Research Council (grant 281600).

Conflict of Interest The authors declare no conflict of interest.

References

Aachoui Y, Leaf IA, Hagar JA, Fontana MF, Campos CG, Zak DE, Tan MH, Cotter PA, Vance RE, Aderem A, Miao EA (2013) Caspase-11 protects against bacteria that escape the vacuole. Science 339(6122):975–978

Aksentijevich I, Nowak M, Mallah M, Chae JJ, Watford WT, Hofmann SR, Stein L, Russo R, Goldsmith D, Dent P, Rosenberg HF, Austin F, Remmers EF, Balow JE Jr, Rosenzweig S, Komarow H, Shoham NG, Wood G, Jones J, Mangra N, Carrero H, Adams BS, Moore TL, Schikler K, Hoffman H, Lovell DJ, Lipnick R, Barron K, O'Shea JJ, Kastner DL, Goldbach-Mansky R (2002) De novo CIAS1 mutations, cytokine activation, and evidence for genetic heterogeneity in patients with neonatal-onset multisystem inflammatory disease (NOMID): a new member of the expanding family of pyrin-associated autoinflammatory diseases. Arthritis Rheum 46(12):3340–3348

Allen IC, Scull MA, Moore CB, Holl EK, McElvania-TeKippe E, Taxman DJ, Guthrie EH, Pickles RJ, Ting JP (2009) The NLRP3 inflammasome mediates in vivo innate immunity to influenza A virus through recognition of viral RNA. Immunity 30(4):556–565

Alnemri ES (2010) Sensing cytoplasmic danger signals by the inflammasome. J Clin Immunol 30 (4):512–519

Amer A, Franchi L, Kanneganti TD, Body-Malapel M, Ozoren N, Brady G, Meshinchi S, Jagirdar R, Gewirtz A, Akira S, Nunez G (2006) Regulation of *Legionella* phagosome maturation and infection through flagellin and host Ipaf. J Biol Chem 281(46):35217–35223

Baker PJ, Boucher D, Bierschenk D, Tebartz C, Whitney PG, D'Silva DB, Tanzer MC, Monteleone M, Robertson AA, Cooper MA, Alvarez-Diaz S, Herold MJ, Bedoui S,

Schroder K, Masters SL (2015) NLRP3 inflammasome activation downstream of cytoplasmic LPS recognition by both caspase-4 and caspase-5. Eur J Immunol 45(10):2918–2926

Balow JE Jr, Shelton DA, Orsborn A, Mangelsdorf M, Aksentijevich I, Blake T, Sood R, Gardner D, Liu R, Pras E, Levy EN, Centola M, Deng Z, Zaks N, Wood G, Chen X, Richards N, Shohat M, Livneh A, Pras M, Doggett NA, Collins FS, Liu PP, Rotter JI, Fischel-Ghodsian N, Gumucio D, Richards RI, Kastner DL (1997) A high-resolution genetic map of the familial Mediterranean fever candidate region allows identification of haplotype-sharing among ethnic groups. Genomics 44(3):280–291

Bann JG (2012) Anthrax toxin protective antigen–insights into molecular switching from prepore to pore. Protein Sci 21(1):1–12

Battistini A (2009) Interferon regulatory factors in hematopoietic cell differentiation and immune regulation. J Interferon Cytokine Res 29(12):765–780

Bauernfeind FG, Horvath G, Stutz A, Alnemri ES, MacDonald K, Speert D, Fernandes-Alnemri T, Wu J, Monks BG, Fitzgerald KA, Hornung V, Latz E (2009) Cutting edge: NF-kappaB activating pattern recognition and cytokine receptors license NLRP3 inflammasome activation by regulating NLRP3 expression. J Immunol 183(2):787–791

Boehm T (2012) Evolution of vertebrate immunity. Curr Biol 22(17):R722–R732

Boyden ED, Dietrich WF (2006) Nalp1b controls mouse macrophage susceptibility to anthrax lethal toxin. Nat Genet 38(2):240–244

Broz P, Newton K, Lamkanfi M, Mariathasan S, Dixit VM, Monack DM (2010a) Redundant roles for inflammasome receptors NLRP3 and NLRC4 in host defense against Salmonella. J Exp Med 207(8):1745–1755

Broz P, von Moltke J, Jones JW, Vance RE, Monack DM (2010b) Differential requirement for caspase-1 autoproteolysis in pathogen-induced cell death and cytokine processing. Cell Host Microbe 8(6):471–483

Broz P, Ruby T, Belhocine K, Bouley DM, Kayagaki N, Dixit VM, Monack DM (2012) Caspase-11 increases susceptibility to Salmonella infection in the absence of caspase-1. Nature 490(7419):288–291

Canna SW, de Jesus AA, Gouni S, Brooks SR, Marrero B, Liu Y, DiMattia MA, Zaal KJ, Sanchez GA, Kim H, Chapelle D, Plass N, Huang Y, Villarino AV, Biancotto A, Fleisher TA, Duncan JA, O'Shea JJ, Benseler S, Grom A, Deng Z, Laxer RM, Goldbach-Mansky R (2014) An activating NLRC4 inflammasome mutation causes autoinflammation with recurrent macrophage activation syndrome. Nat Genet 46(10):1140–1146

Casson CN, Copenhaver AM, Zwack EE, Nguyen HT, Strowig T, Javdan B, Bradley WP, Fung TC, Flavell RA, Brodsky IE, Shin S (2013) Caspase-11 activation in response to bacterial secretion systems that access the host cytosol. PLoS Pathog 9(6):e1003400

Chae JJ, Wood G, Masters SL, Richard K, Park G, Smith BJ, Kastner DL (2006) The B30.2 domain of pyrin, the familial Mediterranean fever protein, interacts directly with caspase-1 to modulate IL-1beta production. Proc Natl Acad Sci USA 103(26):9982–9987

Chavarria-Smith J, Vance RE (2013) Direct proteolytic cleavage of NLRP1B is necessary and sufficient for inflammasome activation by anthrax lethal factor. PLoS Pathog 9(6):e1003452

Chisholm ST, Coaker G, Day B, Staskawicz BJ (2006) Host-microbe interactions: shaping the evolution of the plant immune response. Cell 124(4):803–814

Choubey D (2012) DNA-responsive inflammasomes and their regulators in autoimmunity. Clin Immunol 142(3):223–231

Coll RC, Robertson AA, Chae JJ, Higgins SC, Munoz-Planillo R, Inserra MC, Vetter I, Dungan LS, Monks BG, Stutz A, Croker DE, Butler MS, Haneklaus M, Sutton CE, Nunez G, Latz E, Kastner DL, Mills KH, Masters SL, Schroder K, Cooper MA, O'Neill LA (2015) A small-molecule inhibitor of the NLRP3 inflammasome for the treatment of inflammatory diseascs. Nat Med 21(3):248–255

Consortium FF (1997) A candidate gene for familial Mediterranean fever. Nat Genet 17(1):25–31

Cookson BT, Brennan MA (2001) Pro-inflammatory programmed cell death. Trends Microbiol 9 (3):113–114

Cyster JG (2010) B cell follicles and antigen encounters of the third kind. Nat Immunol 11 (11):989–996

de Gassart A, Martinon F (2015) Pyroptosis: caspase-11 unlocks the gates of death. Immunity 43 (5):835–837

Dinarello CA (2009) Immunological and inflammatory functions of the interleukin-1 family. Annu Rev Immunol 27:519–550

Dostert C, Petrilli V, Van Bruggen R, Steele C, Mossman BT, Tschopp J (2008) Innate immune activation through Nalp3 inflammasome sensing of asbestos and silica. Science 320 (5876):674–7

Eisenbarth SC, Colegio OR, O'Connor W, Sutterwala FS, Flavell RA (2008) Crucial role for the Nalp3 inflammasome in the immunostimulatory properties of aluminium adjuvants. Nature 453 (7198):1122–1126

Fernandes-Alnemri T, Yu JW, Datta P, Wu J, Alnemri ES (2009) AIM2 activates the inflammasome and cell death in response to cytoplasmic DNA. Nature 458(7237):509–513

Fernandes-Alnemri T, Yu JW, Juliana C, Solorzano L, Kang S, Wu J, Datta P, McCormick M, Huang L, McDermott E, Eisenlohr L, Landel CP, Alnemri ES (2010) The AIM2 inflammasome is critical for innate immunity to *Francisella tularensis*. Nat Immunol 11 (5):385–393

Finger JN, Lich JD, Dare LC, Cook MN, Brown KK, Duraiswami C, Bertin J, Gough PJ (2012) Autolytic proteolysis within the function to find domain (FIIND) is required for NLRP1 inflammasome activity. J Biol Chem 287(30):25030–25037

Fink SL, Bergsbaken T, Cookson BT (2008) Anthrax lethal toxin and Salmonella elicit the common cell death pathway of caspase-1-dependent pyroptosis via distinct mechanisms. Proc Natl Acad Sci USA 105(11):4312–4317

Franchi L, Kanneganti TD, Dubyak GR, Nunez G (2007) Differential requirement of P2X7 receptor and intracellular K^+ for caspase-1 activation induced by intracellular and extracellular bacteria. J Biol Chem 282(26):18810–18818

Frew BC, Joag VR, Mogridge J (2012) Proteolytic processing of Nlrp1b Is required for inflammasome activity. PLoS Pathog 8(4):e1002659

Gavrilin MA, Abdelaziz DH, Mostafa M, Abdulrahman BA, Grandhi J, Akhter A, Abu Khweek A, Aubert DF, Valvano MA, Wewers MD, Amer AO (2012) Activation of the pyrin inflammasome by intracellular *Burkholderia cenocepacia*. J Immunol 188(7):3469–3477

Gross O, Poeck H, Bscheider M, Dostert C, Hannesschlager N, Endres S, Hartmann G, Tardivel A, Schweighoffer E, Tybulewicz V, Mocsai A, Tschopp J, Ruland J (2009) Syk kinase signalling couples to the Nlrp3 inflammasome for anti-fungal host defence. Nature 459(7245):433–436

Gu Y, Kuida K, Tsutsui H, Ku G, Hsiao K, Fleming MA, Hayashi N, Higashino K, Okamura H, Nakanishi K, Kurimoto M, Tanimoto T, Flavell RA, Sato V, Harding MW, Livingston DJ, Su MS (1997) Activation of interferon-gamma inducing factor mediated by interleukin-1beta converting enzyme. Science 275(5297):206–209

Guey B, Bodnar M, Manié SN, Tardivel A, Petrilli V (2014) Caspase-1 autoproteolysis is differentially required for NLRP1b and NLRP3 inflammasome function. Proc Natl Acad Sci USA 111(48):17254–17259

Guo C, Chen M, Fa Z, Lu A, Fang W, Sun B, Chen C, Liao W, Meng G (2014) Acapsular *Cryptococcus neoformans* activates the NLRP3 inflammasome. Microbes Infect 16(10):845–854

Gurung P, Malireddi RK, Anand PK, Demon D, Vande Walle L, Liu Z, Vogel P, Lamkanfi M, Kanneganti TD (2012) Toll or interleukin-1 receptor (TIR) domain-containing adaptor inducing interferon-beta (TRIF)-mediated caspase-11 protease production integrates toll-like receptor 4 (TLR4) protein- and Nlrp3 inflammasome-mediated host defense against enteropathogens. J Biol Chem 287(41):34474–34483

Hagar JA, Powell DA, Aachoui Y, Ernst RK, Miao EA (2013) Cytoplasmic LPS activates caspase-11: implications in TLR4-independent endotoxic shock. Science 341(6151): 1250–1253

Halle A, Hornung V, Petzold GC, Stewart CR, Monks BG, Reinheckel T, Fitzgerald KA, Latz E, Moore KJ, Golenbock DT (2008) The NALP3 inflammasome is involved in the innate immune response to amyloid-beta. Nat Immunol 9(8):857–865

Hansen JD, Vojtech LN, Laing KJ (2011) Sensing disease and danger: a survey of vertebrate PRRs and their origins. Dev Comp Immunol 35(9):886–897

He Y, Zeng MY, Yang D, Motro B, Nunez G (2016) NEK7 is an essential mediator of NLRP3 activation downstream of potassium efflux. Nature 530(7590):354–7

Hise AG, Tomalka J, Ganesan S, Patel K, Hall BA, Brown GD, Fitzgerald KA (2009) An essential role for the NLRP3 inflammasome in host defense against the human fungal pathogen *Candida albicans*. Cell Host Microbe 5(5):487–497

Hoffman HM, Mueller JL, Broide DH, Wanderer AA, Kolodner RD (2001) Mutation of a new gene encoding a putative pyrin-like protein causes familial cold autoinflammatory syndrome and Muckle–Wells syndrome. Nat Genet 29(3):301–305

Hornung V, Bauernfeind F, Halle A, Samstad EO, Kono H, Rock KL, Fitzgerald KA, Latz E (2008) Silica crystals and aluminum salts activate the NALP3 inflammasome through phagosomal destabilization. Nat Immunol 9(8):847–856

Hornung V, Ablasser A, Charrel-Dennis M, Bauernfeind F, Horvath G, Caffrey DR, Latz E, Fitzgerald KA (2009) AIM2 recognizes cytosolic dsDNA and forms a caspase-1-activating inflammasome with ASC. Nature 458(7237):514–518

Hulbert SH, Webb CA, Smith SM, Sun Q (2001) Resistance gene complexes: evolution and utilization. Annu Rev Phytopathol 39(1):285–312

Iyer SS, He Q, Janczy JR, Elliott EI, Zhong Z, Olivier AK, Sadler JJ, Knepper-Adrian V, Han R, Qiao L, Eisenbarth SC, Nauseef WM, Cassel SL, Sutterwala FS (2013) Mitochondrial cardiolipin is required for Nlrp3 inflammasome activation. Immunity 39(2):311–323

Jin Y, Mailloux CM, Gowan K, Riccardi SL, LaBerge G, Bennett DC, Fain PR, Spritz RA (2007) NALP1 in vitiligo-associated multiple autoimmune disease. N Engl J Med 356(12):1216–1225

Jin T, Perry A, Jiang J, Smith P, Curry JA, Unterholzner L, Jiang Z, Horvath G, Rathinam VA, Johnstone RW, Hornung V, Latz E, Bowie AG, Fitzgerald KA, Xiao TS (2012) Structures of the HIN domain: DNA complexes reveal ligand binding and activation mechanisms of the AIM2 inflammasome and IFI16 receptor. Immunity 36(4):561–571

Jones JW, Kayagaki N, Broz P, Henry T, Newton K, O'Rourke K, Chan S, Dong J, Qu Y, Roose-Girma M, Dixit VM, Monack DM (2010) Absent in melanoma 2 is required for innate immune recognition of *Francisella tularensis*. Proc Natl Acad Sci U S A 107(21):9771–9776

Juliana C, Fernandes-Alnemri T, Kang S, Farias A, Qin F, Alnemri ES (2012) Non-transcriptional priming and deubiquitination regulate NLRP3 inflammasome activation. J Biol Chem 287 (43):36617–36622

Kanneganti TD (2010) Central roles of NLRs and inflammasomes in viral infection. Nat Rev Immunol 10(10):688–698

Kanneganti TD, Body-Malapel M, Amer A, Park JH, Whitfield J, Franchi L, Taraporewala ZF, Miller D, Patton JT, Inohara N, Nunez G (2006) Critical role for Cryopyrin/Nalp3 in activation of caspase-1 in response to viral infection and double-stranded RNA. J Biol Chem 281 (48):36560–36568

Kanneganti TD, Lamkanfi M, Nunez G (2007) Intracellular NOD-like receptors in host defense and disease. Immunity 27(4):549–559

Katsnelson MA, Rucker LG, Russo HM, Dubyak GR (2015) K^+ efflux agonists induce NLRP3 inflammasome activation independently of Ca^{2+} signaling. J Immunol 194(8):3937–3952

Kayagaki N, Warming S, Lamkanfi M, Vande Walle L, Louie S, Dong J, Newton K, Qu Y, Liu J, Heldens S, Zhang J, Lee WP, Roose-Girma M, Dixit VM (2011) Non-canonical inflammasome activation targets caspase-11. Nature 479(7371):117–121

Kayagaki N, Wong MT, Stowe IB, Ramani SR, Gonzalez LC, Akashi-Takamura S, Miyake K, Zhang J, Lee WP, Muszynski A, Forsberg LS, Carlson RW, Dixit VM (2013) Noncanonical inflammasome activation by intracellular LPS independent of TLR4. Science 341(6151): 1246–1249

Kayagaki N, Stowe IB, Lee BL, O'Rourke K, Anderson K, Warming S, Cuellar T, Haley B, Roose-Girma M, Phung QT, Liu PS, Lill JR, Li H, Wu J, Kummerfeld S, Zhang J, Lee WP, Snipas SJ, Salvesen GS, Morris LX, Fitzgerald L, Zhang Y, Bertram EM, Goodnow CC, Dixit VM (2015) Caspase-11 cleaves gasdermin D for non-canonical inflammasome signalling. Nature 526(7575):666–671

Kim EK, Choi EJ (2010) Pathological roles of MAPK signaling pathways in human diseases. Biochim Biophys Acta (BBA) Mol Basis Dis 1802(4):396–405

Kitamura A, Sasaki Y, Abe T, Kano H, Yasutomo K (2014) An inherited mutation in NLRC4 causes autoinflammation in human and mice. J Exp Med 211(12):2385–2396

Knodler LA, Crowley SM, Sham HP, Yang H, Wrande M, Ma C, Ernst RK, Steele-Mortimer O, Celli J, Vallance BA (2014) Noncanonical inflammasome activation of caspase-4/caspase-11 mediates epithelial defenses against enteric bacterial pathogens. Cell Host Microbe 16(2):249–256

Koch U, Radtke F (2011) Mechanisms of T cell development and transformation. Annu Rev Cell Dev Biol 27:539–562

Kofoed EM, Vance RE (2011) Innate immune recognition of bacterial ligands by NAIPs determines inflammasome specificity. Nature 477(7366):592–595

Kortmann J, Brubaker SW, Monack DM (2015) Cutting edge: inflammasome activation in primary human macrophages is dependent on flagellin. J Immunol 195(3):815–819

Kumar H, Kumagai Y, Tsuchida T, Koenig PA, Satoh T, Guo Z, Jang MH, Saitoh T, Akira S, Kawai T (2009) Involvement of the NLRP3 inflammasome in innate and humoral adaptive immune responses to fungal beta-glucan. J Immunol 183(12):8061–8067

Laing KJ, Purcell MK, Winton JR, Hansen JD (2008) A genomic view of the NOD-like receptor family in teleost fish: identification of a novel NLR subfamily in zebrafish. BMC Evol Biol 8:42

Lamkanfi M (2011) Emerging inflammasome effector mechanisms. Nat Rev Immunol 11(3):213–220

Lamkanfi M, Dixit VM (2014) Mechanisms and functions of inflammasomes. Cell 157(5):1013–1022

Lamkanfi M, Mueller JL, Vitari AC, Misaghi S, Fedorova A, Deshayes K, Lee WP, Hoffman HM, Dixit VM (2009) Glyburide inhibits the Cryopyrin/Nalp3 inflammasome. J Cell Biol 187 (1):61–70

Lamkanfi M, Sarkar A, Vande Walle L, Vitari AC, Amer AO, Wewers MD, Tracey KJ, Kanneganti TD, Dixit VM (2010) Inflammasome-dependent release of the alarmin HMGB1 in endotoxemia. J Immunol 185(7):4385–4392

Lange C, Hemmrich G, Klostermeier UC, Lopez-Quintero JA, Miller DJ, Rahn T, Weiss Y, Bosch TC, Rosenstiel P (2011) Defining the origins of the NOD-like receptor system at the base of animal evolution. Mol Biol Evol 28(5):1687–1702

Lara-Tejero M, Sutterwala FS, Ogura Y, Grant EP, Bertin J, Coyle AJ, Flavell RA, Galan JE (2006) Role of the caspase-1 inflammasome in Salmonella typhimurium pathogenesis. J Exp Med 203(6):1407–1412

Lee GS, Subramanian N, Kim AI, Aksentijevich I, Goldbach-Mansky R, Sacks DB, Germain RN, Kastner DL, Chae JJ (2012) The calcium-sensing receptor regulates the NLRP3 inflammasome through Ca^{2+} and cAMP. Nature 492(7427):123–127

Levandowski CB, Mailloux CM, Ferrara TM, Gowan K, Ben S, Jin Y, McFann KK, Holland PJ, Fain PR, Dinarello CA, Spritz RA (2013) NLRP1 haplotypes associated with vitiligo and autoimmunity increase interleukin-1β processing via the NLRP1 inflammasome. Proc Natl Acad Sci USA 110(8):2952–2956

Levinsohn JL, Newman ZL, Hellmich KA, Fattah R, Getz MA, Liu S, Sastalla I, Leppla SH, Moayeri M (2012) Anthrax lethal factor cleavage of Nlrp1 is required for activation of the inflammasome. PLoS Pathog 8(3):e1002638

Lin KM, Hu W, Troutman TD, Jennings M, Brewer T, Li X, Nanda S, Cohen P, Thomas JA, Pasare C (2014) IRAK-1 bypasses priming and directly links TLRs to rapid NLRP3 inflammasome activation. Proc Natl Acad Sci USA 111(2):775–780

Lopez-Castejon G, Luheshi NM, Compan V, High S, Whitehead RC, Flitsch S, Kirov A, Prudovsky I, Swanton E, Brough D (2013) Deubiquitinases regulate the activity of caspase-1 and interleukin-1beta secretion via assembly of the inflammasome. J Biol Chem 288(4):2721–2733

Lotze MT, Zeh HJ, Rubartelli A, Sparvero LJ, Amoscato AA, Washburn NR, DeVera ME, Liang X, Tör M, Billiar T (2007) The grateful dead: damage-associated molecular pattern molecules and reduction/oxidation regulate immunity. Immunol Rev 220(1):60–81

Magitta NF, Boe Wolff AS, Johansson S, Skinningsrud B, Lie BA, Myhr KM, Undlien DE, Joner G, Njolstad PR, Kvien TK, Forre O, Knappskog PM, Husebye ES (2009) A coding polymorphism in NALP1 confers risk for autoimmune Addison's disease and type 1 diabetes. Genes Immun 10(2):120–124

Mariathasan S, Newton K, Monack DM, Vucic D, French DM, Lee WP, Roose-Girma M, Erickson S, Dixit VM (2004) Differential activation of the inflammasome by caspase-1 adaptors ASC and Ipaf. Nature 430(6996):213–218

Mariathasan S, Weiss DS, Newton K, McBride J, O'Rourke K, Roose-Girma M, Lee WP, Weinrauch Y, Monack DM, Dixit VM (2006) Cryopyrin activates the inflammasome in response to toxins and ATP. Nature 440(7081):228–232

Martinon F, Burns K, Tschopp J (2002) The inflammasome: a molecular platform triggering activation of inflammatory caspases and processing of proIL-β. Mol Cell 10(2):417–426

Martinon F, Petrilli V, Mayor A, Tardivel A, Tschopp J (2006) Gout-associated uric acid crystals activate the NALP3 inflammasome. Nature 440(7081):237–241

Martinon F, Mayor A, Tschopp J (2009) The inflammasomes: guardians of the body. Annu Rev Immunol 27:229–265

Masters SL, Simon A, Aksentijevich I, Kastner DL (2009) Horror autoinflammaticus: the molecular pathophysiology of autoinflammatory disease (*). Annu Rev Immunol 27:621–668

Masters SL, Gerlic M, Metcalf D, Preston S, Pellegrini M, O'Donnell JA, McArthur K, Baldwin TM, Chevrier S, Nowell CJ, Cengia LH, Henley KJ, Collinge JE, Kastner DL, Feigenbaum L, Hilton DJ, Alexander WS, Kile BT, Croker BA (2012) NLRP1 inflammasome activation induces pyroptosis of hematopoietic progenitor cells. Immunity 37(6):1009–1023

Matusiak M, Van Opdenbosch N, Vande Walle L, Sirard JC, Kanneganti TD, Lamkanfi M (2015) Flagellin-induced NLRC4 phosphorylation primes the inflammasome for activation by NAIP5. Proc Natl Acad Sci USA 112(5):1541–1546

McNeela EA, Burke A, Neill DR, Baxter C, Fernandes VE, Ferreira D, Smeaton S, El-Rachkidy R, McLoughlin RM, Mori A, Moran B, Fitzgerald KA, Tschopp J, Petrilli V, Andrew PW, Kadioglu A, Lavelle EC (2010) Pneumolysin activates the NLRP3 inflammasome and promotes proinflammatory cytokines independently of TLR4. PLoS Pathog 6(11):e1001191

Miao EA, Alpuche-Aranda CM, Dors M, Clark AE, Bader MW, Miller SI, Aderem A (2006) Cytoplasmic flagellin activates caspase-1 and secretion of interleukin 1beta via Ipaf. Nat Immunol 7(6):569–575

Miao EA, Ernst RK, Dors M, Mao DP, Aderem A (2008) Pseudomonas aeruginosa activates caspase 1 through Ipaf. Proc Natl Acad Sci USA 105(7):2562–2567

Miao EA, Leaf IA, Treuting PM, Mao DP, Dors M, Sarkar A, Warren SE, Wewers MD, Aderem A (2010a) Caspase-1-induced pyroptosis is an innate immune effector mechanism against intracellular bacteria. Nat Immunol 11(12):1136–1142

Miao EA, Mao DP, Yudkovsky N, Bonneau R, Lorang CG, Warren SE, Leaf IA, Aderem A (2010b) Innate immune detection of the type III secretion apparatus through the NLRC4 inflammasome. Proc Natl Acad Sci USA 107(7):3076–3080

Motta VN, Markle JG, Gulban O, Mortin-Toth S, Liao KC, Mogridge J, Steward CA, Danska JS (2015) Identification of the inflammasome Nlrp1b as the candidate gene conferring diabetes risk at the Idd4.1 locus in the nonobese diabetic mouse. J Immunol 194(12):5663–5673

Munoz-Planillo R, Kuffa P, Martinez-Colon G, Smith BL, Rajendiran TM, Nunez G (2013) $K^{(+)}$ efflux is the common trigger of NLRP3 inflammasome activation by bacterial toxins and particulate matter. Immunity 38(6):1142–1153

Nakahira K, Haspel JA, Rathinam VA, Lee SJ, Dolinay T, Lam HC, Englert JA, Rabinovitch M, Cernadas M, Kim HP, Fitzgerald KA, Ryter SW, Choi AM (2011) Autophagy proteins regulate innate immune responses by inhibiting the release of mitochondrial DNA mediated by the NALP3 inflammasome. Nat Immunol 12(3):222–230

Ng J, Hirota SA, Gross O, Li Y, Ulke-Lemee A, Potentier MS, Schenck LP, Vilaysane A, Seamone ME, Feng H, Armstrong GD, Tschopp J, Macdonald JA, Muruve DA, Beck PL (2010) *Clostridium difficile* toxin-induced inflammation and intestinal injury are mediated by the inflammasome. Gastroenterology 139(2):542–552, 552 e1-3

Nystrom S, Antoine DJ, Lundback P, Lock JG, Nita AF, Hogstrand K, Grandien A, Erlandsson-Harris H, Andersson U, Applequist SE (2013) TLR activation regulates damage-associated molecular pattern isoforms released during pyroptosis. EMBO J 32 (1):86–99

Ozen S, Bilginer Y (2014) A clinical guide to autoinflammatory diseases: familial Mediterranean fever and next-of-kin. Nat Rev Rheumatol 10(3):135–147

Palm NW, Medzhitov R (2009a) Pattern recognition receptors and control of adaptive immunity. Immunol Rev 227(1):221–233

Palm NW, Medzhitov R (2009b) Immunostimulatory activity of haptenated proteins. Proc Natl Acad Sci USA 106(12):4782–4787

Perregaux D, Gabel CA (1994) Interleukin-1 beta maturation and release in response to ATP and nigericin. Evidence that potassium depletion mediated by these agents is a necessary and common feature of their activity. J Biol Chem 269(21):15195–15203

Perregaux DG, Gabel CA (1998) Human monocyte stimulus-coupled IL-1β posttranslational processing: modulation via monovalent cations. Am J Physiol Cell Physiol 275(6):C1538–C1547

Petrilli V, Papin S, Dostert C, Mayor A, Martinon F, Tschopp J (2007) Activation of the NALP3 inflammasome is triggered by low intracellular potassium concentration. Cell Death Differ 14 (9):1583–1589

Py BF, Kim MS, Vakifahmetoglu-Norberg H, Yuan J (2013) Deubiquitination of NLRP3 by BRCC3 critically regulates inflammasome activity. Mol Cell 49(2):331–338

Qu Y, Misaghi S, Izrael-Tomasevic A, Newton K, Gilmour LL, Lamkanfi M, Louie S, Kayagaki N, Liu J, Komuves L, Cupp JE, Arnott D, Monack D, Dixit VM (2012) Phosphorylation of NLRC4 is critical for inflammasome activation. Nature 490(7421):539–542

Rajan JV, Rodriguez D, Miao EA, Aderem A (2011) The NLRP3 inflammasome detects encephalomyocarditis virus and vesicular stomatitis virus infection. J Virol 85(9):4167–4172

Rast JP, Smith LC, Loza-Coll M, Hibino T, Litman GW (2006) Genomic insights into the immune system of the sea urchin. Science 314(5801):952–956

Rathinam VA, Jiang Z, Waggoner SN, Sharma S, Cole LE, Waggoner L, Vanaja SK, Monks BG, Ganesan S, Latz E, Hornung V, Vogel SN, Szomolanyi-Tsuda E, Fitzgerald KA (2010) The AIM2 inflammasome is essential for host defense against cytosolic bacteria and DNA viruses. Nat Immunol 11(5):395–402

Romberg N, Al Moussawi K, Nelson-Williams C, Stiegler AL, Loring E, Choi M, Overton J, Meffre E, Khokha MK, Huttner AJ, West B, Podoltsev NA, Boggon TJ, Kazmierczak BI, Lifton RP (2014) Mutation of NLRC4 causes a syndrome of enterocolitis and autoinflammation. Nat Genet 46(10):1135–1139

Sagulenko V, Thygesen SJ, Sester DP, Idris A, Cridland JA, Vajjhala PR, Roberts TL, Schroder K, Vince JE, Hill JM, Silke J, Stacey KJ (2013) AIM2 and NLRP3 inflammasomes activate both apoptotic and pyroptotic death pathways via ASC. Cell Death Differ 20(9):1149–1160

Said-Sadier N, Padilla E, Langsley G, Ojcius DM (2010) *Aspergillus fumigatus* stimulates the NLRP3 inflammasome through a pathway requiring ROS production and the Syk tyrosine kinase. PLoS ONE 5(4):e10008

Sauer JD, Witte CE, Zemansky J, Hanson B, Lauer P, Portnoy DA (2010) *Listeria monocytogenes* triggers AIM2-mediated pyroptosis upon infrequent bacteriolysis in the macrophage cytosol. Cell Host Microbe 7(5):412–419

Schmid-Burgk JL, Gaidt MM, Schmidt T, Ebert TS, Bartok E, Hornung V (2015) Caspase-4 mediates non-canonical activation of the NLRP3 inflammasome in human myeloid cells. Eur J Immunol 45(10):2911–2917

Schmid-Burgk JL, Chauhan D, Schmidt T, Ebert TS, Reinhardt J, Endl E, Hornung V (2016) A Genome-wide CRISPR (clustered regularly interspaced short palindromic repeats) screen identifies NEK7 as an essential component of NLRP3 inflammasome activation. J Biol Chem 291(1):103–109

Seshadri S, Duncan MD, Hart JM, Gavrilin MA, Wewers MD (2007) Pyrin levels in human monocytes and monocyte-derived macrophages regulate IL-1 processing and release. J Immunol 179(2):1274–1281

Shi J, Zhao Y, Wang Y, Gao W, Ding J, Li P, Hu L, Shao F (2014) Inflammatory caspases are innate immune receptors for intracellular LPS. Nature 514(7521):187–192

Shi J, Zhao Y, Wang K, Shi X, Wang Y, Huang H, Zhuang Y, Cai T, Wang F, Shao F (2015) Cleavage of GSDMD by inflammatory caspases determines pyroptotic cell death. Nature 526 (7575):660–665

Shi H, Wang Y, Li X, Zhan X, Tang M, Fina M, Su L, Pratt D, Bu CH, Hildebrand S, Lyon S, Scott L, Quan J, Sun Q, Russell J, Arnett S, Jurek P, Chen D, Kravchenko VV, Mathison JC, Moresco EM, Monson NL, Ulevitch RJ, Beutler B (2016) NLRP3 activation and mitosis are mutually exclusive events coordinated by NEK7, a new inflammasome component. Nat Immunol 17(3):250–8

Shimada K, Crother TR, Karlin J, Dagvadorj J, Chiba N, Chen S, Ramanujan VK, Wolf AJ, Vergnes L, Ojcius DM, Rentsendorj A, Vargas M, Guerrero C, Wang Y, Fitzgerald KA, Underhill DM, Town T, Arditi M (2012) Oxidized mitochondrial DNA activates the NLRP3 inflammasome during apoptosis. Immunity 36(3):401–414

Spritz RA (2007) The genetics of generalized vitiligo and associated autoimmune diseases. Pigment Cell Res 20(4):271–278

Stein C, Caccamo M, Laird G, Leptin M (2007) Conservation and divergence of gene families encoding components of innate immune response systems in zebrafish. Genome Biol 8(11): R251

Sutterwala FS, Mijares LA, Li L, Ogura Y, Kazmierczak BI, Flavell RA (2007) Immune recognition of *Pseudomonas aeruginosa* mediated by the IPAF/NLRC4 inflammasome. J Exp Med 204(13):3235–3245

Suzuki T, Franchi L, Toma C, Ashida H, Ogawa M, Yoshikawa Y, Mimuro H, Inohara N, Sasakawa C, Nunez G (2007) Differential regulation of caspase-1 activation, pyroptosis, and autophagy via Ipaf and ASC in Shigella-infected macrophages. PLoS Pathog 3(8):e111

Suzuki S, Franchi L, He Y, Munoz-Planillo R, Mimuro H, Suzuki T, Sasakawa C, Nunez G (2014) Shigella type III secretion protein MxiI is recognized by Naip2 to induce Nlrc4 inflammasome activation independently of Pkcdelta. PLoS Pathog 10(2):e1003926

Takeuchi O, Akira S (2010) Pattern recognition receptors and inflammation. Cell 140(6):805–820

Tenthorey JL, Kofoed EM, Daugherty MD, Malik HS, Vance RE (2014) Molecular basis for specific recognition of bacterial ligands by NAIP/NLRC4 inflammasomes. Mol Cell 54(1):17–29

Touitou I, Lesage S, McDermott M, Cuisset L, Hoffman H, Dode C, Shoham N, Aganna E, Hugot JP, Wise C, Waterham H, Pugnere D, Demaille J, Sarrauste de Menthiere C (2004) Infevers: an evolving mutation database for auto-inflammatory syndromes. Hum Mutat 24 (3):194–198

Tsuchiya K, Hara H, Kawamura I, Nomura T, Yamamoto T, Daim S, Dewamitta SR, Shen Y, Fang R, Mitsuyama M (2010) Involvement of absent in melanoma 2 in inflammasome activation in macrophages infected with *Listeria monocytogenes*. J Immunol 185(2):1186–1195

Vajjhala PR, Kaiser S, Smith SJ, Ong QR, Soh SL, Stacey KJ, Hill JM (2014) Identification of multifaceted binding modes for pyrin and ASC pyrin domains gives insights into pyrin inflammasome assembly. J Biol Chem 289(34):23504–23519

Vallabhapurapu S, Karin M (2009) Regulation and function of NF-kappaB transcription factors in the immune system. Annu Rev Immunol 27:693–733
Van Gorp H, Kuchmiy A, Van Hauwermeiren F, Lamkanfi M (2014) NOD-like receptors interfacing the immune and reproductive systems. FEBS J 281(20):4568–4582
Van Opdenbosch N, Gurung P, Vande Walle L, Fossoul A, Kanneganti TD, Lamkanfi M (2014) Activation of the NLRP1b inflammasome independently of ASC-mediated caspase-1 autoproteolysis and speck formation. Nat Commun 5:3209
Vande Walle L, Van Opdenbosch N, Jacques P, Fossoul A, Verheugen E, Vogel P, Beyaert R, Elewaut D, Kanneganti TD, van Loo G, Lamkanfi M (2014) Negative regulation of the NLRP3 inflammasome by A20 protects against arthritis. Nature 512(7512):69–73
Wang S, Miura M, Jung Y-K, Zhu H, Li E, Yuan J (1998) Murine caspase-11, an ICE-interacting protease, is essential for the activation of ICE. Cell 92(4):501–509
Wickliffe KE, Leppla SH, Moayeri M (2008) Anthrax lethal toxin-induced inflammasome formation and caspase-1 activation are late events dependent on ion fluxes and the proteasome. Cell Microbiol 10(2):332–343
Williams TM, Leeth RA, Rothschild DE, Coutermarsh-Ott SL, McDaniel DK, Simmons AE, Heid B, Cecere TE, Allen IC (2015) The NLRP1 inflammasome attenuates colitis and colitis-associated tumorigenesis. J Immunol 194(7):3369–3380
Wilmanski JM, Petnicki-Ocwieja T, Kobayashi KS (2008) NLR proteins: integral members of innate immunity and mediators of inflammatory diseases. J Leukoc Biol 83(1):13–30
Wu J, Fernandes-Alnemri T, Alnemri ES (2010) Involvement of the AIM2, NLRC4, and NLRP3 inflammasomes in caspase-1 activation by *Listeria monocytogenes*. J Clin Immunol 30(5): 693–702
Xu H, Yang J, Gao W, Li L, Li P, Zhang L, Gong YN, Peng X, Xi JJ, Chen S, Wang F, Shao F (2014) Innate immune sensing of bacterial modifications of Rho GTPases by the Pyrin inflammasome. Nature 513(7517):237–241
Yaron JR, Gangaraju S, Rao MY, Kong X, Zhang L, Su F, Tian Y, Glenn HL, Meldrum DR (2015) $K(^{+})$ regulates $Ca(^{2+})$ to drive inflammasome signaling: dynamic visualization of ion flux in live cells. Cell Death Dis 6:e1954
Yu JR, Leslie KS (2011) Cryopyrin-associated periodic syndrome: an update on diagnosis and treatment response. Curr Allergy Asthma Rep 11(1):12–20
Yu JW, Fernandes-Alnemri T, Datta P, Wu J, Juliana C, Solorzano L, McCormick M, Zhang Z, Alnemri ES (2007) Pyrin activates the ASC pyroptosome in response to engagement by autoinflammatory PSTPIP1 mutants. Mol Cell 28(2):214–227
Zamboni DS, Kobayashi KS, Kohlsdorf T, Ogura Y, Long EM, Vance RE, Kuida K, Mariathasan S, Dixit VM, Flavell RA, Dietrich WF, Roy CR (2006) The Birc1e cytosolic pattern-recognition receptor contributes to the detection and control of *Legionella pneumophila* infection. Nat Immunol 7(3):318–325
Zhao Y, Yang J, Shi J, Gong YN, Lu Q, Xu H, Liu L, Shao F (2011) The NLRC4 inflammasome receptors for bacterial flagellin and type III secretion apparatus. Nature 477(7366):596–600
Zurawek M, Fichna M, Januszkiewicz-Lewandowska D, Gryczynska M, Fichna P, Nowak J (2010) A coding variant in NLRP1 is associated with autoimmune Addison's disease. Hum Immunol 71(5):530–534

Structural Mechanisms in NLR Inflammasome Assembly and Signaling

Zehan Hu and Jijie Chai

Abstract Inflammasomes are multimeric protein complexes that mediate the activation of inflammatory caspases. One central component of inflammasomes is nucleotide-binding domain (NBD)- and leucine-rich repeat (LRR)-containing proteins (NLRs) that can function as pattern recognition receptors (PRRs). In resting cells, NLR proteins exist in an auto-inhibited, monomeric, and ADP-bound state. Perception of microbial or damage-associated signals results in NLR oligomerization, thus recruiting inflammatory caspases directly or through the adaptor molecule apoptosis-associated speck-like protein containing a caspase recruitment domain (ASC). The assembled NLR inflammasomes serve as dedicated machinery to facilitate the activation of the inflammatory caspases. Here, we review current understanding of the structures of NLR inflammasomes with an emphasis on the molecular mechanisms of their assembly and activation. We also discuss implications of the self-propagation model derived from the NAIP–NLRC4 inflammasomes for the activation of other NLR inflammasomes and a potential role of the C-terminal LRR domain in the activation of an NLR protein.

Contents

Z. Hu (✉) · J. Chai (✉)
Innovation Center for Structural Biology, Tsinghua-Peking Joint Center for Life Sciences, School of Life Sciences, Tsinghua University, Beijing 100084, China
e-mail: zehan.hu@gmail.com

J. Chai
e-mail: chaijj@tsinghua.edu.cn

S. Backert (ed.), *Inflammasome Signaling and Bacterial Infections*,
Current Topics in Microbiology and Immunology 397,
DOI 10.1007/978-3-319-41171-2_2

1 Introduction

In the evolutionary "arms race," vertebrates from the primitive lamprey to humans have developed an arsenal of weapons conferring protection against invading pathogens, including bacteria, viruses, fungi, and parasites (Boehm et al. 2012; Flajnik and Du Pasquier 2004; Kimbrell and Beutler 2001). The first line of host defense, the innate immune system in mammals, relies on recognition of evolutionarily conserved pathogen components (pathogen-associated molecular patterns, PAMPs) or endogenous danger-associated molecular patterns (DAMPs) by an array of germline-encoded pattern recognition receptors (PRRs) (Janeway and Medzhitov 2002; Medzhitov and Janeway 1997). PRR-mediated signaling cascades get the innate immune system in gear, including production of antimicrobial proteins (Mukherjee and Hooper 2015), secretion of cytokines, and pyroptosis (Takeuchi and Akira 2010).

Thus far, several types of PRRs have been identified, including toll-like receptors (TLRs), C-type lectin receptors (CLRs), retinoic acid-inducible gene 1 (RIG-I)-like receptors (RLRs), absent in melanoma 2 (AIM2)-like receptors (ALRs), and nucleotide-binding domain (NBD)- and leucine-rich repeat (LRR)-containing proteins (NLRs). Activation of some NLRs or ALRs often results in the formation of high molecular weight cytosolic protein complexes, termed inflammasomes, that serve as a platform for recruitment of the pro-inflammatory caspases (Lamkanfi and Dixit 2014; Martinon et al. 2009; von Moltke et al. 2013). Recruitment of the pro-inflammatory caspases by the complexes in many cases is through the adaptor molecule apoptosis-associated speck-like protein containing a caspase recruitment domain (ASC). Among the pro-inflammatory caspases, caspase-1 is best-characterized, and its activation by inflammasomes leads to proteolytic processing of pro-interleukin 1β (pro-IL-1β) and pro-IL-18, and cell death.

Multiple inflammasomes have been identified, denoted by the sensor PRR protein within the inflammasome, such as AIM2 inflammasome (Burckstummer et al. 2009; Fernandes-Alnemri et al. 2009; Hornung et al. 2009; Roberts et al. 2009), NLRP1 inflammasome (Martinon et al. 2002), NLRP3 inflammasome (Agostini et al. 2004), and NAIP–NLRC4 inflammasome (Kofoed and Vance 2011; Mariathasan et al. 2004; Zamboni et al. 2006; Zhao et al. 2011). Other NLR family members including NLRP2 (Minkiewicz et al. 2013), NLRP6 (Elinav et al. 2011), NLRP7 (Khare et al. 2012), NLRP12 (Vladimer et al. 2012), and the ALR protein interferon gamma-inducible protein 16 (IFI16) (Kerur et al. 2011; Unterholzner et al. 2010) may also assemble into inflammasomes, but more studies are needed to confirm their in vivo roles. More recently, a non-canonical inflammasome formed by direct recognition of cytosolic lipopolysaccharide (LPS) by caspase-11 in mice (caspase-4 and caspase-5 in human) was found to mediate pyroptotic cell death (Hagar et al. 2013; Kayagaki et al. 2013; Shi et al. 2014).

During the past few years, rapid progress has been made toward structural and biochemical understanding of inflammasomes including their activation and regulation mechanisms. Several excellent reviews have summarized structural studies of the ALR inflammasomes (Shaw and Liu 2014; Xiao 2015). In this review, we focus on the recent structural elucidations of NLR inflammasome assembly and signaling.

2 Overviews of NOD-Like Receptors

NLR family members are typically characteristic of a tripartite structure comprising a varied N-terminal effector domain, a central nucleotide-binding and oligomerization domain (NOD), and a C-terminal LRR domain. Depending on the N-terminal effector domain, NLR family can be further divided into several subfamilies, such as NLRC with an N-terminal caspase recruitment domain (CARD), NLRP with an N-terminal pyrin domain (PYD), and NAIP with three tandem N-terminal baculovirus inhibitor of apoptosis (BIR) protein repeat domains. The NOD module, which can be further divided into an NBD and a helical domain 1 (HD1) followed by a winged helical domain (WHD), is an ADP-/ATP-binding motif conserved in the signal transduction ATPases with numerous domains (STAND) subfamily, including the apoptotic protein apoptotic peptidase-activating factor-1 (Apaf-1) in mammals (Danot et al. 2009). Similar tripartite structural organization is also present in the plant disease resistance proteins (NLR-type receptors, also known as R proteins) (Maekawa et al. 2011; Jones and Dangl 2006). There are 22 NLR family members in humans and at least 34 in mice. In unchallenged cells, NLR proteins are maintained inactive through their C-terminal LRR domains, as several LRR-truncated NLRs were constitutively active in inducing downstream signaling (Hu et al. 2013; Kofoed and Vance 2011; Poyet et al. 2001). Consistently, unchecked inflammasome activation has been shown to associate with many severe auto-inflammatory diseases (Davis et al. 2011; Lamkanfi and Dixit 2012; Strowig et al. 2012; Wen et al. 2012). Activation of an NLR protein is believed to accompany with the exchange of ADP for ATP (Danot et al. 2009; Duncan et al. 2007).

NAIP–NLRC4, NLRP1, and NLRP3 are subsets of the NLR family which are best understood for their ability to assemble into inflammasomes. The NAIP–NLRC4 inflammasome plays a critical role in host defense against facultative intracellular pathogens such as *Salmonella typhimurium*, *Shigella flexneri*, and *Legionella pneumophila* (Lamkanfi and Dixit 2012). NAIPs with seven homologs (Naip1–7) in mice and one in humans are responsible for the direct recognition of PAMPs from these pathogens. In mice, Naip5 and Naip6 act as an intracellular sensor for bacterial flagellin, whereas Naip1 and Naip2 recognize the needle protein and the rod protein from the type III secretion system (T3SS) of bacteria, respectively (Franchi et al. 2006; Kofoed and Vance 2011; Lightfield et al. 2008; Miao et al. 2006, 2010; Yang et al. 2013; Zhao et al. 2011). Upon recognition by their cognate NAIP proteins, these bacterial components stimulate NAIP–NLRC4 association, resulting in assembly of the NAIP–NLRC4 inflammasomes (Halff et al. 2012; Kofoed and Vance 2011; Zhao et al. 2011).

NLRP1 is the first PRR shown to form inflammasomes (Martinon et al. 2002), though the physiological activation signals were initially unknown. Humans have a single *NLRP1* gene, while mice have three paralogs (*Nlrp1a*, *Nlrp1b*, and *Nlrp1c*) exhibiting high allelic variations. The Nlrp1b inflammasome is a major component of host defense against *Bacillus anthracis*, mediating inflammatory response to the anthrax lethal toxin (LeTx) (Boyden and Dietrich 2006; Moayeri et al. 2010; Terra et al. 2010). LeTx is a two-component toxin composed of protective antigen (PA) and lethal factor (LF). LF is a metalloprotease, and its activity is required for the activation of Nlrp1b inflammasome by cleaving a position at the N-terminal side of a responsive rat Nlrp1b allele (Chavarria-Smith and Vance 2013; Fink et al. 2008; Hellmich et al. 2012; Levinsohn et al. 2012).

The NLRP3 inflammasome can be activated by many stimuli, including components from bacterial, viral, and fungal pathogens (Dostert et al. 2008; Eisenbarth et al. 2008; Ichinohe et al. 2009; Joly and Sutterwala 2010; Kanneganti et al. 2006; Mariathasan et al. 2006; Martinon et al. 2006). It is widely accepted that the activation of NLRP3 inflammasome requires two agonist-induced signals with one priming NF-κB-mediated expression of NLRP3 and inflammatory cytokines (Bauernfeind et al. 2009), and the other one triggering the activation and assembly of NLRP3 inflammasome. Given the large diversity of the agonists, it is believed that all these agonists may cause some converged effects in hosts monitored by the NLRP3 inflammasome (Lamkanfi and Dixit 2014; Tschopp and Schroder 2010). But recent studies appear to support potassium efflux as a common step essential for NLRP3 inflammasome activation induced by many stimuli (He et al. 2016; Munoz-Planillo et al. 2013; Petrilli et al. 2007).

3 Auto-Inhibition Mechanism of NLRC4

Structural studies of mouse Nlrc4 in its active and inactive forms provide insight into the autoinhibition mechanism of NLR family protein (Hu et al. 2013, 2015; Zhang et al. 2015). As observed in other STAND members (Danot et al. 2009), the auto-inhibited Nlrc4 (Hu et al. 2013) is monomeric and ADP-bound (Fig. 1a). The structure of $Nlrc4^{NOD}$ largely resembles that of $Apaf\text{-}1^{NOD}$ (Fig. 1b, c) (Chai and Shi 2013; Reubold et al. 2011; Riedl et al. 2005). All the ADP-interacting residues of Nlrc4 are from the conserved NBD and HD1 except $Nlrc4^{H443}$ from the WHD that forms a single hydrogen bond with the β-phosphate group of the bound ADP (Fig. 1c). Structural comparison between the inactive (crystal structure) and active

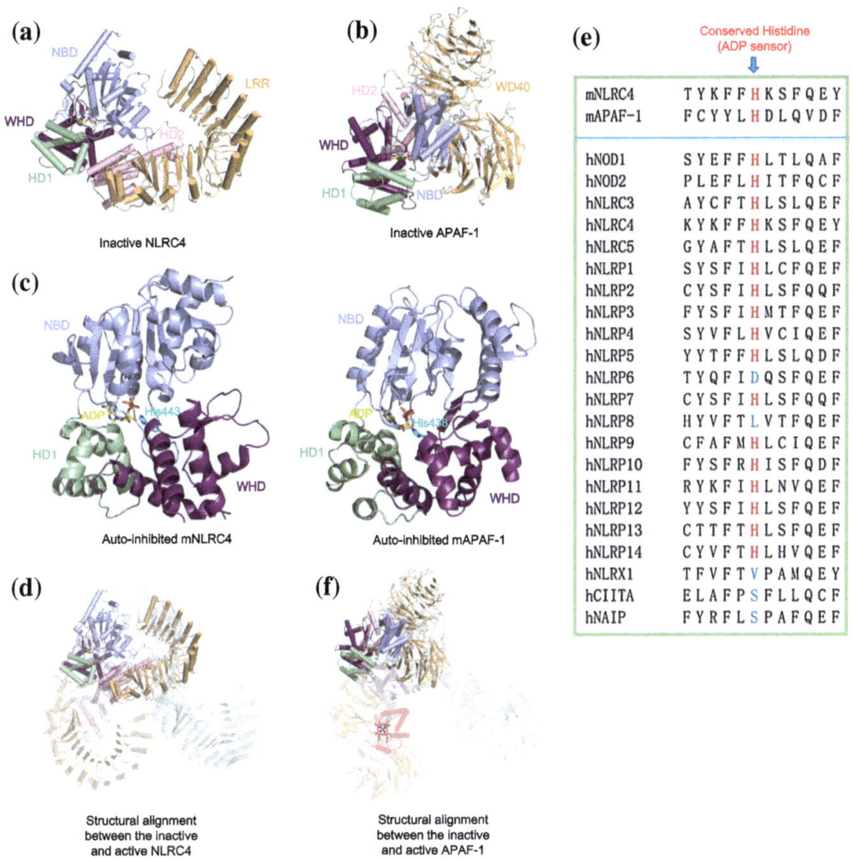

Fig. 1 Auto-inhibition mechanism of NLRC4. **a** The overall structure of auto-inhibited Nlrc4$^{\Delta CARD}$ (PDB code: 4KXF) shown in cartoon. Color codes for domains are indicated. **b** The overall structure of auto-inhibited Apaf-1 (PDB code: 3SFZ) shown in cartoon. Color codes for domains are indicated. The NBD domains of Nlrc4 and Apaf-1 are shown in the same orientation in **a** and **b**. **c** Histidine from the WHD of Nlrc4 and Apaf-1 interact with the bound ADP in the auto-inhibited state. The bound ADP molecules are shown in *yellow* and *stick*, and the side chains of the histidine are shown in cyan and stick. **d** Structural alignment between the inactive Nlrc4 and active Nlrc4. The NBD of auto-inhibited Nlrc4 (in cartoon) is aligned with that of one active Nlrc4 protomer (in transparent cartoon) from a lateral dimer (EMDB code: EMD-3141). The other protomer is shown in *gray* and transparent cartoon. e Most NLR proteins have a conserved histidine residue (highlighted in *red*) as the ADP sensor. **f** Structural alignment between the inactive Apaf-1 and active Apaf-1. The NBD of auto-inhibited Apaf-1 (in cartoon) is aligned with that of one active Apaf-1 protomer (in transparent cartoon) from a lateral dimer (PDB code: 3JBT)

(cryo-EM) Nlrc4 (Hu et al. 2015; Zhang et al. 2015) showed that its WHD, HD2, and LRR as a whole undergo striking structural re-organization relative to the NBD and HD1 on activation (Fig. 1d). This observation indicates that the hydrogen bond formed between Nlrc4^{H443} and ADP is specific for the inactive Nlrc4. Disruption of the specific interaction is expected to favor conformational changes in the WHD.

On the other hand, attenuation of ADP binding caused by loss of the hydrogen bond would promote the ATP-binding activity of Nlrc4 because of a stable ratio between ADP and ATP in cells. Both of the effects would facilitate the activation of Nlrc4. Indeed, the $Nlrc4^{H443L}$ mutation led to ligand- and Naip-independent activation of Nlrc4 in 293T cells (Hu et al. 2013). In strong support of a role played by the single hydrogen bond in Nlrc4 autoinhibition, the $NLRC4^{H443P}$ mutation was recently identified in patients with the familial cold auto-inflammatory syndrome (FCAS) characterized by the constitutive activation of the NAIP–NLRC4 inflammasomes (Kitamura et al. 2014). Furthermore, mice expressing this Nlrc4 mutant also developed dermatitis and arthritis caused by excessive IL-1β-mediated production of IL-17A from neutrophils (Kitamura et al. 2014). $Nlrc4^{H443}$ is highly conserved in most of the NLRs from both mammals (Fig. 1e) and plants, suggesting a conserved role played by the conserved histidine in the autoinhibition of NLR proteins. Consistent with this idea, mutations of the "MHD" motif ("H" is equivalent to $Nlrc4^{H443}$) in plant NLRs also resulted in their constitutive activation (Takken et al. 2006; van der Biezen and Jones 1998; Williams et al. 2011). In fact, the histidine is also conserved in the apoptotic protein Apaf-1 and forms a hydrogen bond with the bound ADP (Fig. 1c). But further studies are needed to test whether this residue is important for Apaf-1 autoinhibition.

In the cryo-EM structures (Hu et al. 2015; Zhang et al. 2015), lateral dimerization of Nlrc4 is largely mediated by packing of one side of the NBD from one protomer against the opposite side of the NBD from the other protomer (Fig. 1d), as observed in the Apaf-1 apoptosome (Fig. 1f) (Zhou et al. 2015). Structural superposition of one protomer from the lateral dimer with the inactive Nlrc4 showed that the LRR domain of an inactive Nlrc4 occupies the position of the other protomer (Fig. 1d). This observation indicates the LRR domain keeps Nlrc4 in inactive by sequestering it in a monomeric state. Interestingly, the two C-terminal WD40 domains of the inactive Apaf-1 also sterically occlude the monomeric Apaf-1 from oligomerization (Fig. 1f). HD2, also known as the arm domain, exists in all the mammalian NLRs and some other STAND proteins. Packing against one side of NBD in the closed form of Nlrc4 (Fig. 1a), HD2 is positioned to completely overlap with WHD in the active form of Nlrc4 that interacts with the HD1 from an adjacent protomer (Fig. 1d). Thus, HD2 exerts its inhibitory effects by steric blockage of both the WHD-HD1 and NBD-NBD interfaces in the oligomeric Nlrc4. In summary, several domain–domain interactions that may act cooperatively or even synergistically are involved in Nlrc4 autoinhibition. Given the striking effect generated by the $Nlrc4^{H443L}$ or $Nlrc4^{H443P}$ mutation (Hu et al. 2013; Kitamura et al. 2014), perturbation of the hydrogen formed between this residue and ADP could be a critical step to initiate Nlrc4 activation.

4 Mechanism of NAIP–NLRC4 Inflammasome Activation and Assembly

Two recent cryo-EM studies by our and Wu's groups provide the first glimpse of the mechanism underlying NAIP–NLRC4 inflammasome activation and assembly (Hu et al. 2015; Zhang et al. 2015). Biochemical studies from both groups showed that the Naip2–Nlrc4 complex induced by PrgJ (a rod protein of type III secretion system from *Salmonella typhimurium*) was substoichiometric with a dominating mount of Nlrc4. Further study using NTA-nanogold labeling EM indicated that only one Naip2 molecule was incorporated into the complex. In the cryo-EM structures, the complex formed a wheel-like architecture containing 10 or 11 protomers (Fig. 2a), reminiscent of the structure of Apaf-1 apoptosome (Yuan and Akey 2013; Zhou et al. 2015). Thus, the stoichiometry between Naip2 and Nlrc4 in the complex is 1:9 or 1:10. In addition to the wheel-like structures, the complex containing Naip2/5 and the CARD-deleted Nlrc4 (Nlrc4$^{\Delta CARD}$) also formed unclosed and twisted structures with variable protomers (Fig. 2b). For each type of the partially oligomerized particles, 2D class averages showed that the protomer at one end differed from the remaining copies, indicating that it corresponds to Naip2. This observation further supports the data from the nanogold labeling study and indicates an unidirectional propagation process for the assembly of the wheel-like structures.

Capturing of the partial oligomers starting with Naip2 suggests that Nlrc4 is capable of self-activation for assembly of the wheel-like structures. This conclusion is further supported by structural and biochemical data (Hu et al. 2015; Zhang et al. 2015). As discussed above, one oligomerization surface of Nlrc4 is completely blocked by the C-terminal LRR domain (Fig. 1d). This oligomerization surface (catalytic surface; Fig. 2c) is therefore activation-created because it can be formed only after structural re-organization during the Nlrc4 activation. In contrast, the other oligomerization surface is largely solvent-exposed and pre-exists in the inactive Nlrc4 (Fig. 2d). The two oligomerization surfaces are clearly discernible in the partially oligomerized structures with Naip2 contacting the receptor surface of Nlrc4 (Fig. 2b). These data led to a model of Nlrc4 self-activation in which Nlrc4, once activated by Naip2/5, unmasks its catalytic surface to interact with the solvent-exposed receptor surface of an inactive Nlrc4, consequently activating its. This would result in self-propagation of the active conformation of Nlrc4 and eventual assembly of the wheel-like structures (Fig. 2e). This model predicts that an Nlrc4 mutant with an impaired catalytic surface would not be able to self-activate, but can still be activated by a Naip protein because of its intact receptor surface. In strong support of this conclusion, the Nlrc4^{R288A} mutation located at the catalytic surface abolished Nlrc4-mediated caspase-1 activation, but the mutant protein still formed a flagellin-induced heterodimer with Naip5 (Hu et al. 2015). Furthermore, the mutation terminated oligomerization of wild-type Nlrc4 induced by flagellin and Naip5, forming partial oligomers as observed with the PrgJ–Naip2–Nlrc4$^{\Delta CARD}$ complex (Hu et al. 2015). Further evidence for the model comes from the

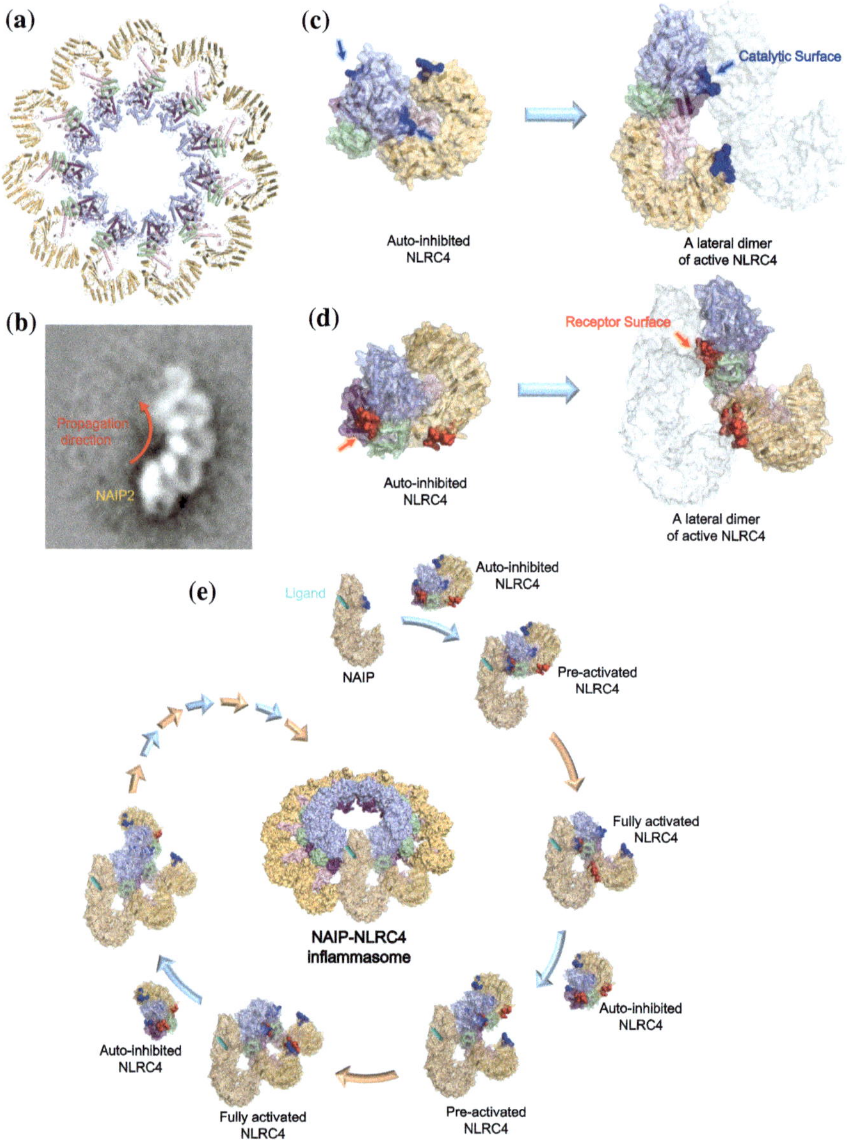

observation that the constitutively active Nlrc4^{H443L} mutant activated the wild-type Nlrc4 independent of Naip2/5 (Hu et al. 2015).

The self-activation model explains why only one Naip molecule is incorporated into the Naip–Nlrc4 inflammasomes. A Naip protein is needed to activate Nlrc4, exposing the catalytic receptor of Nlrc4 to initiate self-activation. Both the wheel-like and unclosed structures demonstrated that Naip2 occupies an equivalent position to Nlrc4, suggesting that Naip2 can use a similar mechanism to that of

◀ **Fig. 2** Assembly of NAIP–NLRC4 inflammasome through self-propagation. **a** The overall structure of oligomerized Nlrc4$^{\Delta CARD}$ (EMDB code: EMD-3141) shown in cartoon. Structural domains are labeled in the same color as indicated in Fig. 1a. **b** A representative 2D class average image of the partially oligomerized PrgJ–Naip2–Nlrc4$^{\Delta CARD}$ complex using negative-stain EM. The protomer at one end corresponding to Naip2 is indicated. The *red arrow* indicates the propagation of the partial oligmerized complex. **c** Structural remodeling creates the catalytic surface for oligomerization during the activation of Nlrc4. Structures are shown in surface, and the catalytic surface is shown in *blue* and indicated by *arrows*. The region involved in the LRR–LRR interaction is also shown in *blue*. **d** The receptor surface is largely solvent-exposed and pre-exists in the inactive Nlrc4. Structures are shown in surface, and the receptor surface is shown in *red* and indicated by *arrows*. The region involved in the LRR–LRR interaction is also shown in *red*. **e** A schematic diagram for the ligand-induced and self-propagated activation mechanism of NAIP–NLRC4 inflammasome

Nlrc4 self-activation for the activation of Nlrc4. Indeed, the residues from the catalytic surface of Nlrc4 are highly conserved among all the Naip members (Hu et al. 2015; Zhang et al. 2015). Mutations of the critical residues from the catalytic surface of Naip5 abrogated flagellin-induced interaction with Nlrc4, contrasting with Nlrc4^{R288A} from the catalytic surface of Nlrc4. In contrast, those from the receptor surface of Nlrc4 are highly variable in the Naip proteins. Thus, the receptor surface from a Naip protein does not match the catalytic surface of Nlrc4 or its own. This ensures that Naip members can neither self-oligomerize nor be further recruited into an existing Naip–Nlrc4 complex by an activated Nlrc4, resulting in incorporation of only one Naip molecule into one Naip–Nlrc4 inflammasome.

5 A Positive Role of the C-Terminal LRR Domain in the Activation of NAIP–NLRC4 Inflammasomes?

As mentioned above, the receptor surface of a Naip protein does not match the catalytic surface of Nlrc4. Then an ensuing question is how the Naip–Nlrc4 inflammasomes are closed to form wheel-like structures? The underlying mechanism remains unclear, but the Nlrc4^{CARD} clearly has an important role in this process, because its deletion resulted in the formation of unclosed oligomers that were not found with the full-length Nlrc4 (Hu et al. 2015; Zhang et al. 2015). The Nlrc4^{CARD} contributes to the closure of the inflammasomes likely through the formation of a circular structure formed in the hub of the wheel-like structure (Hu et al. 2015). Additionally, the ligands of a Naip could also have a role in the closure of the Naip–Nlrc4 inflammasomes. Clearly, higher-resolution structures of Naip–Nlrc4 inflammasomes are expected to address this question.

Available structural, biochemical, and cell-based data strongly support the notion that the C-terminal LRR domain is important to keep an NLR autoinhibited. Unexpectedly, however, the Nlrc4^{LRR} and Naip2^{LRR} also contribute to the assembly of Naip–Nlrc4 inflammasomes through electrostatic complementarity between two consecutive LRRs (Fig. 2c, d). This is in contrast to the WD40 domains of Apaf-1 in the formation of the apoptosome (Fig. 1f). It currently remains unknown whether the LRR–LRR interaction has a role in the full activation of the Naip–Nlrc4 inflammasomes. But the NOD module, while important to mediate Nlrc4 oligomerization, is not sufficient for assembly of the wheel-like structures of Naip–Nlrc4 inflammasomes as evidenced by the data from $\mathrm{Nlrc4^{CARD}}$ deletion (Hu et al. 2015; Zhang et al. 2015). Structurally, other domains, if presented to a proper position, could serve a similar role to an LRR domain in sequestering an NLR protein in a monomeric state, as only marginal interactions are formed between $\mathrm{Nlrc4^{LRR}}$ and $\mathrm{Nlrc4^{NBD}}$ in an inactive Nlrc4 (Fig. 1a). Mechanistically, the horse-shoed LRR domain appears optimal to act as a structural fold for closure of the inflammasomes because of its curvature. In agreement with this possibility, a more recent study showed that fusion of a 76-kDa protein to the C-terminus of Naip5 resulted in the formation of a helical structure of the flagellin–Naip5–Nlrc4 complex (Diebolder et al. 2015). Thus, it appears that the $\mathrm{Naip^{LRR}}$ and $\mathrm{Nlrc4^{LRR}}$ might have a dual role in the full activation of the Naip–Nlrc4 inflammasomes. Interestingly, a similar role has been proposed for the PYD of AIM2, which can exert its auto-inhibitory effect in the resting state of AIM2 and also stabilize the dsDNA-binding conformation of AIM2 via intermolecular PYD–PYD interaction upon activation (Xiao 2015).

A potential role of the $\mathrm{Nlrc4^{LRR}}$ in inflammasome activation is not necessarily contradictory with the observation that its deletion led to constitutively active Nlrc4 in caspase-1 activation (Hu et al. 2013; Kofoed and Vance 2011; Poyet et al. 2001). One plausible reason for this may be that the Nlrc4 mutant is only partially active because some of the mutant protein is unable to assemble into wheel-like structures required for caspase-1 activation. Quantitative assays of the caspase-1 activating activity of wild-type Nlrc4 and the mutant would be useful to test this possibility. Alternatively or additionally, it is also conceivable that unclosed oligomers formed by the Nlrc4 mutant can still bring caspase-1 molecules into proximity for activation. This is particularly possible in the presence of ASC, because overexpression of the $\mathrm{Nlrc4^{CARD}}$ alone induced ASC into filaments to activate caspase-1 (Cai et al. 2014; Zhang et al. 2015). Whether this is associated with ASC-promoted caspase-1 activation by the Naip–Nlrc4 inflammasomes (Broz et al. 2010) remains unknown. But many interesting questions could be raised by this possibility. For example, what are the functional differences between the closed and unclosed Naip–Nlrc4 inflammasomes? Is it a possible way of regulating the activity of Naip–Nlrc4 inflammasomes in vivo?

6 Activation of Other NLR Inflammasomes

A number of single gain-of-function mutations in Nlrc4 including the disease-related ones have been identified (Canna et al. 2014; Kitamura et al. 2014; Romberg et al. 2014). All these mutations were mapped to destabilize the inactive conformation of Nlrc4, resulting in the activation of caspase-1. The self-propagation property of Nlrc4 suggests that a small population of fully activated Nlrc4 molecules resulting from the mutations can induce assembly of immunoactive Nlrc4 homo-oligomers by activating the mutation-destabilized inactive Nlrc4. A similar mechanism in principle could be utilized by the gain-of-function and disease-associated mutations for the activation of NLRP3 inflammasome (Schroder and Tschopp 2010). Consistently, modeling studies suggested that some of the NLRP3 mutations directly lead to relief of autoinhibition and constitutive activation of NLRP3 (Albrecht et al. 2003; Chai and Shi 2013). However, further studies are required to determine whether NLRP3 or other immune NLR proteins also possesses self-propagation activity. But such an activity would endow them with high sensitivity to PAMPs or DAMPs and efficiency for signaling, allowing quick response of hosts to invading pathogens. The NOD module should be responsible for the self-propagation activity of Nlrc4, as the Nlrc4$^{\Delta CARD}$ mutant still formed Naip-induced oligomers with multiple Nlrc4 s. Given the highly conserved NOD among NLRs, it appears reasonable to assume that other NLRs may also have a similar activity to Nlrc4. Naips and Apaf-1, however, are known not to be the case. The two buried oligomerization surfaces in an inactive Apaf-1 rule out the possibility of self-activation. No structural information is available for a Naip protein yet. But it is expected to follow the conserved oligomerization mode of the STAND proteins, in which one side of the NBD from the one protomer stacking against the opposite side of the NBD from the other protomer in a lateral dimer. Structure-based sequence alignment showed that the two oligomerization surfaces of a Naip protein do not match each other for homo-oligomerization, thus precluding it from homo-oligomerization.

The self-propagation activity of Nlrc4 is induced by a Naip protein. However, other inflammasomes containing two different NLRs have not been biochemically demonstrated. Thus, for an NLR to self-propagate, either an as-yet identified paired NLR(s) similar to Naips is needed or the single NLR can act as both the sensor and the self-propagating protomer. To probe the former possibility, sequence alignment could provide useful information, as homology between Naips and Nlrc4 is important for their heterodimerization. Some tangent evidence from the studies of Nlrp1b activation seems consistent with the latter possibility. Data from several studies indicated that mouse and rat Nlrp1b sense the protease activity of lethal toxin by acting as substrates to activate the Nlrp1b inflammasome (Chavarria-Smith and Vance 2013; Hellmich et al. 2012; Levinsohn et al. 2012). Cleavage of Nlrp1b by LeTx is required for caspase-1 activation in non-macrophage cells, and a strict correlation between cleavage sensitivity and the activation of Nlrp1b inflammasome has been observed (Chavarria-Smith and Vance 2013). However, only a very small fraction of Nlrp1b was cleaved within 90 min, and at this time point, mouse macrophages showed robust immune responses (Boyden and Dietrich 2006; Chavarria-Smith and Vance 2013; Muehlbauer et al. 2007). While several mechanisms can be formulated to explain these results, self-activation would be the most straightforward one.

7 Insights into the Assembly of ASC-Dependent Inflammasomes

ASC as a core component of inflammasomes contains an N-terminal PYD and C-terminal CARD (de Alba 2009). The ASC^{PYD} can interact with those from almost all the PYD-containing inflammasomes, and the ASC^{CARD} can recruit the downstream pro-caspase-1 through homotypic CARD–CARD interactions (Poyet et al. 2001). In the CARD-containing inflammasomes, such as the Naip–Nlrc4 and NLRP1b inflammasomes, the CARDs from them are believed to mediate the direct recruitment of pro-caspase-1 for activation. In the absence of ASC, the Naip–Nlrc4 and NLRP1b inflammasomes induce caspase-1-mediated pyroptosis, but interestingly fail to induce caspase-1 auto-proteolysis (Broz et al. 2010; Van Opdenbosch et al. 2014). In the presence of ASC, Nlrc4-mediated caspase-1 proteolytic processing is significantly promoted (Broz et al. 2010).

Recent studies provide significant insight into the mechanism underlying assembly of ASC-dependent inflammasomes (Fig. 3) (Cai et al. 2014; Lu et al. 2014; Sborgi et al. 2015). The upstream sensors NLRs or ALRs form a high-order oligomer upon the recognition of their ligands, leading to the clustering of their PYDs or CARDs (Hu et al. 2015; Jin et al. 2012; Morrone et al. 2015; Zhang et al. 2015). The clustered PYDs or CARDs can serve as a platform to recruit the adaptor ASC via PYD–PYD or CARD–CARD interactions and nucleate the self-propagated filament formation of monomeric ASC^{PYD}. Structural studies of the ASC^{PYD} filament showed a cylinder-like structure with the ASC^{PYD} packing densely in a spiral (Lu et al. 2014; Sborgi et al. 2015). The ASC^{CARD}, which localizes at the outer layer of the ASC^{PYD} filament, forms another platform to nucleate the formation of caspase-1 filaments through CARD–CARD interactions (Lu et al. 2014). The highly polymerized caspase-1 could further lead to the proximity-induced caspase-1 auto-processing, resulting in the full activation of ASC-dependent inflammasomes. Consistently, the reconstituted $AIM2^{PYD}$–ASC–caspase-1 ternary complex showed a one-by-one increased stoichiometry, with $AIM2^{PYD}$ under stoichiometric to ASC and ASC under stoichiometric to caspase-1 (Lu et al. 2014). EM studies of the AIM2–ASC complex revealed star-shaped structures in which $AIM2^{PYD}$ and ASC form the central hub, whereas caspase-1 forms multiple filaments extended radially from the center (Lu et al. 2014). These results support the observations ASC formed micron-sized punctum or specks associated with upstream sensors and downstream caspases upon inflammasome activation (Masumoto et al. 1999) and a rapid "all-or-none" response generated by ASC oligomerization.

Fig. 3 Schematic diagram for the assembly of ASC-dependent inflammasomes. ▶ Upstream-sensing proteins (NLRs or AIM2) oligomerize upon activation, forming a platform to recruit the adaptor protein ASC. The nucleated ASC can promote the ASC filament formation through a self-propagated manner, leading to the clustering of the CARD of ASC, which can further promote caspase-1 filament formation and proximity-induced activation

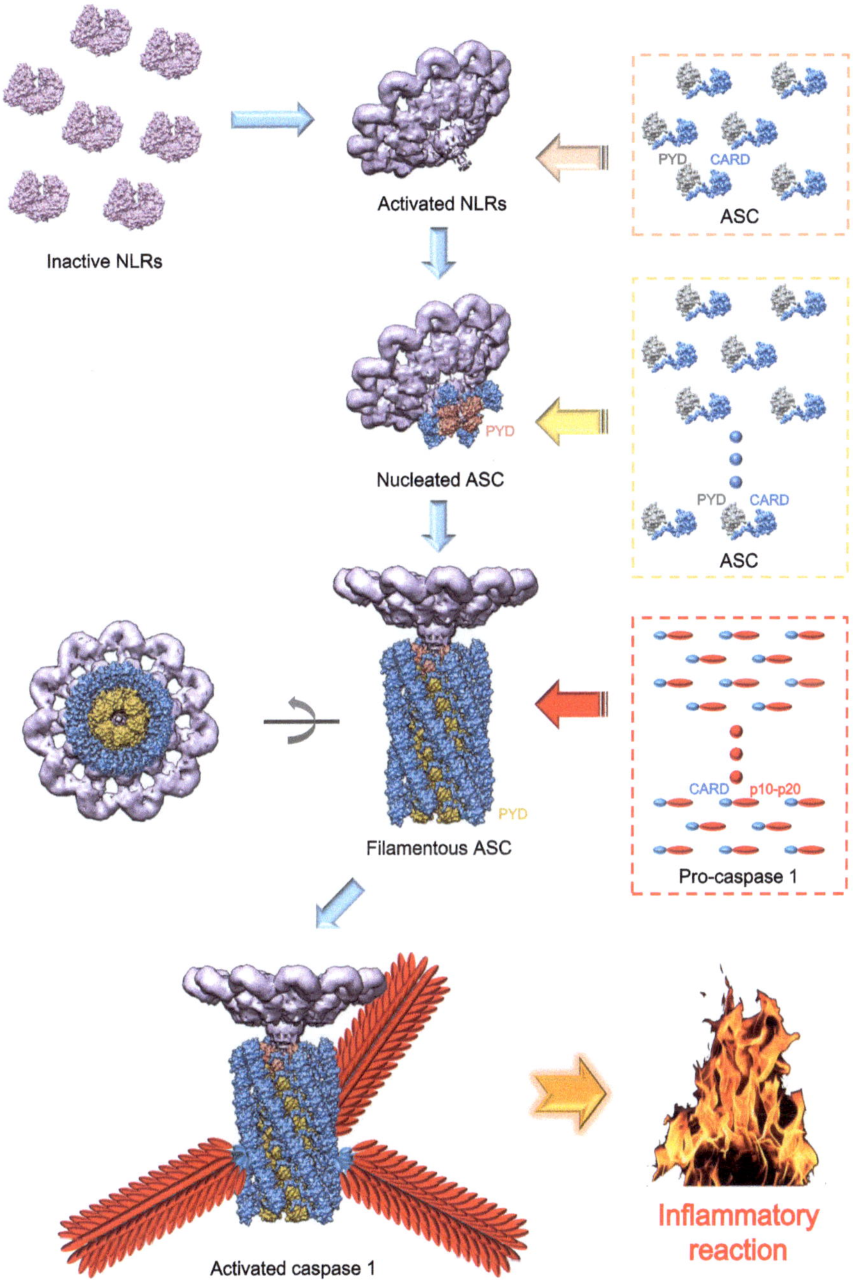
Inactive NLRs
Activated NLRs
PYD
CARD
ASC
Nucleated ASC
PYD
PYD
CARD
ASC
Filamentous ASC
PYD
CARD
p10-p20
Pro-caspase 1
Activated caspase 1
Inflammatory
reaction

8 Concluding Remarks

Although significant progress has been made toward inflammasomes during the past few years, our understanding on their assembly and activation is far from being complete. At present, only a few inflammasomes have been successfully purified or reconstituted using homogeneous recombinant proteins, which is an important step in understanding the structural and functional mechanism of inflammasomes. Ligand-induced oligomerization of an NLR is widely believed to be an important step for the formation of NLR inflammasomes. Future structural studies are required to investigate whether a wheel-like structure is also formed in other NLR inflammasomes. A few NLRs have been shown to directly recognize their cognate ligands, but the underlying structural mechanisms are largely unknown. A recent study demonstrated that the structural determinants for the recognition of bacterial PAMPs by Naips lie in the central NOD module rather than the C-terminal LRR domain as previously hypothesized (Tenthorey et al. 2014). Whether this holds true with ligand recognition by other NLRs is unknown. Moreover, the possibility still remains that the LRR also has a role in ligand recognition by Naips. Recent studies showed that a member of NIMA-related kinases (NEK7) interacts with NLRP3 and acts as an essential component downstream of potassium efflux for NLRP3 inflammasome assembly and activation, which interestingly do not require the kinase activity of NEK7 (He et al. 2016; Schmid-Burgk et al. 2016; Shi et al. 2015). How the interaction with NEK7 activates NLRP3 remains unknown. ATP binding was shown to be important for the activation of several NLRs, such as NLRP3 (Duncan et al. 2007), human NLRP1 (Faustin et al. 2007), and NOD1/2 (Strober et al. 2006), but mutation of the ATP-binding motif in mouse Nlrp1b resulted in constitutive activation of the Nlrp1b inflammasome (Liao and Mogridge 2013). Therefore, the differential roles of ATP in the assembly of NLR inflammasomes still remain to be further scrutinized.

Formation of filamentous structures by ASC is emerging as a general mechanism for the assembly and signal transduction of the ASC-dependent inflammasomes (Kagan et al. 2014; Lu and Wu 2015). The PYDs of an oligomerized NLR or ALR should function to seed the filaments, but the underlying structural mechanism is currently unclear. Given the high diversity of the ASC-dependent inflammasomes, it is hard to imagine that PYDs form a conserved structure in the oligomerized NLRs or ALRs for the seeding activity. Another challenge is that how ASC filaments activate caspase-1. Caspase-1 recruitment by the ASC filaments through the homotypic CARD–CARD interaction presumably brings caspase-1 molecules into close proximity for activation. But the model does not reveal the molecular mechanism underlying caspase-1 activation. The pro-inflammatory caspases, mouse caspase-11 and human caspase-4/5, directly recognize cytoplasmic LPS, forming non-canonical inflammasomes (Shi et al. 2014). Structural studies in conjunction with other biophysical approaches are needed to elucidate how these caspases recognize LPS and consequently become activated. In contrast to other NLRs, activated NOD1/2 recruits receptor-interacting serine/threonine protein kinase 2

(RIPK2) through homotypic CARD–CARD interactions, resulting in close proximity and activation of RIPK2-IκB kinase (IKK) (Caruso et al. 2014; Strober et al. 2006). Future studies are required to determine whether these two NLRs use similar principles to other NLR for kinase activation.

References

Agostini L, Martinon F, Burns K, McDermott MF, Hawkins PN, Tschopp J (2004) NALP3 forms an IL-1beta-processing inflammasome with increased activity in Muckle-Wells autoinflammatory disorder. Immunity 20(3):319–325. doi:10.1016/S1074-7613(04)00046-9

Albrecht M, Domingues FS, Schreiber S, Lengauer T (2003) Structural localization of disease-associated sequence variations in the NACHT and LRR domains of PYPAF1 and NOD2. FEBS Lett 554(3):520–528. doi:10.1016/S0014-5793(03)01222-5

Bauernfeind FG, Horvath G, Stutz A, Alnemri ES, MacDonald K, Speert D, Fernandes-Alnemri T, Wu J, Monks BG, Fitzgerald KA, Hornung V, Latz E (2009) Cutting edge: NF-kappaB activating pattern recognition and cytokine receptors license NLRP3 inflammasome activation by regulating NLRP3 expression. J Immunol 183(2):787–791. doi:10.4049/jimmunol.0901363

Boehm T, Iwanami N, Hess I (2012) Evolution of the immune system in the lower vertebrates. Annu Rev Genomics Hum Genet 13:127–149. doi:10.1146/annurev-genom-090711-163747

Boyden ED, Dietrich WF (2006) Nalp1b controls mouse macrophage susceptibility to anthrax lethal toxin. Nat Genet 38(2):240–244. doi:10.1038/ng1724

Broz P, von Moltke J, Jones JW, Vance RE, Monack DM (2010) Differential requirement for Caspase-1 autoproteolysis in pathogen-induced cell death and cytokine processing. Cell Host Microbe 8(6):471–483. doi:10.1016/j.chom.2010.11.007

Burckstummer T, Baumann C, Bluml S, Dixit E, Durnberger G, Jahn H, Planyavsky M, Bilban M, Colinge J, Bennett KL, Superti-Furga G (2009) An orthogonal proteomic-genomic screen identifies AIM2 as a cytoplasmic DNA sensor for the inflammasome. Nat Immunol 10(3):266–272. doi:10.1038/ni.1702

Cai X, Chen J, Xu H, Liu S, Jiang QX, Halfmann R, Chen ZJ (2014) Prion-like polymerization underlies signal transduction in antiviral immune defense and inflammasome activation. Cell 156(6):1207–1222. doi:10.1016/j.cell.2014.01.063

Canna SW, de Jesus AA, Gouni S, Brooks SR, Marrero B, Liu Y, DiMattia MA, Zaal KJ, Sanchez GA, Kim H, Chapelle D, Plass N, Huang Y, Villarino AV, Biancotto A, Fleisher TA, Duncan JA, O'Shea JJ, Benseler S, Grom A, Deng Z, Laxer RM, Goldbach-Mansky R (2014) An activating NLRC4 inflammasome mutation causes autoinflammation with recurrent macrophage activation syndrome. Nat Genet 46(10):1140–1146. doi:10.1038/ng.3089

Caruso R, Warner N, Inohara N, Núñez G (2014) NOD1 and NOD2: Signaling, host defense, and inflammatory disease. Immunity 41(6):898–908. doi:10.1016/j.immuni.2014.12.010

Chai J, Shi Y (2013) Apoptosome and inflammasome: conserved machineries for caspase activation. Nat Sci Rev 1(1):101–118. doi:10.1093/nsr/nwt025

Chavarria-Smith J, Vance RE (2013) Direct proteolytic cleavage of NLRP1B is necessary and sufficient for inflammasome activation by anthrax lethal factor. PLoS Pathog 9(6):e1003452. doi:10.1371/journal.ppat.1003452

Danot O, Marquenet E, Vidal-Ingigliardi D, Richet E (2009) Wheel of life, wheel of death: a mechanistic insight into signaling by STAND proteins. Structure 17(2):172–182. doi:10.1016/j.str.2009.01.001

Davis BK, Wen H, Ting JP (2011) The inflammasome NLRs in immunity, inflammation, and associated diseases. Annu Rev Immunol 29:707–735. doi:10.1146/annurev-immunol-031210-101405

de Alba E (2009) Structure and interdomain dynamics of apoptosis-associated speck-like protein containing a CARD (ASC). J Biol Chem 284(47):32932–32941. doi:10.1074/jbc.M109.024273
Diebolder CA, Halff EF, Koster AJ, Huizinga EG, Koning RI (2015) Cryoelectron tomography of the NAIP5/NLRC4 inflammasome: implications for NLR activation. Structure 23(12):2349–2357. doi:10.1016/j.str.2015.10.001
Dostert C, Petrilli V, Van Bruggen R, Steele C, Mossman BT, Tschopp J (2008) Innate immune activation through Nalp3 inflammasome sensing of asbestos and silica. Science 320 (5876):674–677. doi:10.1126/science.1156995
Duncan JA, Bergstralh DT, Wang Y, Willingham SB, Ye Z, Zimmermann AG, Ting JP (2007) Cryopyrin/NALP3 binds ATP/dATP, is an ATPase, and requires ATP binding to mediate inflammatory signaling. Proc Natl Acad Sci USA 104(19):8041–8046. doi:10.1073/pnas.0611496104
Eisenbarth SC, Colegio OR, O'Connor W, Sutterwala FS, Flavell RA (2008) Crucial role for the Nalp3 inflammasome in the immunostimulatory properties of aluminium adjuvants. Nature 453 (7198):1122–1126. doi:10.1038/nature06939
Elinav E, Strowig T, Kau AL, Henao-Mejia J, Thaiss CA, Booth CJ, Peaper DR, Bertin J, Eisenbarth SC, Gordon JI, Flavell RA (2011) NLRP6 inflammasome regulates colonic microbial ecology and risk for colitis. Cell 145(5):745–757. doi:10.1016/j.cell.2011.04.022
Faustin B, Lartigue L, Bruey JM, Luciano F, Sergienko E, Bailly-Maitre B, Volkmann N, Hanein D, Rouiller I, Reed JC (2007) Reconstituted NALP1 inflammasome reveals two-step mechanism of caspase-1 activation. Mol Cell 25(5):713–724. doi:10.1016/j.molcel.2007.01.032
Fernandes-Alnemri T, Yu JW, Datta P, Wu J, Alnemri ES (2009) AIM2 activates the inflammasome and cell death in response to cytoplasmic DNA. Nature 458(7237):509–513. doi:10.1038/nature07710
Fink SL, Bergsbaken T, Cookson BT (2008) Anthrax lethal toxin and Salmonella elicit the common cell death pathway of caspase-1-dependent pyroptosis via distinct mechanisms. Proc Natl Acad Sci USA 105(11):4312–4317. doi:10.1073/pnas.0707370105
Flajnik MF, Du Pasquier L (2004) Evolution of innate and adaptive immunity: can we draw a line? Trends Immunol 25(12): 640–644. doi:10.1016/j.it.2004.10.001
Franchi L, Amer A, Body-Malapel M, Kanneganti TD, Ozoren N, Jagirdar R, Inohara N, Vandenabeele P, Bertin J, Coyle A, Grant EP, Nunez G (2006) Cytosolic flagellin requires Ipaf for activation of caspase-1 and interleukin 1beta in salmonella-infected macrophages. Nat Immunol 7(6):576–582. doi:10.1038/ni1346
Hagar JA, Powell DA, Aachoui Y, Ernst RK, Miao EA (2013) Cytoplasmic LPS activates caspase-11: implications in TLR4-independent endotoxic shock. Science 341(6151):1250–1253. doi:10.1126/science.1240988
Halff EF, Diebolder CA, Versteeg M, Schouten A, Brondijk TH, Huizinga EG (2012) Formation and structure of a NAIP5-NLRC4 inflammasome induced by direct interactions with conserved N- and C-terminal regions of flagellin. J Biol Chem 287(46):38460–38472. doi:10.1074/jbc.M112.393512
He Y, Zeng MY, Yang D, Motro B, Núñez G (2016) NEK7 is an essential mediator of NLRP3 activation downstream of potassium efflux. Nature 530(7590):354–357. doi:10.1038/nature16959
Hellmich KA, Levinsohn JL, Fattah R, Newman ZL, Maier N, Sastalla I, Liu S, Leppla SH, Moayeri M (2012) Anthrax lethal factor cleaves mouse nlrp1b in both toxin-sensitive and toxin-resistant macrophages. PLoS One 7(11):e49741. doi:10.1371/journal.pone.0049741
Hornung V, Ablasser A, Charrel-Dennis M, Bauernfeind F, Horvath G, Caffrey DR, Latz E, Fitzgerald KA (2009) AIM2 recognizes cytosolic dsDNA and forms a caspase-1-activating inflammasome with ASC. Nature 458(7237):514–518. doi:10.1038/nature07725
Hu Z, Yan C, Liu P, Huang Z, Ma R, Zhang C, Wang R, Zhang Y, Martinon F, Miao D, Deng H, Wang J, Chang J, Chai J (2013) Crystal structure of NLRC4 reveals its autoinhibition mechanism. Science 341(6142):172–175. doi:10.1126/science.1236381

Hu Z, Zhou Q, Zhang C, Fan S, Cheng W, Zhao Y, Shao F, Wang HW, Sui SF, Chai J (2015) Structural and biochemical basis for induced self-propagation of NLRC4. Science 350 (6259):399–404. doi:10.1126/science.aac5489

Ichinohe T, Lee HK, Ogura Y, Flavell R, Iwasaki A (2009) Inflammasome recognition of influenza virus is essential for adaptive immune responses. J Exp Med 206(1):79–87. doi:10.1084/jem.20081667

Janeway CJ, Medzhitov R (2002) Innate immune recognition. Annu Rev Immunol 20:197–216. doi:10.1146/annurev.immunol.20.083001.084359

Jin T, Perry A, Jiang J, Smith P, Curry JA, Unterholzner L, Jiang Z, Horvath G, Rathinam VA, Johnstone RW, Hornung V, Latz E, Bowie AG, Fitzgerald KA, Xiao TS (2012) Structures of the HIN domain:DNA complexes reveal ligand binding and activation mechanisms of the AIM2 inflammasome and IFI16 receptor. Immunity 36(4):561–571. doi:10.1016/j.immuni.2012.02.014

Joly S, Sutterwala FS (2010) Fungal pathogen recognition by the NLRP3 inflammasome. Virulence 1(4):276–280. doi:10.4161/viru.1.4.11482

Jones JD, Dangl JL (2006) The plant immune system. Nature 444(7117):323–329. doi:10.1038/nature05286

Kagan JC, Magupalli VG, Wu H (2014) SMOCs: supramolecular organizing centres that control innate immunity. Nat Rev Immunol 14(12):821–826. doi:10.1038/nri3757

Kanneganti TD, Ozoren N, Body-Malapel M, Amer A, Park JH, Franchi L, Whitfield J, Barchet W, Colonna M, Vandenabeele P, Bertin J, Coyle A, Grant EP, Akira S, Nunez G (2006) Bacterial RNA and small antiviral compounds activate caspase-1 through cryopyrin/Nalp3. Nature 440(7081):233–236. doi:10.1038/nature04517

Kayagaki N, Wong MT, Stowe IB, Ramani SR, Gonzalez LC, Akashi-Takamura S, Miyake K, Zhang J, Lee WP, Muszynski A, Forsberg LS, Carlson RW, Dixit VM (2013) Noncanonical inflammasome activation by intracellular LPS independent of TLR4. Science 341(6151):1246–1249. doi:10.1126/science.1240248

Kerur N, Veettil MV, Sharma-Walia N, Bottero V, Sadagopan S, Otageri P, Chandran B (2011) IFI16 acts as a nuclear pathogen sensor to induce the inflammasome in response to Kaposi Sarcoma-associated herpesvirus infection. Cell Host Microbe 9(5):363–375. doi:10.1016/j.chom.2011.04.008

Khare S, Dorfleutner A, Bryan NB, Yun C, Radian AD, de Almeida L, Rojanasakul Y, Stehlik C (2012) An NLRP7-containing inflammasome mediates recognition of microbial lipopeptides in human macrophages. Immunity 36(3):464–476. doi:10.1016/j.immuni.2012.02.001

Kimbrell DA, Beutler B (2001) The evolution and genetics of innate immunity. Nat Rev Genet 2 (4):256–267. doi:10.1038/35066006

Kitamura A, Sasaki Y, Abe T, Kano H, Yasutomo K (2014) An inherited mutation in NLRC4 causes autoinflammation in human and mice. J Exp Med 211(12):2385–2396. doi:10.1084/jem.20141091

Kofoed EM, Vance RE (2011) Innate immune recognition of bacterial ligands by NAIPs determines inflammasome specificity. Nature 477(7366):592–595. doi:10.1038/nature10394

Lamkanfi M, Dixit VM (2012) Inflammasomes and their roles in health and disease. Annu Rev Cell Dev Biol 28:137–161. doi:10.1146/annurev-cellbio-101011-155745

Lamkanfi M, Dixit VM (2014) Mechanisms and functions of inflammasomes. Cell 157(5):1013–1022. doi:10.1016/j.cell.2014.04.007

Levinsohn JL, Newman ZL, Hellmich KA, Fattah R, Getz MA, Liu S, Sastalla I, Leppla SH, Moaycri M (2012) Anthrax lethal factor cleavage of Nlrp1 is required for activation of the inflammasome. PLoS Pathog 8(3):e1002638. doi:10.1371/journal.ppat.1002638

Liao KC, Mogridge J (2013) Activation of the Nlrp1b inflammasome by reduction of cytosolic ATP. Infect Immun 81(2):570–579. doi:10.1128/IAI.01003-12

Lightfield KL, Persson J, Brubaker SW, Witte CE, von Moltke J, Dunipace EA, Henry T, Sun YH, Cado D, Dietrich WF, Monack DM, Tsolis RM, Vance RE (2008) Critical function for Naip5 in inflammasome activation by a conserved carboxy-terminal domain of flagellin. Nat Immunol 9(10):1171–1178. doi:10.1038/ni.1646

Lu A, Wu H (2015) Structural mechanisms of inflammasome assembly. FEBS J 282(3):435–444. doi:10.1111/febs.13133

Lu A, Magupalli VG, Ruan J, Yin Q, Atianand MK, Vos MR, Schroder GF, Fitzgerald KA, Wu H, Egelman EH (2014) Unified polymerization mechanism for the assembly of ASC-dependent inflammasomes. Cell 156(6):1193–1206. doi:10.1016/j.cell.2014.02.008

Maekawa T, Kufer TA, Schulze-Lefert P (2011) NLR functions in plant and animal immune systems: so far and yet so close. Nat Immunol 12(9):817–826. doi:10.1038/ni.2083

Mariathasan S, Newton K, Monack DM, Vucic D, French DM, Lee WP, Roose-Girma M, Erickson S, Dixit VM (2004) Differential activation of the inflammasome by caspase-1 adaptors ASC and Ipaf. Nature 430(6996):213–218. doi:10.1038/nature02664

Mariathasan S, Weiss DS, Newton K, McBride J, O'Rourke K, Roose-Girma M, Lee WP, Weinrauch Y, Monack DM, Dixit VM (2006) Cryopyrin activates the inflammasome in response to toxins and ATP. Nature 440(7081):228–232. doi:10.1038/nature04515

Martinon F, Burns K, Tschopp J (2002) The inflammasome: a molecular platform triggering activation of inflammatory caspases and processing of proIL-beta. Mol Cell 10(2):417–426. doi:10.1016/S1097-2765(02)00599-3

Martinon F, Petrilli V, Mayor A, Tardivel A, Tschopp J (2006) Gout-associated uric acid crystals activate the NALP3 inflammasome. Nature 440(7081):237–241. doi:10.1038/nature04516

Martinon F, Mayor A, Tschopp J (2009) The inflammasomes: guardians of the body. Annu Rev Immunol 27:229–265. doi:10.1146/annurev.immunol.021908.132715

Masumoto J, Taniguchi S, Ayukawa K, Sarvotham H, Kishino T, Niikawa N, Hidaka E, Katsuyama T, Higuchi T, Sagara J (1999) ASC, a novel 22-kDa protein, aggregates during apoptosis of human promyelocytic leukemia HL-60 cells. J Biol Chem 274(48):33835–33838. doi:10.1074/jbc.274.48.33835

Medzhitov R, Janeway CJ (1997) Innate immunity: the virtues of a nonclonal system of recognition. Cell 91(3):295–298. doi:10.1016/S0092-8674(00)80412-2

Miao EA, Alpuche-Aranda CM, Dors M, Clark AE, Bader MW, Miller SI, Aderem A (2006) Cytoplasmic flagellin activates caspase-1 and secretion of interleukin 1beta via Ipaf. Nat Immunol 7(6):569–575. doi:10.1038/ni1344

Miao EA, Mao DP, Yudkovsky N, Bonneau R, Lorang CG, Warren SE, Leaf IA, Aderem A (2010) Innate immune detection of the type III secretion apparatus through the NLRC4 inflammasome. Proc Natl Acad Sci USA 107(7):3076–3080. doi:10.1073/pnas.0913087107

Minkiewicz J, de Rivero VJ, Keane RW (2013) Human astrocytes express a novel NLRP2 inflammasome. Glia 61(7):1113–1121. doi:10.1002/glia.22499

Moayeri M, Crown D, Newman ZL, Okugawa S, Eckhaus M, Cataisson C, Liu S, Sastalla I, Leppla SH (2010) Inflammasome sensor Nlrp1b-dependent resistance to anthrax is mediated by caspase-1, IL-1 signaling and neutrophil recruitment. PLoS Pathog 6(12):e1001222. doi:10.1371/journal.ppat.1001222

Morrone SR, Matyszewski M, Yu X, Delannoy M, Egelman EH, Sohn J (2015) Assembly-driven activation of the AIM2 foreign-dsDNA sensor provides a polymerization template for downstream ASC. Nat Commun 6:7827. doi:10.1038/ncomms8827

Muehlbauer SM, Evering TH, Bonuccelli G, Squires RC, Ashton AW, Porcelli SA, Lisanti MP, Brojatsch J (2007) Anthrax lethal toxin kills macrophages in a strain-specific manner by apoptosis or caspase-1-mediated necrosis. Cell Cycle 6(6):758–766. doi:10.4161/cc.6.6.3991

Mukherjee S, Hooper LV (2015) Antimicrobial defense of the intestine. Immunity 42(1):28–39. doi:10.1016/j.immuni.2014.12.028

Munoz-Planillo R, Kuffa P, Martinez-Colon G, Smith BL, Rajendiran TM, Nunez G (2013) K(+) efflux is the common trigger of NLRP3 inflammasome activation by bacterial toxins and particulate matter. Immunity 38(6):1142–1153. doi:10.1016/j.immuni.2013.05.016

Petrilli V, Papin S, Dostert C, Mayor A, Martinon F, Tschopp J (2007) Activation of the NALP3 inflammasome is triggered by low intracellular potassium concentration. Cell Death Differ 14 (9):1583–1589. doi:10.1038/sj.cdd.4402195

Poyet JL, Srinivasula SM, Tnani M, Razmara M, Fernandes-Alnemri T, Alnemri ES (2001) Identification of Ipaf, a human caspase-1-activating protein related to Apaf-1. J Biol Chem 276 (30):28309–28313. doi:10.1074/jbc.C100250200

Reubold TF, Wohlgemuth S, Eschenburg S (2011) Crystal structure of full-length Apaf-1: how the death signal is relayed in the mitochondrial pathway of apoptosis. Structure 19(8):1074–1083. doi:10.1016/j.str.2011.05.013

Riedl SJ, Li W, Chao Y, Schwarzenbacher R, Shi Y (2005) Structure of the apoptotic protease-activating factor 1 bound to ADP. Nature 434(7035):926–933. doi:10.1038/nature03465

Roberts TL, Idris A, Dunn JA, Kelly GM, Burnton CM, Hodgson S, Hardy LL, Garceau V, Sweet MJ, Ross IL, Hume DA, Stacey KJ (2009) HIN-200 proteins regulate caspase activation in response to foreign cytoplasmic DNA. Science 323(5917):1057–1060. doi:10.1126/science.1169841

Romberg N, Al MK, Nelson-Williams C, Stiegler AL, Loring E, Choi M, Overton J, Meffre E, Khokha MK, Huttner AJ, West B, Podoltsev NA, Boggon TJ, Kazmierczak BI, Lifton RP (2014) Mutation of NLRC4 causes a syndrome of enterocolitis and autoinflammation. Nat Genet 46(10):1135–1139. doi:10.1038/ng.3066

Sborgi L, Ravotti F, Dandey VP, Dick MS, Mazur A, Reckel S, Chami M, Scherer S, Huber M, Bockmann A, Egelman EH, Stahlberg H, Broz P, Meier BH, Hiller S (2015) Structure and assembly of the mouse ASC inflammasome by combined NMR spectroscopy and cryo-electron microscopy. Proc Natl Acad Sci USA 112(43):13237–13242. doi:10.1073/pnas.1507579112

Schmid-Burgk JL, Chauhan D, Schmidt T, Ebert TS, Reinhardt J, Endl E, Hornung V (2016) A Genome-wide CRISPR (Clustered Regularly Interspaced Short Palindromic Repeats) Screen Identifies NEK7 as an Essential Component of NLRP3 Inflammasome Activation. J Biol Chem 291(1):103–109. doi:10.1074/jbc.C115.700492

Schroder K, Tschopp J (2010) The inflammasomes. Cell 140(6):821–832. doi:10.1016/j.cell.2010.01.040

Shaw N, Liu ZJ (2014) Role of the HIN domain in regulation of innate immune responses. Mol Cell Biol 34(1):2–15. doi:10.1128/MCB.00857-13

Shi J, Zhao Y, Wang Y, Gao W, Ding J, Li P, Hu L, Shao F (2014) Inflammatory caspases are innate immune receptors for intracellular LPS. Nature 514(7521):187–192. doi:10.1038/nature13683

Shi H, Wang Y, Li X, Zhan X, Tang M, Fina M, Su L, Pratt D, Bu CH, Hildebrand S, Lyon S, Scott L, Quan J, Sun Q, Russell J, Arnett S, Jurek P, Chen D, Kravchenko VV, Mathison JC, Moresco EM, Monson NL, Ulevitch RJ, Beutler B (2015) NLRP3 activation and mitosis are mutually exclusive events coordinated by NEK7, a new inflammasome component. Nat Immunoldoi. doi:10.1038/ni.3333

Strober W, Murray PJ, Kitani A, Watanabe T (2006) Signalling pathways and molecular interactions of NOD1 and NOD2. Nat Rev Immunol 6(1):9–20. doi:10.1038/nri1747

Strowig T, Henao-Mejia J, Elinav E, Flavell R (2012) Inflammasomes in health and disease. Nature 481(7381):278–286. doi:10.1038/nature10759

Takeuchi O, Akira S (2010) Pattern recognition receptors and inflammation. Cell 140(6):805–820. doi:10.1016/j.cell.2010.01.022

Takken FL, Albrecht M, Tameling WI (2006) Resistance proteins: molecular switches of plant defence. Curr Opin Plant Biol 9(4):383–390. doi:10.1016/j.pbi.2006.05.009

Tenthorey JL, Kofoed EM, Daugherty MD, Malik HS, Vance RE (2014) Molecular basis for specific recognition of bacterial ligands by NAIP/NLRC4 inflammasomes. Mol Cell 54(1):17–29. doi:10.1016/j.molcel.2014.02.018

Terra JK, Cote CK, France B, Jenkins AL, Bozue JA, Welkos SL, LeVine SM, Bradley KA (2010) Cutting edge: resistance to Bacillus anthracis infection mediated by a lethal toxin sensitive allele of Nalp1b/Nlrp1b. J Immunol 184(1):17–20. doi:10.4049/jimmunol.0903114

Tschopp J, Schroder K (2010) NLRP3 inflammasome activation: the convergence of multiple signalling pathways on ROS production? Nat Rev Immunol 10(3):210–215. doi:10.1038/nri2725

Unterholzner L, Keating SE, Baran M, Horan KA, Jensen SB, Sharma S, Sirois CM, Jin T, Latz E, Xiao TS, Fitzgerald KA, Paludan SR, Bowie AG (2010) IFI16 is an innate immune sensor for intracellular DNA. Nat Immunol 11(11):997–1004. doi:10.1038/ni.1932

van der Biezen EA, Jones JD (1998) The NB-ARC domain: a novel signalling motif shared by plant resistance gene products and regulators of cell death in animals. Curr Biol 8(7):R226–R227. doi:10.1016/S0960-9822(98)70145-9

Van Opdenbosch N, Gurung P, Vande WL, Fossoul A, Kanneganti TD, Lamkanfi M (2014) Activation of the NLRP1b inflammasome independently of ASC-mediated caspase-1 autoproteolysis and speck formation. Nat Commun 5:3209. doi:10.1038/ncomms4209

Vladimer GI, Weng D, Paquette SW, Vanaja SK, Rathinam VA, Aune MH, Conlon JE, Burbage JJ, Proulx MK, Liu Q, Reed G, Mecsas JC, Iwakura Y, Bertin J, Goguen JD, Fitzgerald KA, Lien E (2012) The NLRP12 inflammasome recognizes Yersinia pestis. Immunity 37(1):96–107. doi:10.1016/j.immuni.2012.07.006

von Moltke J, Ayres JS, Kofoed EM, Chavarria-Smith J, Vance RE (2013) Recognition of bacteria by inflammasomes. Annu Rev Immunol 31:73–106. doi:10.1146/annurev-immunol-032712-095944

Wen H, Ting JP, O'Neill LA (2012) A role for the NLRP3 inflammasome in metabolic diseases–did Warburg miss inflammation? Nat Immunol 13(4):352–357. doi:10.1038/ni.2228

Williams SJ, Sornaraj P, DeCourcy-Ireland E, Menz RI, Kobe B, Ellis JG, Dodds PN, Anderson PA (2011) An autoactive mutant of the M flax rust resistance protein has a preference for binding ATP, whereas wild-type M protein binds ADP. Mol Plant Microbe Interact 24(8):897–906. doi:10.1094/MPMI-03-11-0052

Xiao TS (2015) The nucleic acid-sensing inflammasomes. Immunol Rev 265(1):103–111. doi:10.1111/imr.12281

Yang J, Zhao Y, Shi J, Shao F (2013) Human NAIP and mouse NAIP1 recognize bacterial type III secretion needle protein for inflammasome activation. Proc Natl Acad Sci USA 110 (35):14408–14413. doi:10.1073/pnas.1306376110

Yuan S, Akey CW (2013) Apoptosome structure, assembly, and procaspase activation. Structure 21(4):501–515. doi:10.1016/j.str.2013.02.024

Zamboni DS, Kobayashi KS, Kohlsdorf T, Ogura Y, Long EM, Vance RE, Kuida K, Mariathasan S, Dixit VM, Flavell RA, Dietrich WF, Roy CR (2006) The Birc1e cytosolic pattern-recognition receptor contributes to the detection and control of legionella pneumophila infection. Nat Immunol 7(3):318–325. doi:10.1038/ni1305

Zhang L, Chen S, Ruan J, Wu J, Tong AB, Yin Q, Li Y, David L, Lu A, Wang WL, Marks C, Ouyang Q, Zhang X, Mao Y, Wu H (2015) Cryo-EM structure of the activated NAIP2-NLRC4 inflammasome reveals nucleated polymerization. Science 350(6259):404–409. doi:10.1126/science.aac5789

Zhao Y, Yang J, Shi J, Gong YN, Lu Q, Xu H, Liu L, Shao F (2011) The NLRC4 inflammasome receptors for bacterial flagellin and type III secretion apparatus. Nature 477(7366):596–600. doi:10.1038/nature10510

Zhou M, Li Y, Hu Q, Bai XC, Huang W, Yan C, Scheres SH, Shi Y (2015) Atomic structure of the apoptosome: mechanism of cytochrome c- and dATP-mediated activation of Apaf-1. Genes Dev 29(22):2349–2361. doi:10.1101/gad.272278.115

Salmonella and the Inflammasome: Battle for Intracellular Dominance

Shauna M. Crowley, Leigh A. Knodler and Bruce A. Vallance

Abstract Inflammasomes are macromolecular cytoplasmic complexes that act as signaling platforms for the activation of inflammatory caspases. Their activation triggers the processing and secretion of the pro-inflammatory cytokines IL-1β and IL-18, as well as the induction of a specialized form of inflammatory cell death termed pyroptosis. Here, we review the mechanisms of inflammasome activation triggered by the intracellular pathogen *Salmonella enterica* serovar Typhimurium. We highlight the different inflammasome subfamilies utilized by macrophages, neutrophils, dendritic cells, and intestinal epithelial cells response to a *Salmonella* infection as well as the *Salmonella* ligands that trigger each inflammasome's formation. We also discuss the evasion strategies utilized by *Salmonella* to avoid inflammasome detection. Overall, inflammasomes play a key and multilayered role at distinct stages of host cell defense against *Salmonella* infection.

Contents

S.M. Crowley · B.A. Vallance (✉)
Division of Gastroenterology, Department of Pediatrics, Child and Family Research Institute, University of British Columbia, Vancouver, BC V6H 3V4, Canada
e-mail: bvallance@cw.bc.ca

L.A. Knodler
Paul G. Allen School for Global Animal Health, College of Veterinary Medicine, Washington State University, Pullman, WA 99164-7090, USA

S. Backert (ed.), *Inflammasome Signaling and Bacterial Infections*,
Current Topics in Microbiology and Immunology 397,
DOI 10.1007/978-3-319-41171-2_3

1 Introduction

1.1 Salmonella

Salmonella is a genus of pathogenic, motile, Gram-negative bacteria that are part of the Enterobacteriaceae family. It is divided into two species: *Salmonella bongori* and *Salmonella enterica*, which both cause enteric disease in a wide range of animals. The *S. enterica* serovars are clinically important due to their significant impact on human morbidity and mortality worldwide (Crump et al. 2004, 2008). *S. enterica* serovar Typhimurium is associated with self-limiting gastroenteritis in immunocompetent individuals and bacteraemia in the immunocompromised, whereas *S. enterica* serovar Typhi causes typhoid fever which is characterized by fever and abdominal pain. Both serovars have a considerable worldwide impact (Majowicz et al. 2010; Kirk et al. 2015; Havelaar et al. 2015). Approximately 21.7 million cases of typhoid fever are diagnosed annually, leading to ~433,000 deaths per year and non-typhoid *Salmonella* causes approximately 93.8 million cases of gastroenteritis annually, leading to ~155,000 deaths per year.

In addition to its significant clinical impact, *S.* Typhimurium is widely considered an important model organism. A broad range of molecular and cell biology-based tools have been developed to study its pathogenesis. *S.* Typhimurium is also a natural pathogen of mice, which allows for the elucidation of host–pathogen interactions through infection of inbred and immuno-deficient mouse strains. Unfortunately, the study of *S.* Typhi is largely restricted to cell culture-based experiments due to its limited host range as a human-adapted pathogen. Thus, most studies discussed in this chapter have used *S.* Typhimurium (from here denoted as *Salmonella*) to understand its interactions with host cells.

Salmonella infection is contracted through the ingestion of contaminated water or food products. *Salmonella* uses its *Salmonella* pathogenicity island-1 (SPI-1) type three secretion system (T3SS) to breach the intestinal mucosa, preferentially in the terminal ileum. It induces membrane ruffling and direct invasion of a variety of intestinal epithelial cells (IECs) including enterocytes, as well as the microfold (M) cells typically found overlying Peyer's Patches. Macropinocytic uptake of *Salmonella* occurs rapidly, with the invading bacteria initially residing in a phagosomal compartment derived from the endocytic pathway termed the '*Salmonella*-containing vacuole' (SCV) (Steele-Mortimer et al. 1999; Drecktrah et al. 2007). The poor nutritional status and acidic pH in the lumen of the SCV triggers expression of *Salmonella*'s second T3SS, SPI-2 (Beuzón et al. 1999; Cirillo et al. 1998; Löber et al. 2006). Survival within the SCV is possible through the injection of numerous SPI-2 translocated effectors that functionally impact a wide variety of host cellular processes such as prevention of lysosomal fusion with the SCV, modulation of the host cytoskeleton, SCV membrane integrity, as well as immune signaling interference. Recently, it has been recognized that a subpopulation of *Salmonella* escapes the SCV and replicates within the host cell cytosol (Malik-Kale et al. 2012; Knodler et al. 2010). However, the permissiveness of the cytosol to *Salmonella* growth depends on the cell type infected. In cultured IECs, within 15 min of host cell invasion, *Salmonella* can escape the nascent SCV and within 1 h, replicate to reach 20–30 % of the total internalized pathogen population (cytosolic plus vacuolar *Salmonella*) (Birmingham et al. 2006; Cemma et al. 2011; Knodler et al. 2014). Similar levels of cytosolic *Salmonella* were also observed in embryonic fibroblasts by 4 h post-invasion (Birmingham et al. 2006). However, considerably fewer cytosolic *Salmonella* were observed at 1 h post-invasion in bone marrow-derived macrophages (BMDMs), likely due to a more hostile (antimicrobial) environment within these cells (Meunier et al. 2014). This particular intracellular lifestyle exposes *Salmonella* to detection by inflammasomes which have been identified as playing a key role in the early host response to *Salmonella* by both professional immune cells and IECs.

1.2 Inflammasomes

Inflammasomes are macromolecular cytoplasmic complexes that function as platforms for the activation of the inflammatory caspases. These complexes promote the processing of the pro-inflammatory cytokines interleukin-1β (IL-1β) and IL-18, as well as the induction of a specialized form of inflammatory cell death termed pyroptosis (Man and Kanneganti 2015; Storek and Monack 2015; Stowe et al. 2015). Since the initial characterization of inflammasomes in 2002, major strides have been made in this field including the discovery of multiple inflammatory caspases: caspase-1, -4, -5, -11, and -12 (Martinon et al. 2002; Man and Kanneganti 2015). Caspase-1, -4, -5, and -12 are expressed by primates, while caspase-1 and -11 are expressed in mice (caspase-11 is an ortholog of caspase-4 and -5) (Lamkanfi et al. 2002; Stowe et al. 2015). Researchers have also identified several pattern recognition receptors (PRRs) that upon recognizing their respective cytosolic pathogen-associated molecular patterns (PAMPs) or specific cellular stress or danger molecules, trigger the activation of these caspases. The components that comprise the inflammasome are located intracellularly and are only activated when a bacterial pathogen has compromised the sterility of the host's cytoplasm, introducing bacterial products or damaging cellular homeostasis. Modulation of the inflammasome is critical because once it is activated, a chain of events is triggered that primes the cell and surrounding tissues to produce a pro-inflammatory response that can result in the rapid self-destruction of the compromised cell (Fink and Cookson 2006).

Inflammasome activation is classified into two pathways: the canonical inflammasome and the non-canonical inflammasome. The canonical inflammasome consists of three components: a nucleotide-binding oligomerization domain-like (NOD) receptor (NLR) PRR, an adaptor protein bridge, and caspase-1 (Storek and Monack 2015). Upon ligand recognition, the PRR sensor associates with the adaptor protein via their shared domains. This prompts the recruitment of caspase-1 to the adaptor protein-PRR sensor complex through activation of the 'caspase activation and recruitment domain' (CARD) present in both the adaptor protein and caspase-1, which results in the formation of the inflammasome. Caspase-1 exists as a zymogen under homeostatic conditions. The assembly of the inflammasome is presumed to autocatalytically activate caspase-1 through the proteolytic processing of its pro-domain, which in turn cascades to the processing of both pro-IL-1β and pro-IL-18 into their active forms, and the induction of pyroptosis.

The non-canonical inflammasome utilizes caspase-4/5 or -11, and while it has been shown to promote pyroptosis and the secretion of IL-1β and IL-18, its exact functions as well as the interplay between these caspases and caspase-1 remains controversial (Stowe et al. 2015). The unifying feature of the inflammatory caspases is that each contains a CARD at their N-termini that mediates their protein–protein interactions and subsequent activation (Storek and Monack 2015). Surprisingly, a recent report suggested that caspase-11, as well as caspase-4 and -5, do not associate with a lipopolysaccharide (LPS)-sensing PRR, but instead directly bind LPS via their CARDs, effectively making caspase-4/5/11 their own upstream sensors (Shi et al. 2014). This novel finding requires further examination as it pivots away from the typical PRR-centric model, where specialized sensors are required to activate downstream enzymes via a complex signaling pathway, to an atypical response where the sensors themselves detect, and respond to pathogenic stimuli.

Inflammasomes are vital in the restriction of a *Salmonella* infection. Raupach et al. (2006) observed significantly higher *Salmonella* burdens in the Peyer's patches, mesenteric lymph nodes, and spleens of both *Casp1/11*$^{-/-}$ and *Il1β/Il18*$^{-/-}$ mice compared to C57BL/6 mice, five days after oral infection. This chapter will present an overview of the various inflammasomes activated by *Salmonella* and the mechanisms of their activation (summarized in Fig. 1). We will also focus on understanding how *Salmonella*–inflammasome interactions differ among different host cell types (summarized in Fig. 2). The inflammasome literature has largely focused on monocytes and macrophages as the primary effectors of inflammasome-mediated restriction of *Salmonella*. However, recent evidence has revealed numerous cell types utilize inflammasomes to combat intracellular *Salmonella*, and although identical signaling pathways are utilized, the inflammasome-induced phenotype appears individual to each cell type. This chapter will summarize the extensive body of work describing the interactions between *Salmonella* and macrophage-based inflammasomes as well as highlight new studies detailing the role of inflammasome-mediated control of *Salmonella* in dendritic cells (DCs), neutrophils, and IECs.

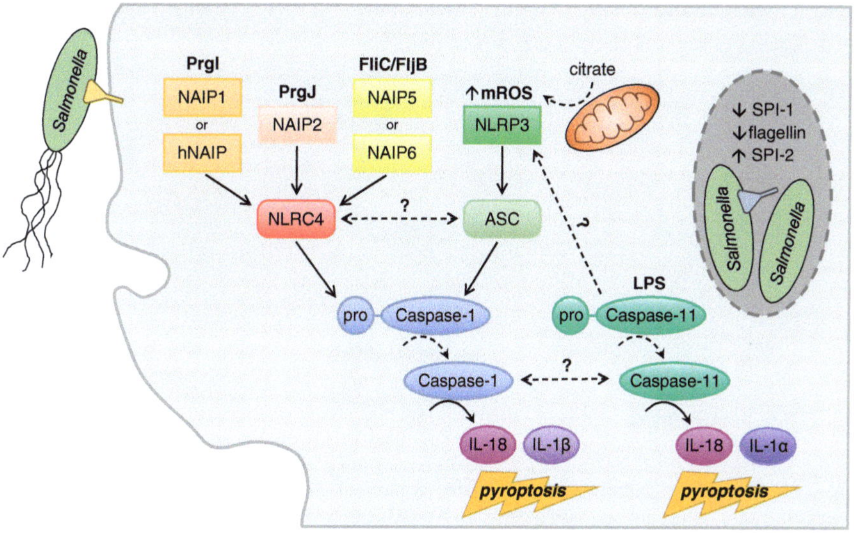

Fig. 1 Proposed function of the inflammasome in response to a *Salmonella* infection. Inflammasome PRRs directly recognize bacterial components as well as sense danger molecules induced by a *Salmonella* infection. The receptors, Naip5 and Naip6, recognize bacterial flagellin (FliC and FljB), while Naip1 and Naip2 recognize the T3SS needle protein (PrgI) and rod protein (PrgJ), respectively. Upon ligand recognition, the Naips interact with Nlrc4 to activate caspase-1 resulting in IL-1β and IL-18 cytokine processing and secretion as well as induction of pyroptosis. Bacterial LPS is directly sensed by caspase-11 to elicit pyroptosis and secrete IL-1α and IL-18. Nlrp3 responds to host cell changes that elevate mitochondrial reactive oxygen (mROS) species through increased *Salmonella* TCA citrate production. *Salmonella* evades cytosolic inflammasome detection in the SCV by decreasing SPI-1 and flagellin expression as well as through SPI-2 translocated effectors. Abbreviations: *Naip1-6* neuronal apoptosis inhibitor protein 1-6; *hNAIP* human neuronal apoptosis inhibitor protein; *Nlrc4* NOD-like receptor family, caspase-activating and recruitment domain containing protein 4; *Nlrp3* NOD-like receptor family, pyrin domain containing 3; *Asc* apoptosis-associated speck-like protein; *LPS* lipopolysaccharide; *IL-1β* interleukin 1 beta; *IL-18* interleukin 18

2 *Salmonella* and the Immune Cell Inflammasome

2.1 Macrophages

After breaching the epithelial barrier, *Salmonella* is engulfed by nearby macrophages or can induce its own uptake via its SPI-1 T3SS. *Salmonella* ultimately activates the nuclear factor kappa-light-chain-enhancer of activated B cells (NF-κB) signaling pathway through detection by Toll-like receptors (TLRs), and/or through SPI-1 translocated effectors activating the mitogen-activated protein kinase (MAPK) pathway (Gurung et al. 2012; Yang et al. 2014; Tam et al. 2008). This triggers secretion of IL-6 and tumor necrosis factor alpha (TNF-α) as well as the expression of pro-IL-1β and pro-IL-18 (LaRock et al. 2015; Man et al. 2013). Once inside the host cell, *Salmonella* dampens the host pro-inflammatory response by

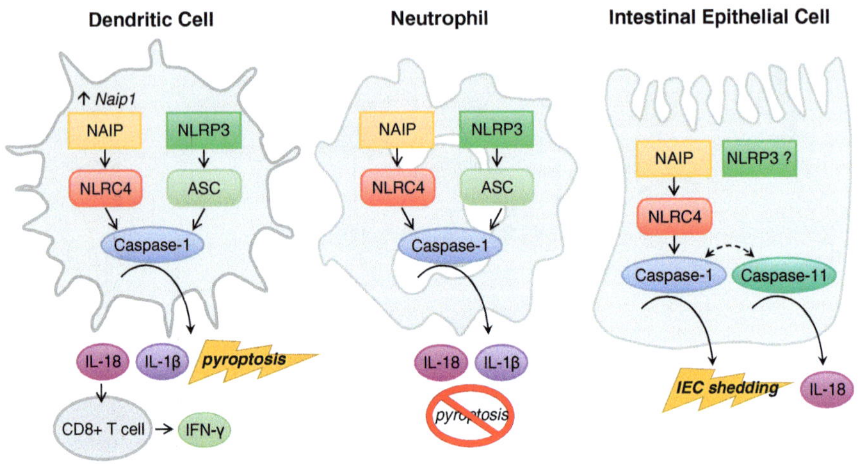

Fig. 2 Differences in inflammasome-mediated control of *Salmonella* in dendritic cells (DCs), neutrophils, and intestinal epithelial cells (IECs). DCs have increased *Naip1* transcript levels and their activation prompts the secretion of IL-18, which directly signals memory CD8+ T cells to secrete IFN-γ. *Salmonella* triggers both Naip and Nlrp3 inflammasome activation in neutrophils resulting in IL-1β and IL-18 secretion. However, neutrophil inflammasome activation does not elicit pyroptosis producing a cellular source of sustained IL-1β. *Salmonella* also activates an epithelial cell-intrinsic inflammasome through the Naip-Nlrc4 axis as well as the non-canonical inflammasome, caspase-11. Activation produces IL-18 secretion as well as IEC extrusion from the epithelium. Abbreviations: *Naip* neuronal apoptosis inhibitor protein; *Nlrc4* NOD-like receptor family, caspase-activating and recruitment domain containing protein 4; *Nlrp3* NOD-like receptor family, pyrin domain containing 3; *Asc* apoptosis-associated speck-like protein; *IL-1β* interleukin 1 beta; *IL-18* interleukin 18; *IFN-γ* interferon-gamma; *IEC* intestinal epithelial cell

residing in the SCV, shielded from inflammasome detection, as well as SPI-2 T3SS effectors. However, the SCV is under constant assault by host molecular pathways attempting to destabilize it as well as induce its fusion with lysosomes (LaRock et al. 2015; Keestra-Gounder et al. 2015). If the SCV is breached and *Salmonella*–inflammasome activating PAMPs become cytoplasmic, this triggers inflammasome formation. The inflammasome can also assemble in response to cellular stress or danger molecules induced by *Salmonella*'s intracellular presence. Once an inflammatory caspase is activated, this triggers processing of the pro-forms of IL-1β and IL-18 as well as rapid cell lysis. Intracellular *Salmonella* is released into the extracellular space, which can benefit both host and pathogen. Extracellular *Salmonella* are especially vulnerable to responding neutrophils, which can rapidly clear the infection through the generation of reactive oxygen species (ROS) (Keestra-Gounder et al. 2015). However, extracellular release can also provide *Salmonella* with basolateral entry to the intestinal epithelium or access to liver and spleen tissues if the infection becomes systemic (Mastroeni et al. 2009). Therefore, inflammatory caspase activation must be tightly regulated and inflammasome formation is likely utilized as a measure of last resort.

Salmonella-induced inflammasome formation and subsequent caspase-1 activation in macrophages is mainly orchestrated by the PRRs: Naip and Nlrp3 (Fig. 1). In mice, Naip has multiple homologues that are able to detect either flagellin (Naip5 and Naip6) or T3SS components PrgJ and PrgI (Naip2, Naip1), while humans express only a single Naip, hNAIP, which responds to PrgI (Storek and Monack 2015; Man and Kanneganti 2015). Once stimulated, Naip associates with the adapter protein Nlrc4 which triggers caspase-1 activation. Intracellular *Salmonella* are also detected by the PRR Nlrp3 which identifies molecules associated with cellular stress or danger such as host cell potassium efflux, calcium influx, extracellular ATP, mitochondrial reactive oxygen species, mitochondrial DNA, and the translocation of cardiolipin from the inner mitochondrial membrane to the outer membrane (Storek and Monack 2015; Man and Kanneganti 2015). Both Naip and Nlrp3 contain different protein sensing, recruitment, and activation domains which impact how they interact with caspase-1. Naip associates with the adaptor Nlrc4 via leucine-rich repeat domains, while Nlrp3 binds Asc via their shared pyrin domains. Both Nlrc4 and Asc associate with caspase-1 via their individual CARD domains, separately forming a functional inflammasome and activating caspase-1.

2.1.1 Naip1-6, hNAIP, and Nlrc4

Salmonella is a motile pathogen and produces six to eight peritrichous flagella in order to move through its environment and reach the site of invasion (McCarter 2006; Fàbrega and Vila 2013). The flagellum consists of a basal body, an anchoring hook component, and a long flagellin filament made up of repeating FliC and FljB structural subunits. In a murine *Salmonella* BMDM infection model, IL-1β secretion by infected cells is dramatically reduced when *Salmonella* lacks FliC or their FliC C-terminus amino acid composition is altered (Franchi et al. 2006; Miao et al. 2006). This sensing takes place via the direct intracellular detection of FliC by the PRRs Naip5 and Naip6 in mice (Lightfield et al. 2008; Zhao et al. 2011; Kofoed and Vance 2011). FljB also induced inflammasome-mediated cytotoxicity at similar levels to a FliC-driven response (Miao et al. 2006). However, there was no significant difference between secreted IL-1β levels comparing an *S.* Typhi *fliC* mutant to a double *S.* Typhimurium *fljBfliC* mutant, indicating FliC alone is sufficient for inflammasome formation (Winter et al. 2015).

Out of the murine Naips, Naip5 and Naip6 have the highest amino acid sequence identity (94.7 %). Naip5 expression in primary macrophages far surpasses that of Naip6, suggesting these receptors may not play equal roles in inflammasome signaling (Wright et al. 2003). Both Naip5 and Naip6 act as co-receptors for Nlrc4 binding, inducing the activation of caspase-1 in BMDMs. Interestingly, there appears to be a functional link between Nlrc4 signaling and actin turnover in macrophages. Actin polymerization is required for Nlrc4-dependent intracellular bacterial burdens, inflammasome assembly, pyroptosis, and IL-1β production (Man et al. 2014a). Man and colleagues report this Nlrc4-induced cellular stiffness restricts intracellular replication and *Salmonella* dissemination throughout tissues. Murine Nlrc4 is also an interesting adaptor because it links caspase-1 activation with the detection of multiple structurally different PAMPs. In addition to associating with the flagellar-detecting Naip5 and Naip6, Nlrc4 also binds to the sensors Naip2 and Naip1 which recognize the T3SS rod and needle apparatus, respectively.

Salmonella is also detected by the inflammasome via its SPI-1 T3SS. The T3SS is a molecular syringe-like apparatus that punctures the host cell membrane, forming a continuous channel between the bacterium and host, allowing bacterial effectors to be pumped directly into the host cell cytoplasm. *Salmonella* contains two T3SS, and their expression is tightly controlled based on the stage of *Salmonella*'s pathogenesis as well as the detection of specific environmental cues from the gut and within host cells. The SPI-1 T3SS is expressed during the initial invasion of host cells, inducing bacterial-mediated endocytosis. Its various translocated effectors such as SopB, SopE, and SopE2 activate the small GTPases, Cdc42 and Rac-1, through their guanine nucleotide exchange factor activity (LaRock et al. 2015). This causes alterations in the actin cytoskeleton and disrupts tight junctions between IECs, producing conditions that facilitate *Salmonella's* invasion of IECs as well as a generalized weakening of the overall epithelial barrier. Once *Salmonella* has undergone endocytosis, it alters its vacuole maturation through its SPI-2 T3SS. While not as abundantly expressed as the SPI-1 T3SS, the SPI-2 T3SS injects effectors across the SCV membrane, yielding a relatively safe replication niche for *Salmonella* by altering the vacuolar membrane and trafficking to limit lysosomal fusion as well as rerouting vital metabolites to be imported into the SCV (Figueira and Holden 2012).

Initially, the discovery that the *Salmonella* SPI-1 T3SS inner rod protein (PrgJ) also triggered a Nlrc4 inflammasome-dependent response was confusing because of the protein structural differences between PrgJ and FliC. Zhao et al. (2011) discovered flagellin and Naip5 coimmunoprecipitated with Nlrc4, but neither Naip1 nor Naip2 coimmunoprecipitated in the presence of flagellin. In a yeast two hybrid assay probing for associations with the *Burkholderia thailandensis* T3SS rod protein BsaK, the authors uncovered an interaction between BsaK and Naip2. They confirmed Naip2 detected the T3SS rod protein, through infection of BMDMs with either *Salmonella* Δ*fliC/fljB* or Δ*fliC/fljB/prgJ* mutants, to remove the confounding caspase-1 activation from flagellin detection by Naip5/6. The presence of PrgJ produced cleaved and activated caspase-1 after 30 min of infection that was eliminated following a siRNA knockdown of *Naip2*. Also, no cleaved caspase-1 was detected from the cells infected with the Δ*fliC/fljB/prgJ* strain. Further studies revealed that the flagellin and T3SS rod protein's ability to stimulate Naip activity was localized to their C-terminal leucine-rich helical hairpin regions (Lightfield et al. 2008; Poyraz et al. 2010; Miao et al. 2011). This is of special interest to studies of *Salmonella* because the amino acid composition of its SPI-1 (PrgJ) and SPI-2 (SsaI) T3SS inner rod proteins differ at these residues. SsaI encodes several amino acid substitutions compared to PrgJ, but most notably it displays a switch from valine to leucine at its C-terminus (Miao et al. 2010). When *Salmonella* grown under SPI-2 inducing conditions were used to infect BMDMs, the infected cells did not secrete IL-1β after 8 h of infection, indicating an absence of inflammasome activation. Interestingly, swapping the eight amino acids at the C-terminus of SsaI with those from PrgJ restored Nlrc4 binding and IL-1β secretion, uncovering a potential evasion strategy utilized by intracellular *Salmonella* to evade the inflammasome.

In humans, only one Naip ortholog exists (hNAIP) and although it has high sequence identity to murine Naip5 (68 % amino acid identity), it does not recognize/respond to flagellin (Zhao et al. 2011; Storek and Monack 2015). In phorbol myristate acetate-stimulated U937 and THP-1 human macrophages, hNAIP was been shown to respond to PrgI (the T3SS needle proteins from SPI-1) but not PrgJ (Yang et al. 2013). This finding was of significant interest because murine Naip activity against the T3SS needle protein had not previously been detected. Yang et al. (2013) proposed that murine Nlrc4 was indeed capable of responding to intracellular PrgI, but the activity of Naip2 and Naip5 shielded its effects. To this end, they delivered the needle protein directly into the cytoplasm of BMDMs utilizing the LFn-PA delivery system (this consists of a fusion of needle proteins to the N-terminal domain of *Bacillus anthracis* lethal factor (LFn) which mediates cytosolic delivery through the anthrax protective antigen (PA) channel) (Yang et al. 2013; von Moltke et al. 2012). This triggered robust caspase-1 activation, which was abolished in *Nlrc4*-deficient BMDMs (Yang et al. 2013). Furthermore, when HEK293T cells were co-transfected with Nlrc4, Naip1, and PrgI, large oligomeric complexes containing all three components could be resolved by native polyacrylamide gels, confirming Naip1 was an active detector of the T3SS needle protein and that it activated caspase-1 through Nlrc4 binding.

This difference between the number of Naips between human and mice as well as the specificity of detecting a single PAMP by hNAIP could reflect a control mechanism for the intracellular sensing of bacteria. In mice, both Naip2 and Naip5 are highly expressed in systemic organs such as the spleen, whereas Naip1 could not be detected (Yaraghi et al. 1999; Wright et al. 2003). These differences in expression could explain the overall responsiveness of murine macrophage inflammasomes to endogenous PrgJ and FliC from *Salmonella*, whereas PrgI must be artificially injected into cells to activate caspase-1 (Yang et al. 2013). However, hNAIP, like Naip1, detects PrgI. Compared to flagellin (which is downregulated after *Salmonella*'s internalization) and the PrgJ inner rod protein (which remains sheathed within the T3SS unit), it is more likely that the host cell cytosol could be exposed to PrgI during an active *Salmonella* infection. Limiting the number of PAMPs that can activate the inflammasome and restricting activation to PAMPs that are only detected once *Salmonella* has successfully infected the host cells may represent a regulatory strategy by the host to limit inflammasome activation as a measure of last resort. Little information is currently available regarding how inflammasome responses are terminated or if they can be terminated at all.

2.1.2 Nlrp3 and Asc

Inflammasome stimulatory PAMPs such as flagellin and T3SS structural proteins are essential bacterial components for the pathogenesis of *Salmonella* and are readily detected by the Naip-Nlrc4 inflammasome. In order to shield itself from constant assault by host cell molecular processes and continue its intracellular life cycle, *Salmonella* can alter its PAMPs by decreasing expression of flagellin and switching from its SPI-1 to its SPI-2 T3SS. To combat this evasion strategy, the host cell utilizes the Nlrp3 inflammasome, which instead of detecting PAMPs, responds to increased levels of host cell-derived stress and danger molecules.

Study of the Nlrp3 inflammasome across a wide range of bacterial infections has uncovered its responsiveness to host cellular stress molecules such as excessive potassium efflux, calcium influx, extracellular ATP, mitochondrial reactive oxygen species, mitochondrial DNA, and the translocation of cardiolipin from the inner mitochondrial membrane to the outer membrane. The mechanism by which Nlrp3 can detect such a wide range of structurally and chemically unrelated stimuli is still under investigation. In the case of *Salmonella* interactions with macrophages, Nlrp3 is responsible for activation of the inflammasome independently of SPI-1 and SPI-2 T3SS and tied to alterations in host cell most likely induced from intracellular *Salmonella* metabolism (Broz et al. 2012; Gurung et al. 2012; Rathinam et al. 2012; Sanman et al. 2016). During a screen of *Salmonella* genes that modulate Nlrp3 activation, Wynosky-Dolfi et al. (2014) identified four genes; *acnB*—aconitase, which mediates conversion of citrate to isocitrate in the tricarboxylic acid cycle (TCA); *bcfB*—fimbrial chaperone; *rcsD*—a two component system which controls capsule and flagellar synthesis; and *melB*—a symporter of melibiose and monovalent cations. The authors focused on the link between TCA cycle disruption and Nlrp3 activation by constructing mutants deficient in aconitase or isocitrate dehydrogenase. Mice infected with *acnB* mutants displayed significantly higher serum levels of IL-18 as compared to mice infected with wild-type *Salmonella,* and these levels were reduced in $Casp1/11^{-/-}$ and $Nlrp3^{-/-}$ mice. Mutation of these bacterial TCA cycle enzymes in conjunction with flagellar proteins led to high intracellular citrate levels and elevated levels of mitochondrial ROS, which resulted in rapid Nlrp3-dependent, Nlrc4-independent inflammasome activation. Limiting the resulting mitochondrial ROS (mROS) levels in murine BMDMs through pharmacological inhibitors or through the use of mitochondrial catalase, transgenic mice eliminated Nlrp3 inflammasome activation by the *Salmonella* TCA cycle mutants, indicating that *Salmonella* specifically activated the Nlrp3 inflammasome through mROS production.

Nlrp3-based caspase-1 activation occurs through the binding of these two molecules to the adaptor protein Asc. Moreover, Asc expression is crucial for *Salmonella*-induced IL-1β and IL-18 secretion by murine macrophages (Broz et al. 2010; De Jong et al. 2014). The physiological importance of Asc-containing foci in Naip-Nlrc4 inflammasome activation is still under investigation. The CARD of Nlrc4 directly interacts with the CARD of caspase-1; therefore, T3SS and flagellin–inflammasome signaling should occur independently of Asc foci formation.

However, Asc foci were detected in *Salmonella* infected *Nlrp3*$^{-/-}$-deficient murine BMDMs and this aggregation was found to be dependent on the presence of flagellin (Broz et al. 2010). Proell et al. (2013) explored this phenotype through a bimolecular fluorescent complementation system, which allowed them to visualize Asc foci within a living cell. The CARDs of Nlrc4 and caspase-1 were fused with two complementary fragments of the fluorescent protein Venus. These fragments do not reconstitute spontaneously, but a functional fluorescent signal is produced when Nlrc4 and caspase-1 associate. When the CARDs of Nlrc4 and caspase-1 were co-expressed in HEK293T cells, this produced a diffuse signal throughout the cell. However, when full-length Asc was introduced, tight punctuated foci were formed, suggesting that Naip-Nlrc4-caspase-1 inflammasomes could utilize Asc as a platform for formation. Further support for model is given through co-localization experiments performed by Man et al. (2014b). The authors observed multiple NLRs, such as Nlrp3 and Nlrc4, co-localizing with Asc specks. They propose that *Salmonella* PAMPs activate NLRs which associate forming a nucleation point for the assembly of an Asc foci. This in turn recruits caspase-1, producing a functional inflammasome. There is also the potential that Asc might play a role in inflammatory signal amplification. Once infected macrophages undergo pyroptosis and lyse, researchers found that Asc aggregates appear to remain active in the extracellular space (Storek and Monack 2015). These aggregates can be phagocytosed by nearby cells, leading to the activation of their own inflammasomes and the subsequent cleavage and secretion of IL-1β and IL-18. As a result, Asc foci released by pyroptotic cells may play a key role in amplifying inflammatory signals by providing a platform for inflammasome formation.

2.1.3 Caspase-11

Detection of cytosolic *Salmonella* LPS can also induce inflammasome-mediated cell death through the actions of caspase-11. LPS must enter the cytosol before it can activate the caspase-11 inflammasome, in the case of *Salmonella*, this means the bacterium must be directly exposed to the cytoplasm, not contained inside the SCV. Experimentally, this can be induced through direct microinjection or transfection of purified LPS into the cytoplasm, or the use of a Δ*fljAB*Δ*fliC*Δ*sifA* *Salmonella* mutant, which does not express flagellin and upon infecting a host cell, readily escapes the SCV. SifA is a SPI-2 translocated effector which is vital for maintaining the stability of the SCV (Beuzón et al. 2000; Cirillo et al. 1998). Utilizing this *Salmonella* mutant, Aachoui et al. (2013) demonstrated its enhanced clearance by caspase-11 driven pyroptosis was independent of Nlrp3, Nlrc4, and Asc.

Once caspase-11 has detected LPS, this triggers its proteolytic activation inducing pyroptosis and the cleavage and secretion of IL-1β and IL-18. A group based out of Genentech reported that caspase-1 is dispensable for caspase-11 mediated pyroptosis, but caspase-11 dependent IL-1β processing requires Nlrp3, Asc, and caspase-1 (Kayagaki et al. 2011, 2013; Baker et al. 2015; Rühl and Broz

2015; Schmid-Burgk et al. 2015). The reduced IL-1β secretion by *Casp11*$^{-/-}$ cells is not due to lower expression of pro-IL-1β as both wild-type and *Casp11*$^{-/-}$-deficient BMDMs expressed similar pro-IL-1β levels when infected with *Salmonella* (Broz et al. 2012). Recently Casson et al. (2015) observed that the human ortholog of caspase-11, caspase-4, mediated IL-1α secretion, and cell death in response to *Salmonella* infection of primary human macrophages. They also noted that caspase-5 was not processed during the infection and did not appear to play a role in host restriction of intracellular *Salmonella*.

The interplay between canonical and non-canonical inflammasomes remains poorly understood. Caspase-1 and caspase-11 can interact, but it is unclear whether the two caspases form a catalytically active heterocomplex or whether proximity of the two caspases within a cell is sufficient for caspase-1 and -11 activation. Caspase-11's major role appears to be in the initiation of pyroptosis. Pyroptosis is responsible for the clearance of intracellular bacteria through the host's self--directed destruction of the infected cell. This destroys the protected niche established by the pathogen within the host cytosol, exposing the previously intracellular *Salmonella* to extracellular neutrophil attack. Pyroptosis, rather than the release of IL-1β and IL-18, has been proposed to be the dominant process underlying caspase-11's key role in the LPS-induced lethal sepsis model (Kayagaki et al. 2011). Mice deficient in *Casp1/11*$^{-/-}$, or both *IL-1β/IL-18*$^{-/-}$ were shown to rapidly succumb to high dose injections of LPS, whereas *Casp11*$^{-/-}$ mice remained resistant. However, the mechanism of pyroptosis and the cell types responsible for in vivo lethal sepsis are still under investigation.

Caspase-11 readily detects cytosolic *Salmonella* but it can also detect SCV-protected *Salmonella* once the vacuole is compromised (Broz et al. 2012; Rathinam et al. 2012). Meunier et al. (2014) explored the requirement for Trif-dependent production of type I interferons and caspase-11 driven cell death by Gram-negative bacteria. Through a proteomics-based expression analysis for proteins highly induced upon *Salmonella* infection, they identified the interferon-induced GTPases as the most strongly upregulated proteins. In particular, they observed the large 65–67 kDa guanylate-binding proteins (GBPs) as being highly expressed. They then moved to examine the interaction of these GBPs during a *Salmonella* infection and detected guanylate-binding protein 2 (GBP2) co-localizing with the SCV in murine BMDMs and its expression led to higher cytosolic populations of *Salmonella* (Meunier et al. 2014). The expression of GBP2 was also linked to caspase-11 based cytotoxicity in response to *Salmonella* infection, while overall depletion of the GBPs led to a significant decrease in IL-1β secretion. This suggests the GBPs destabilize the SCV, by an as-yet-unknown mechanism, leading to the release of *Salmonella* into the cytosol, exposing them for detection by both the canonical and non-canonical inflammasomes. The authors also observed that if cytosolic *Salmonella* was targeted by autophagy, this significantly reduced caspase-11 activation (Meunier et al. 2014).

2.1.4 Other Inflammasomes Involved in *Salmonella* Detection

The NLRP6 and NLRP12 inflammasomes have been linked to the maintenance of intestinal homeostasis, primarily in the context of chemical-induced colitis (Zaki et al. 2010; Elinav et al. 2011). Studies have observed that NLRP6 and NLRP12 suppress colon inflammation through inhibition of NF-κB signaling (Zaki et al. 2011; Allen et al. 2012). In terms of *Salmonella* i.p. infection, both *Nlrp6-/-* and *Nlrp12*$^{-/-}$ mice were shown to carry significant lower bacterial burdens in their spleens and livers (Zaki et al. 2014; Anand et al. 2012). Moreover, both *Nlrp6*$^{-/-}$ and *Nlrp12*$^{-/-}$ BMDMs were found to produce elevated levels of TNF-α, IL-6, and KC in response to bacterial infection, but showed no difference when compared to wild-type cells in terms of IL-1β and IL-18 expression. The authors also observed higher numbers of circulating monocytes and neutrophils in this study (Anand et al. 2012). While the ligand that the NLRP6 and NLRP12 inflammasomes detect has not yet been identified, over-expression studies suggest NLRP6 association with Asc and caspase-1 results in IL-1β secretion, and *Salmonella* LPS alone was able to induce NLRP12-mediated inhibition of NF-κB (Grenier et al. 2002; Zaki et al. 2014). Therefore, the NLRP6 and NLRP12 inflammasomes may play immune dampening roles in response to specific stimuli, and this suppression might be exploited by *Salmonella* to modulate host antimicrobial responses as a means to promote their prolonged persistence and survival within infected hosts.

One of the strongest *Salmonella*–inflammasome stimulants is the SPI-1 translocon protein SipB, which activates caspase-1 (Hersh et al. 1999). Deletion of *sipB* causes a severe deficiency in inflammasome activation within murine BMDMs which is independent of the number of intracellular bacteria with these cells (Storek and Monack 2015; Wynosky-Dolfi et al. 2014). The initial characterization of SipB via affinity purification uncovered its interaction with caspase-1, but it is unclear whether this occurs through direct binding, or alternatively, via an interaction with a PAMP or DAMP sensor. During the initial stages of *Salmonella* invasion, SipB associates with the *Salmonella* proteins SipC and SipD to assemble a channel on the distal tip of the needle complex that spans the host cell plasma membrane and facilitates passage of T3SS effectors (Galán 1998). It is also possible that SipB signals through the Nlrp3 inflammasome through cellular stress molecules induced upon host membrane damage. However, this does not explain the direct binding of SipB to caspase-1 and indicates further study is required to uncover the underlying mechanism.

2.2 Dendritic Cells (DCs)

The majority of inflammasome research has centered on murine macrophages; however, recent studies suggest inflammasomes are active in multiple cell types and each cell type may be unique in its cellular response to *Salmonella*–inflammasome activation (Fig. 2).

Both macrophages and dendritic cells are phagocytic cells critical in the immune response to *Salmonella* infection. The primary function of DCs is to process antigen material and present it to naïve T-lymphocytes. Early work by Fink et al. (2008) established that *Salmonella* infection of murine BMDCs triggers a caspase-1 dependent cell death. In an intravenous mouse model of *Salmonella* infection, the authors observed activation of the Naip-Nlrc4-caspase-1 inflammasome (Kupz et al. 2012). Activation prompted the secretion of IL-18, which directly signaled memory $CD8^+$ T cells to secrete IFN-γ. This interaction is important in highlighting that DC-T cell interactions are not limited to simple antigen-specific responses. DCs can also orchestrate innate immune responses through inflammasome-directed cytokine secretion. Interestingly, the majority of murine DC inflammasome signaling appears to utilize Naip5 and Naip1. DC-based Naip2 sensing of PrgJ was not required for the early regulation of innate IFN-γ secretion by memory $CD8^+$ T cells. Yang et al. (2013) observed that primary murine BMDCs and immortalized DC2.4 cells produce higher levels of *Naip1* transcripts compared to BMDMs. When DC2.4 cells were stimulated via transfection with plasmids encoding T3SS needle transcripts, this triggered a more robust caspase-1 activation than in BMDM cells and this activation was silenced following siRNA knockdown of *Nlrc4* or *Naip1*.

There has also been a preliminary study of *Salmonella* infection of human monocyte-derived DCs (Dreher et al. 2002). The resulting inflammatory response was dependent on SipB expression and resulted in caspase-1 activation and the release of IL-18.

2.3 Neutrophils

Neutrophils are recruited in large numbers to the site of *Salmonella* infection where they play a key role in clearing the infection. Also commonly viewed as cellular targets of IL-1β, caspase-1 expression by infected hosts is associated with efficient neutrophil clearance of *Salmonella.* Interestingly, recent evidence has established that neutrophils also express inflammasomes, and their activation plays a vital role in the restriction of *Salmonella* infection (Broz et al. 2012; Chen et al. 2014). Chen et al. (2014) observed that both human and mouse neutrophils contained transcripts for *Nlrp3* and *Nlrc4*, and in human neutrophils, these mRNA levels were significantly greater than those expressed by BMDMs or BMDCs. When murine neutrophils and BMDMs were infected with *Salmonella*, both cell types secreted IL-1β, albeit 5-fold less for neutrophils. The secretion of IL-1β was lacking in cells purified from *caspase-1/11*$^{-/-}$ mice and reduced in *Nlrc4*$^{-/-}$ and *Asc*$^{-/-}$ mice. Interestingly, neutrophils secreted negligible amounts of IL-18. The authors then moved to in vivo experiments where they depleted neutrophils from mice utilizing α-Ly6G antibodies, which did not alter the abundance of other myeloid cells. The authors then challenged isotype control (mock treated) and neutrophil-depleted mice with *Salmonella* and observed that IL-1β levels remained at baseline levels in the neutrophil-depleted mice while control mice displayed 3-fold higher IL-1β

levels after 6 h of infection. Neutrophil depletion did not significantly affect IL-18 levels even after 12 h of *Salmonella* infection. Interestingly, although neutrophils display a potent inflammasome response, they were highly resistant to pyroptosis. When the authors quantified intracellular lactate dehydrogenase released into the supernatant as a measure of in vitro and ex vivo pyroptosis, they reported significantly decreased neutrophil cytotoxicity compared to BMDMs when the cells were infected by *Salmonella*. It appears that neutrophils express a specialized inflammasome, decoupling pyroptosis from Nlrp3 and Nlrc4 signaling. The molecular mechanisms underlying pyroptosis remain elusive and the proteins and proteolytic substrates involved are still under intense investigation. Only HMGB1 and Gasdermin D have been implicated as downstream products from macrophage caspase activation (Kayagaki et al. 2015; Shi et al. 2015). This suggests that inflammasome driven pyroptosis and cytokine activation/secretion can be mechanistically separated, likely through the downregulation or proteolytic removal of a vital substrate required for the lytic cell death pathway. Overall, this suggests an innate defense mechanism for the restriction of *Salmonella* infection. Macrophages secrete IL-1β to recruit neutrophils and then undergo pyroptosis to limit vacuolar replication of *Salmonella*. This releases the bacteria into the extracellular environment where they are more effectively killed by neutrophils through ROS production (Miao et al. 2011).

3 *Salmonella* and the Intestinal Epithelial Inflammasome

Interestingly, the specific PRR responsible for downstream caspase activation appears to be dependent on the route of *Salmonella* infection. In a systemic *Salmonella* infection model, where *Salmonella* primarily encounters myeloid cells, Nlrc4 and Nlrp3 play redundant roles in stimulating caspase-1 activation. Only mice deficient in both Nlrc4 and Nlrp3 demonstrate increased susceptibility to *Salmonella* infection (Broz et al. 2010). However, in a streptomycin (Sm)-pretreatment oral gavage mouse model for *Salmonella* gastroenteritis, where *Salmonella* must first breach the intestinal epithelium to reach systemic organs, this redundancy is not apparent and mice deficient in only Nlrc4 were found to be more susceptible to *Salmonella* infection (Franchi et al. 2012). Wynosky-Dolfi and colleagues observed during *Salmonella* infection that *acnB* mutants (which signals through Nlrp3 by increasing bacterial-induced mitochondrial ROS production) displayed similar levels of splenic colonization as wild-type *Salmonella* following i.p. injection of C57BL/6 mice (Wynosky-Dolfi et al. 2014). In contrast, acnB-deficient bacteria were significantly attenuated in numbers in the spleens of C57BL/6 mice after oral infection. These results suggest that *Salmonella* encounters a unique inflammasome pathway at the intestinal mucosal surface.

Understanding the mechanisms by which IECs defend against enteric pathogens is paramount to our ability to prevent and treat food-borne illnesses. Historically, the role played by IECs in innate defense was defined by their location. A single layer of cells that function as a structural barrier, physically sequestering bacteria and other noxious stimuli within the gut lumen, as a means to prevent their interaction with, and overt stimulation of the underlying immune system. However, we and others have hypothesized that the intestinal epithelium is not a passive barrier, but instead actively responds and defends the host when challenged by pathogenic attack. Recently, our group as well as the group of Dietrich-Hardt published two separate studies characterizing the early stages of *Salmonella* infection and the role played by IEC inflammasomes in promoting protection of the intestinal mucosal barrier (Knodler et al. 2014; Sellin et al. 2014).

Sellin et al. (2014) focused on the initial stages of *Salmonella* infection where they observed that by 6 h post-infection (p.i.), the cecum contained numerous infected enterocytes that contained up to 20 densely packed bacteria, predominantly occupying the supranuclear region of IECs. These populations, which the authors termed 'microcolonies', were still evident until 12 h p.i., at which time infected IECs began to be extruded from the epithelial layer and released into the lumen. By 36 h p.i., the hyper-infected IECs were entirely cleared from the epithelium and only small numbers of IECs containing one to two bacteria remained. Mice deficient in *Naip1-6* or *Nlrc4* displayed much higher densities of *Salmonella* in their intestinal epithelium ($\sim$ 50- to 100-fold higher load), with these loads persisting for up to 36 h p.i., indicating the mechanism responsible for infected IEC extrusion was tied to caspase-1 activation by Nlrc4. However, after 36 h p.i., microcolony-containing IECs for both *Naip1-6*$^{-/-}$ and *Nlrc4*$^{-/-}$ mice were also cleared, indicating that caspase-1 activation is not the only mechanism at play promoting epithelial restriction of *Salmonella*. Finally, the authors demonstrated this microcolony phenotype was reliant on IEC-exclusive Naip-Nlrc4 inflammasome activation through a series of elegant bone marrow transplant studies as well as the use of a *Naip1-6*$^{\Delta IEC-/-}$ mouse.

Our study published in the same issue of Cell Host and Microbe, focused on the role of the non-canonical inflammasome in restricting *Salmonella* burdens (Knodler et al. 2014). We demonstrated that *Casp11*$^{-/-}$ mice carried higher *Salmonella* loads in their cecal tissues at 7 days p.i., but we observed no significant difference in systemic pathogen loads, which is consistent with previous reports (Knodler et al. 2014; Broz et al. 2012). Also we observed the microcolony phenotype in the epithelial cells of *Casp11*$^{-/-}$ mice but at 24 h p.i. (Knodler et al. 2014). Upon staining the ceca of wild-type mice, we found individual intracellular bacteria that were scattered throughout the epithelium. However, *Casp11*$^{-/-}$ mice displayed numerous IECs containing greater than five bacteria per cell. These microcolonies were observed only rarely in wild-type mice, suggesting that epithelial cell sloughing may be delayed in *Casp11*$^{-/-}$ mice. This phenotype extended to other organ sites, as infection of the gallbladder epithelium by *Salmonella* was also marked by heavy intracellular infection of the epithelium that remained intact in *Casp11*$^{-/-}$ mice, whereas infected epithelial cells in wild-type mice were largely shed into the lumen.

Currently a role for the canonical Naip/Nlrc4 and non-canonical caspase-11 inflammasome pathways has been established in driving IEC resistance against

Salmonella invasion (Knodler et al. 2014; Sellin et al. 2014). It also appears, based on our own work as well as DSS-induced colitis reports that an IEC Nlrp3 inflammasome exists (Knodler et al. 2014; Oficjalska et al. 2015). However, the contribution of each separate IEC inflammasome to host protection against *Salmonella* has not yet been established nor has it been elucidated whether each inflammasome operates independently, or in unison, to ultimately clear the infection. Controversy also surrounds the downstream products of inflammatory caspase activation. Caspase-1 and -4/-11 activation ultimately results in IL-1β and IL-18 secretion as well as host cell pyroptosis. There are conflicting reports about which caspase is responsible for which function. Stowe et al. (2015) propose caspase-4/-11 activation alone induces pyroptosis, while others suggest caspase-1 activation also plays a role in pyroptosis (Man and Kanneganti 2015). Some researchers report that caspase-4/-11 induces caspase-1 activation and subsequent processing of IL-1β and IL-18, while others observe caspase-4/-11 direct processing of IL-18 (Stowe et al. 2015; Storek and Monack 2015; Knodler et al. 2014; Man and Kanneganti 2015). Finally, questions still remain concerning the intracellular location of *Salmonella* required for IEC inflammasome activation. Sellin and colleagues observed their IEC microcolonies remained largely contained within the SCV while it has been established that in BMDM *Salmonella* must escape the SCV and replicate in the cytosol to be detected by caspase-11 (Sellin et al. 2014; Aachoui et al. 2013; Knodler et al. 2014).

4 Inflammasome Evasion by *Salmonella*

Once *Salmonella* has breached the host cell membrane, it engages in an intracellular battle for survival with host cell molecular defense pathways. To evade detection by inflammasomes, *Salmonella* can modify its detectable PAMPs through downregulation of PAMP expression, chemical modification of PAMPs, or through the targeting of host cell types with limited inflammasome activity.

FliC, PrgJ and PrgI are potent activators of the Naip-Nlrc4 inflammasome in murine macrophages. To evade detection, *Salmonella* represses expression of all three proteins during its intracellular lifecycle through its PhoP-PhoQ regulatory system (Miao et al. 2006). The PhoP-PhoQ system is expressed when *Salmonella* is intracellular and paramount for the survival of *Salmonella* within macrophages (Valdivia and Falkow 1997; Miller et al. 1989; Heithoff et al. 1999; Mouslim and Groisman 2003). It is also involved in the regulation of SPI-2 expression, alongside other regulatory sensors (Groisman 2001; Lee et al. 2000, Bijlsma and Groisman 2005). PhoP-PhoQ strongly represses *fliC* and SPI-1 expression, decreasing the presence of these inflammasome formation-inducing PAMPs (Miao et al. 2006). When *Salmonella* is engineered to constitutively express flagellin through a plasmid-based expression system where *fliC* expression is driven by *sseJ* promoter (encoded in SPI-2), this significantly increased IL-1β secretion and caspase-1 cleavage in BMDMs compared to wild-type *Salmonella* (Miao et al. 2010).

Constitutive FliC expression also led to the attenuation of systemic *Salmonella* infection in mice. Following challenge with wild-type *Salmonella*, mice succumbed to infection within 6-8 days whereas the P_{SPI2}-*fliC* SPI-2 expressing strain was unable to kill C57BL/6 mice. The hyperactivation of the Naip5/6-Nlrc4 inflammasome is likely responsible for this decrease in pathogenesis and demonstrates that the ability of *Salmonella* to actively regulate PAMP expression plays a key role in its ability to survive in the host intracellular environment.

Another target of the *Salmonella*-induced inflammasome is LPS. Intracellular LPS stimulates non-canonical inflammasome activity via caspase-11. Recent evidence has suggested the lipid A of LPS binds caspase-11 via its CARD (Shi et al. 2015). LPS binding was shown to induce caspase-11 oligomerization, activation and subsequent pyroptosis by the host cell (Kayagaki et al. 2013). Interestingly, caspase-11 demonstrates decreased activity in response to tetra-acylated lipid IVa and penta-acylated LPS (Shi et al. 2015; Hagar et al. 2013; Kayagaki et al. 2013). This is independent of the capacity of caspase-11 to bind to these moieties and instead was linked to an inability of caspase-11 to undergo oligomerization once bound to these forms of LPS (Shi et al. 2015). *Salmonella* can chemically modify its LPS through the PhoP-PhoQ and PmrA-PmrB regulatory systems (Ernst et al. 2001). Interestingly, LPS from *Salmonella phoP* mutants induces higher TNF-α expression from BMDMs compared to wild-type LPS (Guo et al. 1997). Potential modifications of LPS include reduction in O-antigen length, addition of palmitate to lipid A, addition of phosphate, phosphoethanolamine and/or aminoarabinose to lipid A, or the incorporation or removal of myristate from lipid A (Ernst et al. 2001; Raetz et al. 2007). The lipid A from intracellular wild-type *Salmonella* growing inside the murine macrophage cell line RAW264.7 cells has been shown to be hexa-acylated, and heavily modified with 4-amino-4-deoxy-l-arabinose, phosphoethanolamine, 2-hydroxymyristate, and palmitate (Gibbons et al. 2005). Manipulation of the LPS modification protein machinery alters the immunogenicity of *Salmonella* LPS. When *Salmonella*'s lipid A is changed into a penta-acylated form through mutation to both *msbB* and *pagP*, this significantly reduced its ability to stimulate the human monocyte cell line MM-6 to secrete IL-1β and TNF-α compared to both wild-type LPS and single *msbB* mutants, which expressed a mix of both penta- and hexa-acylated LPS forms (Kong et al. 2011). During an oral *Salmonella* infection of BALB/c mice, pooled serum murine IL-1α levels were also decreased in response to infection by the double *msbB* and *pagP* mutants as compared to wild-type *Salmonella,* whereas IFN-γ levels were significantly increased (Kong et al. 2011). This suggests that *Salmonella* can dampen the host inflammatory response by modifying its LPS, which may impact its ability to associate and interact with caspase-11. The non-canonical inflammasome field is still in its infancy and the ability of caspase-11 to bind to different modified forms of *Salmonella* LPS has not been explored. The binding experiments reported by Shi et al. (2015) were conducted with *Salmonella* lysates produced from boiled overnight cultures, and the structure of the *Salmonella* LPS that activated caspase-11 was not determined. It is likely that different chemical modifications to *Salmonella*'s LPS directly impacts the ability of caspase-11 to undergo oligomerization and represents another strategy by which *Salmonella* modifies its inflammasome-inducing PAMPs as a means to promote its intracellular survival.

In addition to the intrinsic modifications *Salmonella* utilizes to evade the inflammasome, *Salmonella* can also target cellular niches that exhibit reduced inflammasome activity. Recently, *Salmonella* was discovered to preferentially survive in macrophages that display an M2 phenotype (Eisele et al. 2013; McCoy et al. 2012). This was tied to the altered metabolisms displayed between M1 and M2 macrophages. In particular, the upregulation of the eukaryotic transcription factor PPARδ in M2 cells, which is involved in sustaining fatty acid metabolism and was shown to directly impact glucose availability to intracellular *Salmonella*, resulting in enhanced bacterial replication (Eisele et al. 2013). However, it is also possible that the persistence of *Salmonella* in this cell type could be tied to the comparatively reduced expression of inflammasome components in M2 macrophages (Storek and Monack 2015). Jourdan et al. (2013) observed that obese ZDF rats display high mRNA transcripts of *Nlrp3*, *Il1β* and *Il18*, high IL-1β and IL-18 protein levels and robust caspase-1 activity. When treated with the cannabinoid inverse agonist JD5037, this induced an M1-to-M2 shift and the authors observed a significant decrease in all three markers of inflammasome function. Another study observed an increase in the number of M2 macrophages in *Nlrp3*$^{-/-}$-deficient mice (Vandanmagsar et al. 2011). These studies suggest that the inflammasomes of M2 macrophages are less active and this could be an additional factor that regulates what makes these cells attractive as a cellular niche for *Salmonella* replication and persistence.

5 Concluding Remarks

Inflammasome research has seen intense and growing interest in the past ten years and is one of the most exciting and rapidly growing areas in immunology. However, the field itself is not without controversy due to complications resulting from the presence of multiple inflammasomes signaling receptors (each with a unique signaling pathway), the potential for cross-talk between inflammasome signaling pathways, and the discovery that inflammasomes display unique effector functions in distinct cell types. Studying the interactions between *Salmonella* and host inflammasomes has unveiled crucial mechanisms by which the host can restrict the growth of this intracellular pathogen while simultaneously unraveling the highly sophisticated molecular mechanisms that drive inflammasome activation. Inflammasomes combat *Salmonella* infection through their actions in multiple cell types; including monocytes, macrophages, DCs, neutrophils, and IECs. Characterizing the cell-specific factors involved in inflammasome regulation and signaling is a field still in its infancy, but could lead to the elucidation of the proteolytic pathways involved in initiating pyroptosis and the factors responsible for the secretion of IL-1β and IL-18. Taken together, we can anticipate many new and exciting findings as we further explore how inflammasomes regulate host defense against *Salmonella*.

References

Aachoui Y, Leaf IA, Hagar JA, Fontana MF, Campos CG, Zak DE, Tan MH, Cotter PA, Vance RE, Aderem A, Miao EA (2013) Caspase-11 protects against bacteria that escape the vacuole. Science 339(6122):975–978

Anand PK, Malireddi RK, Lukens JR, Vogel P, Bertin J, Lamkanfi M, Kanneganti TD (2012) NLRP6 negatively regulates innate immunity and host defence against bacterial pathogens. Nature 488(7411):389–393

Allen IC, Wilson JE, Schneider M, Lich JD, Roberts RA, Arthur JC, Woodford RM, Davis BK, Uronis JM, Herfarth HH, Jobin C, Rogers AB, Ting JP (2012) NLRP12 suppresses colon inflammation and tumorigenesis through the negative regulation of noncanonical NF-κB signaling. Immunity 36(5):742–754

Baker PJ, Boucher D, Bierschenk D, Tebartz C, Whitney PG, D'Silva DB, Tanzer MC, Monteleone M, Robertson AA, Cooper MA, Alvarez-Diaz S, Herold MJ, Bedoui S, Schroder K, Masters SL (2015) Nlrp3 inflammasome activation downstream of cytoplasmic LPS recognition by both caspase-4 and caspase-5. Eur J Immunol 45(10):2918–2926

Beuzón CR, Banks G, Deiwick J, Hensel M, Holden DW (1999) pH-dependent secretion of SseB, a product of the SPI-2 type III secretion system of *Salmonella typhimurium*. Mol Microbiol 33 (4):806–816

Beuzón CR, Méresse S, Unsworth KE, Ruíz-Albert J, Garvis S, Waterman SR, Ryder TA, Boucrot E, Holden DW (2000) *Salmonella* maintains the integrity of its intracellular vacuole through the action of SifA. EMBO J 19(13):3235–3249 (Erratum in: EMBO J 2000 Aug 1;19 (15):4191)

Bijlsma JJ, Groisman EA (2005) The PhoP/PhoQ system controls the intramacrophage type three secretion system of *Salmonella enterica*. Mol Microbiol 57(1):85–96

Birmingham CL, Smith AC, Bakowski MA, Yoshimori T, Brumell JH (2006) Autophagy controls *Salmonella* infection in response to damage to the *Salmonella*-containing vacuole. J Biol Chem 281(16):11374–11383

Broz P, Newton K, Lamkanfi M, Mariathasan S, Dixit VM, Monack DM (2010) Redundant roles for inflammasome receptors Nlrp3 and Nlrc4 in host defense against *Salmonella*. J Exp Med 207(8):1745–1755

Broz P, Ruby T, Belhocine K, Bouley DM, Kayagaki N, Dixit VM, Monack DM (2012) Caspase-11 increases susceptibility to *Salmonella* infection in the absence of caspase-1. Nature 490(7419):288–291

Cemma M, Kim PK, Brumell JH (2011) The ubiquitin-binding adaptor proteins p62/SQSTM1 and NDP52 are recruited independently to bacteria-associated microdomains to target *Salmonella* to the autophagy pathway. Autophagy 7(3):341–345

Chen KW, Groß CJ, Sotomayor FV, Stacey KJ, Tschopp J, Sweet MJ, Schroder K (2014) The neutrophil Nlrc4 inflammasome selectively promotes IL-1β maturation without pyroptosis during acute *Salmonella* challenge. Cell Rep 8(2):570–582

Casson CN, Yu J, Reyes VM, TAschuk FO, Yadav A, Copenhaver AM, Nguyen HT, Collman RG, Shin S (2015) Human caspase-4 mediates non-canonical inflammasome activation against Gram-negative bacterial pathogens. Proc Natl Acad Sci USA 112(21): 6688–6693

Cirillo DM, Valdivia RH, Monack DM, Falkow S (1998) Macrophage-dependent induction of the *Salmonella* pathogenicity island 2 type III secretion system and its role in intracellular survival. Mol Microbiol 30(1):175–188

Crump JA, Luby SP, Mintz ED (2004) The global burden of typhoid fever. Bull World Health Organ 82:346–353

Crump JA, Ram PK, Gupta SK, Miller MA, Mintz ED (2008) Part I. Analysis of data gaps pertaining to *Salmonella enterica* serotype Typhi infections in low and medium human development index countries, 1984–2005. Epidemiol Infect 136:436–448

De Jong HK, Koh GC, van Lieshout MH, Roelofs JJ, van Dissel JT, van der Poll T, Wiersinga WJ (2014) Limited role for Asc and Nlrp3 during in vivo *Salmonella* Typhimurium infection. BMC Immunol 15:30

Drecktrah D, Knodler LA, Howe D, Steele-Mortimer O (2007) *Salmonella* trafficking is defined by continuous dynamic interactions with the endolysosomal system. Traffic 8(3):212–225

Dreher D, Kok M, Obregon C, Kiama SG, Gehr P, Nicod LP (2002) *Salmonella* virulence factor SipB induces activation and release of IL-18 in human dendritic cells. J Leukoc Biol 72(4):743–751

Eisele NA, Ruby T, Jacobson A, Manzanillo PS, Cox JS, Lam L, Mukundan L, Chawla A, Monack DM (2013) *Salmonella* require the fatty acid regulator PPARδ for the establishment of a metabolic environment essential for long-term persistence. Cell Host Microbe 14(2):171–182

Elinav E, Strowig T, Kau AL, Henao-Mejia J, Thaiss CA, Booth CJ, Peaper DR, Bertin J, Eisenbarth SC, Gordon JI, Flavell RA (2011) NLRP6 inflammasome regulates colonic microbial ecology and risk for colitis. Cell 145(5):745–757

Ernst RK, Guina T, Miller SI (2001) *Salmonella typhimurium* outer membrane remodeling: Role in resistance to host innate immunity. Microbes Infect 3(14–15):1327–1334

Fàbrega A, Vila J (2013) *Salmonella enterica* serovar Typhimurium skills to succeed in the host: virulence and regulation. Clin Microbiol Rev 26(2):308–341

Figueira R, Holden DW (2012) Functions of the *Salmonella* pathogenicity island 2 (SPI-2) type III secretion system effectors. Microbiology 158(Pt 5):1147–1161

Fink SL, Bergsbaken T, Cookson BT (2008) Anthrax lethal toxin and *Salmonella* elicit the common cell death pathway of caspase-1-dependent pyroptosis via distinct mechanisms. Proc Natl Acad Sci USA 105(11):4312–4317

Fink SL, Cookson BT (2006) Caspase-1-dependent pore formation during pyroptosis leads to osmotic lysis of infected host macrophages. Cell Microbiol 8(11):1812–1825

Franchi L, Amer A, Body-Malapel M, Kanneganti TD, Ozören N, Jagirdar R, Inohara N, Vandenabeele P, Bertin J, Coyle A, Grant EP, Núñez G (2006) Cytosolic flagellin requires Ipaf for activation of caspase-1 and interleukin 1beta in *Salmonella*-infected macrophages. Nat Immunol 7(6):576–582

Franchi L, Kamada N, Nakamura Y, Burberry A, Kuffa P, Suzuki S, Shaw MH, Kim YG, Núñez G (2012) Nlrc4-driven production of IL-1β discriminates between pathogenic and commensal bacteria and promotes host intestinal defense. Nat Immunol 13(5):449–456

Galán JE (1998) Interactions of *Salmonella* with host cells: encounters of the closest kind. Proc Natl Acad Sci USA 95(24):14006–14008

Gibbons HS, Kalb SR, Cotter RJ, Raetz CR (2005) Role of Mg2+ and pH in the modification of *Salmonella* lipid A after endocytosis by macrophage tumour cells. Mol Microbiol 55(2):425–440

Keestra-Gounder AM, Tsolis RM, Bäumler AJ (2015) Now you see me, now you don't: The interaction of *Salmonella* with innate immune receptors. Nat Rev Microbiol 13(4):206–216

Kirk MD, Pires SM, Black RE, Caipo M, Crump JA, Devleesschauwer B, Döpfer D, Fazil A, Fischer-Walker CL, Hald T, Hall AJ, Keddy KH, Lake RJ, Lanata CF, Torgerson PR, Havelaar AH, Angulo FJ (2015) World health organization estimates of the global and regional disease burden of 22 foodborne bacterial, protozoal, and viral diseases, 2010: a data synthesis. PLoS Med 12(12):e1001921

Grenier JM, Wang L, Manji GA, Huang WJ, Al-Garawi A, Kelly R, Carlson A, Merriam S, Lora JM, Briskin M, DiStefano PS, Bertin J (2002) Functional screening of five PYPAF family members identifies PYPAF5 as a novel regulator of NF-κB and caspase-1. FEBS Lett 530(1–3):73–78

Groisman EA (2001) The pleiotropic two-component regulatory system PhoP-PhoQ. J Bacteriol 183(6):1835–1842

Guo L, Lim KB, Gunn JS, Bainbridge B, Darveau RP, Hackett M, Miller SI (1997) Regulation of lipid A modifications by *Salmonella typhimurium* virulence genes phoP-phoQ. Science 276(5310):250–253

Gurung P, Malireddi RK, Anand PK, Demon D, Vande Walle L, Liu Z, Vogel P, Lamkanfi M, Kanneganti TD (2012) Toll or interleukin-1 receptor (TIR) domain-containing adaptor inducing interferon-β (TRIF)-mediated caspase-11 protease production integrates Toll-like receptor 4 (TLR4) protein- and Nlrp3 inflammasome-mediated host defense against enteropathogens. J Biol Chem 287(41):34474–34483

Hagar JA, Powell DA, Aachoui Y, Ernst RK, Miao EA (2013) Cytoplasmic LPS activates caspase-11: implications in TLR4-independent endotoxic shock. Science 341(6151): 1250–1253

Havelaar AH, Kirk MD, Torgerson PR, Gibb HJ, Hald T, Lake RJ, Praet N, Bellinger DC, de Silva NR, Gargouri N, Speybroeck N, Cawthorne A, Mathers C, Stein C, Angulo FJ, Devleesschauwer B (2015) World health organization foodborne disease burden epidemiology reference group. World health organization global estimates and regional comparisons of the burden of foodborne disease in 2010. PLoS Med 12(12):e1001923

Heithoff DM, Conner CP, Hentschel U, Govantes F, Hanna PC, Mahan MJ (1999) Coordinate intracellular expression of *Salmonella* genes induced during infection. J Bacteriol 181(3): 799–807

Hersh D, Monack DM, Smith MR, Ghori N, Falkow S, Zychlinsky A (1999) The *Salmonella* invasin SipB induces macrophage apoptosis by binding to caspase-1. Proc Natl Acad Sci USA 96(5):2396–2401

Jourdan T, Godlewski G, Cinar R, Bertola A, Szanda G, Liu J, Tam J, Han T, Mukhopadhyay B, Skarulis MC, Ju C, Aouadi M, Czech MP, Kunos G (2013) Activation of the Nlrp3 inflammasome in infiltrating macrophages by endocannabinoids mediates beta cell loss in type 2 diabetes. Nat Med 19(9):1132–1140

Kayagaki N, Warming S, Lamkanfi M, Vande Walle L, Louie S, Dong J, Newton K, Qu Y, Liu J, Heldens S, Zhang J, Lee WP, Roose-Girma M, Dixit VM (2011) Non-canonical inflammasome activation targets caspase-11. Nature 479(7371):117–121

Kayagaki N, Wong MT, Stowe IB, Ramani SR, Gonzalez LC, Akashi-Takamura S, Miyake K, Zhang J, Lee WP, Muszyński A, Forsberg LS, Carlson RW, Dixit VM (2013) Non-canonical inflammasome activation by intracellular LPS independent of TLR4. Science 341(6151): 1246–1249

Kayagaki N, Stowe IB, Lee BL, O'Rourke K, Anderson K, Warming S, Cuellar T, Haley B, Roose-Girma M, Phung QT, Liu PS, Lill JR, Li H, Wu J, Kummerfeld S, Zhang J, Lee WP, Snipas SJ, Salvesen GS, Morris LX, Fitzgerald L, Zhang Y, Bertram EM, Goodnow CC, Dixit VM (2015) Caspase-11 cleaves gasdermin D for non-canonical inflammasome signalling. Nature 526(7575):666–671

Knodler LA, Vallance BA, Celli J, Winfree S, Hansen B, Montero M, Steele-Mortimer O (2010) Dissemination of invasive *Salmonella* via bacterial-induced extrusion of mucosal epithelia. Proc Natl Acad Sci USA 107(41):17733–17738

Knodler LA, Crowley SM, Sham HP, Yang H, Wrande M, Ma C, Ernst RK, Steele-Mortimer O, Celli J, Vallance BA (2014) Non-canonical inflammasome activation of caspase-4/caspase-11 mediates epithelial defenses against enteric bacterial pathogens. Cell Host Microbe 16(2): 249–256

Kofoed EM, Vance RE (2011) Innate immune recognition of bacterial ligands by Naips determines inflammasome specificity. 477(7366):592–595

Kong Q, Six DA, Liu Q, Gu L, Roland KL, Raetz CR, Curtiss R (2011) Palmitoylation state impacts induction of innate and acquired immunity by the *Salmonella enterica* serovar *typhimurium* msbB mutant. Infect Immun 79(12):5027–5038 (Erratum in: Infect Immun. 2012 Mar;80(3):1302)

Kupz A, Guarda G, Gebhardt T, Sander LE, Short KR, Diavatopoulos DA, Wijburg OL, Cao H, Waithman JC, Chen W, Fernandez-Ruiz D, Whitney PG, Heath WR, Curtiss R 3rd, Tschopp J, Strugnell RA, Bedoui S (2012) Nlrc4 inflammasomes in dendritic cells regulate noncognate effector function by memory $CD8^+$ T cells. Nat Immunol 13(2):162–169

Lamkanfi M, Declercq W, Kalai M, Saelens X, Vandenabeele P (2002) Alice in caspase land. A phylogenetic analysis of caspases from worm to man. Cell Death Differ 9(4):358–361

LaRock DL, Chaudhary A, Miller SI (2015) *Salmonellae* interactions with host processes. Nat Rev Microbiol 13(4):191–205

Lee AK1, Detweiler CS, Falkow S (2000) OmpR regulates the two-component system SsrA-ssrB in *Salmonellae* pathogenicity island 2. J Bacteriol 182(3):771–781

Lightfield KL, Persson J, Brubaker SW, Witte CE, von Moltke J, Dunipace EA, Henry T, Sun YH, Cado D, Dietrich WF, Monack DM, Tsolis RM, Vance RE (2008) Critical function for Naip5 in inflammasome activation by a conserved carboxy-terminal domain of flagellin. Nat Immunol 9(10):1171–1178

Löber S, Jäckel D, Kaiser N, Hensel M (2006) Regulation of *Salmonella* pathogenicity island 2 genes by independent environmental signals. Int J Med Microbiol 296(7):435–447

Majowicz SE, Musto J, Scallan E, Angulo FJ, Kirk M, O'Brien SJ, Jones TF, Fazil A, Hoekstra RM (2010) The global burden of nontyphoidal *Salmonella* gastroenteritis. Clin Infect Dis 50(6):882–889

Malik-Kale P, Winfree S, Steele-Mortimer O (2012) The bimodal lifestyle of intracellular *Salmonella* in epithelial cells: replication in the cytosol obscures defects in vacuolar replication. PLoS One 7(6):e38732

Man SM, Tourlomousis P, Hopkins L, Monie TP, Fitzgerald KA, Bryant CE (2013) *Salmonella* infection induces recruitment of Caspase-8 to the inflammasome to modulate IL-1β production. J Immunol 191(10):5239–5246

Man SM, Ekpenyong A, Tourlomousis P, Achouri S, Cammarota E, Hughes K, Rizzo A, Ng G, Wright JA, Cicuta P, Guck JR, Bryant CE (2014a) Actin polymerization as a key innate immune effector mechanism to control *Salmonella* infection. Proc Natl Acad Sci USA 111 (49):17588–17593

Man SM, Hopkins LJ, Nugent E, Cox S, Glück IM, Tourlomousis P, Wright JA, Cicuta P, Monie TP, Bryant CE (2014b) Inflammasome activation causes dual recruitment of Nlrc4 and Nlrp3 to the same macromolecular complex. Proc Natl Acad Sci USA 111(20):7403–7408

Man SM, Kanneganti TD (2015) Regulation of inflammasome activation. Immunol Rev 265 (1):6–21

Martinon F, Burns K, Tschopp J (2002) The inflammasome: A molecular platform triggering activation of inflammatory caspases and processing of proIL-beta. Mol Cell 10(2):417–426

Mastroeni P, Grant A, Restif O, Maskell D (2009) A dynamic view of the spread and intracellular distribution of *Salmonella enterica*. Nat Rev Microbiol 7(1):73–80

McCarter LL (2006) Regulation of flagella. Curr Opin Microbiol 2:180–186

McCoy MW, Moreland SM, Detweiler CS (2012) Hemophagocytic macrophages in murine typhoid fever have an anti-inflammatory phenotype. Infect Immun 80(10):3642–3649

Meunier E, Dick MS2, Dreier RF, Schürmann N, Kenzelmann Broz D, Warming S, Roose-Girma M, Bumann D, Kayagaki N, Takeda K, Yamamoto M, Broz P (2014) Caspase-11 activation requires lysis of pathogen-containing vacuoles by IFN-induced GTPases. Nature 509(7500): 366–370

Miao EA, Alpuche-Aranda CM, Dors M, Clark AE, Bader MW, Miller SI, Aderem A (2006) Cytoplasmic flagellin activates caspase-1 and secretion of interleukin 1beta via Ipaf. Nat Immunol 7(6):569–575

Miao EA, Leaf IA, Treuting PM, Mao DP, Dors M, Sarkar A, Warren SE, Wewers MD, Aderem A (2011) Caspase-1-induced pyroptosis is an innate immune effector mechanism against intracellular bacteria. Nat Immunol 11(12):1136–1142

Miller SI, Kukral AM, Mekalanos JJ (1989) A two-component regulatory system (phoP phoQ) controls *Salmonella typhimurium* virulence. Proc Natl Acad Sci USA 86(13):5054–5058

Mouslim C, Groisman EA (2003) Control of the *Salmonella ugd* gene by three two-component regulatory systems. Mol Microbiol 47(2):335–344

Oficjalska K, Raverdeau M, Aviello G, Wade SC, Hickey A, Sheehan KM, Corr SC, Kay EW, O'Neill LA, Mills KH, Creagh EM (2015) Protective role for caspase-11 during acute experimental murine colitis. J Immunol 194(3):1252–1260

Poyraz O, Schmidt H, Seidel K, Delissen F, Ader C, Tenenboim H, Goosmann C, Laube B, Thünemann AF, Zychlinsky A, Baldus M, Lange A, Griesinger C, Kolbe M (2010) Protein refolding is required for assembly of the type three secretion needle. Nat Struct Mol Biol 17 (7):788–792

Proell M, Gerlic M, Mace PD, Reed JC, Riedl SJ (2013) The CARD plays a critical role in Asc foci formation and inflammasome signalling. Biochem J 449(3):613–621

Rathinam VA, Vanaja SK, Waggoner L, Sokolovska A, Becker C, Stuart LM, Leong JM, Fitzgerald KA (2012) TRIF licenses caspase-11-dependent Nlrp3 inflammasome activation by gram-negative bacteria. Cell 150(3):606–619

Raupach B, Peuschel SK, Monack DM, Zychlinsky A (2006) Caspase-1-mediated activation of interleukin-1beta (IL-1beta) and IL-18 contributes to innate immune defenses against *Salmonella enterica* serovar Typhimurium infection. Infect Immun 74(8):4922–4926

Raetz CR, Reynolds CM, Trent MS, Bishop RE (2007) Lipid a modification systems in gram-negative bacteria. Annu Rev Biochem 76:295–329

Rühl S, Broz P (2015) Caspase-11 activates a canonical Nlrp3 inflammasome by promoting K(+) efflux. Eur J Immunol 45(10):2927–2936

Sanman LE, Qian Y, Eisele NA, Ng TM, van der Linden WA, Monack DM, Weerapana E, Bogyo M (2016) Disruption of glycolytic flux is a signal for inflammasome signaling and pyroptotic cell death. Elife 5(pii):e13663

Schmid-Burgk JL, Gaidt MM, Schmidt T, Ebert TS, Bartok E, Hornung V (2015) Caspase-4 mediates non-canonical activation of the Nlrp3 inflammasome in human myeloid cells. Eur J Immunol 45(10):2911–2917

Sellin ME, Müller AA, Felmy B, Dolowschiak T, Diard M, Tardivel A, Maslowski KM, Hardt WD (2014) Epithelium-intrinsic Naip/Nlrc4 inflammasome drives infected enterocyte expulsion to restrict *Salmonella* replication in the intestinal mucosa. Cell Host Microbe 16 (2):237–248

Steele-Mortimer O, Méresse S, Gorvel JP, Toh BH, Finlay BB (1999) Biogenesis of *Salmonella typhimurium*-containing vacuoles in epithelial cells involves interactions with the early endocytic pathway. Cell Microbiol 1(1):33–49

Shi J, Zhao Y, Wang Y, Gao W, Ding J, Li P, Hu L, Shao F (2014) Inflammatory caspases are innate immune receptors for intracellular LPS. Nature 514(7521):187–192

Shi J, Zhao Y, Wang K, Shi X, Wang Y, Huang H, Zhuang Y, Cai T, Wang F, Shao F (2015) Cleavage of GSDMD by inflammatory caspases determines pyroptotic cell death. Nature 526 (7575):660–665

Storek KM, Monack DM (2015) Bacterial recognition pathways that lead to inflammasome activation. Immunol Rev 265(1):112–129

Stowe I, Lee B, Kayagaki N (2015) Caspase-11: arming the guards against bacterial infection. Immunol Rev 265(1):75–84

Tam MA, Rydström A, Sundquist M, Wick MJ (2008) Early cellular responses to *Salmonella* infection: dendritic cells, monocytes, and more. Immunol Rev 225:140–162

Valdivia RH, Falkow S (1997) Fluorescence-based isolation of bacterial genes expressed within host cells. Science 277(5334):2007–2011

Vandanmagsar B, Youm YH, Ravussin A, Galgani JE, Stadler K, Mynatt RL, Ravussin E, Stephens JM, Dixit VD (2011) The Nlrp3 inflammasome instigates obesity-induced inflammation and insulin resistance. Nat Med 17(2):179–188

von Moltke J, Trinidad NJ, Moayeri M, Kintzer AF, Wang SB, van Rooijen N, Brown CR, Krantz BA, Leppla SH, Gronert K, Vance RE (2012) Rapid induction of inflammatory lipid mediators by the inflammasome *in vivo*. Nature 490(7418):107–111

Winter SE, Winter MG, Atluri V, Poon V, Romão EL, Tsolis RM, Bäumler AJ (2015) The flagellar regulator TviA reduces pyroptosis by *Salmonella enterica* serovar typhi. Infect Immun 83(4):1546–1555

Wright EK, Goodart SA, Growney JD, Hadinoto V, Endrizzi MG, Long EM, Sadigh K, Abney AL, Bernstein-Hanley I, Dietrich WF (2003) Naip5 affects host susceptibility to the intracellular pathogen Legionella pneumophila. Curr Biol 13(1):27–36

Wynosky-Dolfi MA, Snyder AG, Philip NH, Doonan PJ, Poffenberger MC, Avizonis D, Zwack EE, Riblett AM, Hu B, Strowig T, Flavell RA, Jones RG, Freedman BD, Brodsky IE

(2014) Oxidative metabolism enables *Salmonella* evasion of the Nlrp3 inflammasome. J Exp Med 211(4):653–668

Yang J, Zhao Y, Shi J, Shao F (2013) Human Naip and mouse Naip1 recognize bacterial type III secretion needle protein for inflammasome activation. Proc Natl Acad Sci USA 110 (35):14408–14413

Yang J, Zhang E, Liu F, Zhang Y, Zhong M, Li Y, Zhou D, Chen Y, Cao Y, Xiao Y, He B, Yang Y, Sun Y, Lu M, Yan H (2014) Flagellins of *Salmonella Typhi* and nonpathogenic *Escherichia coli* are differentially recognized through the Nlrc4 pathway in macrophages. J Innate Immun 6(1):47–57

Yaraghi Z, Diez E, Gros P, MacKenzie A (1999) cDNA cloning and the 5′ genomic organization of Naip2, a candidate gene for murine Legionella resistance. Mamm Genome 10(7):761–763

Zaki MH, Vogel P, Body-Malapel M, Lamkanfi M, Kanneganti TD (2010) IL-18 production downstream of the Nlrp3 inflammasome confers protection against colorectal tumor formation. J Immunol 185(8):4912–4920

Zaki MH, Vogel P, Malireddi RK, Body-Malapel M, Anand PK, Bertin J, Green DR, Lamkanfi M, Kanneganti TD (2011) The NOD-like receptor NLRP12 attenuates colon inflammation and tumorigenesis. Cancer Cell 20(5):649–660

Zaki MH, Man SM, Vogel P, Lamkanfi M, Kanneganti TD (2014) *Salmonella* exploits NLRP12-dependent innate immune signaling to suppress host defenses during infection. Proc Natl Acad Sci USA 111(1):385–390

Zhao Y, Yang J, Shi J, Gong YN, Lu Q, Xu H, Liu L, Shao F (2011) The Nlrc4 inflammasome receptors for bacterial flagellin and type III secretion apparatus. Nature 477(7366):596–600

Activation and Evasion of Inflammasomes by *Yersinia*

Naomi H. Philip, Erin E. Zwack and Igor E. Brodsky

Abstract The innate immune system plays an essential role in initiating the early response against microbial infection, as well as instructing and shaping subsequent responses. Microbial pathogens are enormously diverse in terms of the niches they occupy, their metabolic properties and requirements, and the cellular pathways that they target. Nevertheless, innate sensing of pathogens triggers a relatively stereotyped set of responses that involve transcriptional induction of key inflammatory mediators, as well as post-translational assembly and activation of a multiprotein inflammatory complex termed 'the inflammasome.' Along with classical Pattern Recognition Receptors, the inflammasome activation pathway has emerged as a key regulator of tissue homeostasis and immune defense. Components of the inflammasome generally exist within the cell in a soluble, monomeric state, and oligomerize in response to diverse enzymatic activities associated with infection or cellular stress. Inflammasome assembly triggers activation of the pro-enzyme caspase-1, resulting in the cleavage of caspase-1 targets. The most extensively studied targets are the cytokines of the IL-1 family, but the recent discovery of Gasdermin D as a novel target of caspase-1 and the related inflammatory caspase, caspase-11, has begun to mechanistically define the links between caspase-1 activation and cell death. Cell death is a hallmark of macrophage infection by many pathogens, including the gram-negative bacterial pathogens of the genus *Yersinia*. Intriguingly, the activities of the *Yersinia*-secreted effector proteins and the type III secretion system (T3SS) itself have been linked to both inflammasome activation

N.H. Philip · E.E. Zwack · I.E. Brodsky (✉)
Department of Pathobiology, University of Pennsylvania School of Veterinary Medicine, Philadelphia, PA 19104, USA
e-mail: ibrodsky@vet.upenn.edu

N.H. Philip · I.E. Brodsky
Institute for Immunology, University of Pennsylvania Perelman School of Medicine, Philadelphia, PA 19104, USA

N.H. Philip
Immunology Graduate Group, Philadelphia, PA 19104, USA

E.E. Zwack
Cell and Molecular Biology Graduate Group, Philadelphia, PA 19104, USA

S. Backert (ed.), *Inflammasome Signaling and Bacterial Infections*,
Current Topics in Microbiology and Immunology 397,
DOI 10.1007/978-3-319-41171-2_4

and evasion during infection. The balance between these activating and inhibitory activities shapes the outcome of *Yersinia* infection. Here, we describe the current state of knowledge on interactions between *Yersinia* and the inflammasome system, with the goal of integrating these findings within the general framework of inflammasome responses to microbial pathogens.

Contents

1 Introduction

1.1 Evolutionary Relationships Among the Pathogenic Yersinia

Cell death is an evolutionarily conserved immune response to microbial infection, as it prevents pathogen replication and can provide pro-inflammatory signals necessary for an effective immune response (Campisi et al. 2014; Munoz-Pinedo 2012). Extensive death of immune cells has been a well-appreciated hallmark of infection by pathogenic bacteria of the *Yersinia* species. The three pathogenic *Yersiniae, Y. pestis* (*Yp*), *Y. pseudotuberculosis* (*Yptb*), and *Y. enterocolitica* (*Ye*), all harbor a 70 kB virulence plasmid encoding a conserved type III secretion system (T3SS) and virulence factors, termed the *Yersinia* outer proteins (Yops) (Viboud and Bliska 2005; Wren 2003). While environmental *Yersinia* species that do not possess the T3SS are capable of causing disease in fish, for example, *Y. ruckeri*, all mammalian pathogenic *Yersinia* require the T3SS and associated Yops to cause disease. The T3SS provides multiple functions that enable *Yersinia* to evade or modulate host responses, including blockade of phagocytosis, interference with oxidative burst in neutrophils, and disruption of host signaling pathways to suppress innate and adaptive immune responses (Cornelis 2006). The mammalian pathogenic

Yersinia have therefore long been thought to replicate extracellularly in infected lymphoid organs, although it is possible that an intracellular stage exists during the *Yersinia* life cycle in vivo (Grabenstein et al. 2004). In this review, we focus on the interactions between *Yersinia* virulence factors and host cell death machinery. While a great deal of work has been done in this area, recent advances in understanding distinct pathways and mechanisms of cell death in the context of infection and inflammation have enabled further advances in understanding how *Yersinia* interference with these pathways might alter the balance between pathogenesis and host defense.

Yp was identified as the causative agent of plague in 1894 independently by Alexander Yersin and Shibasaburo Kitasato, who came from the Pasteur and Koch schools of microbiology, respectively (Rosenberg 1968). Transmission typically occurs either through the bite of an infected flea or the inhalation of *Yp* containing droplets from an individual infected with the pneumonic form of plague (Wren 2003). Currently, wild animal populations in the western USA provide a reservoir for *Yp* in the USA, and a number of cases have occurred recently in National Parks (Ben Ari et al. 2008; Kwit et al. 2015). The appearance of multidrug resistance and the potential use of *Yp* as a bioterrorism agent continue to draw interest in understanding the biology of *Yersinia* infection. Inside the mammalian host, *Yp* has a preference for leukocytes (Balada-Llasat and Mecsas 2006; Maldonado-Arocho et al. 2013; Marketon et al. 2005). During the bubonic form of the plague, bacteria traffic to and replicate in the lymph node, resulting in inflammation and swelling of the lymph node. *Yp* can also spread to the lungs (pneumonic plague) or systemically (septicemic plague). *Yp* evolved directly from *Yptb* through a combination of gene acquisition and gene loss that allowed for transmission by fleas, a change in virulence, and restriction of lifestyle (Achtman et al. 1999; Chain et al. 2004; Hinnebusch et al. 2002; Zimbler et al. 2015).

While *Yp* is a vector-borne disease that often spreads to systemic tissues and sites, both *Yptb* and *Ye* are transmitted through the oral–fecal route (Wren 2003). All pathogenic *Yersinia* share a tropism for lymph nodes and leukocytes (Durand et al. 2010; Koberle et al. 2009). In contrast to *pestis*, the diseases caused by the gastro-intestinal *Yersinia* species are generally self-limiting in immunocompetent individuals. Interestingly, *Ye* is found as a commensal in over 80 % of commercially farmed pigs and is the most common cause of food-borne *Yersinia* infection (Bhaduri et al. 2005; Wesley et al. 2008).

Yptb in contrast, is a natural pathogen of sylvatic rodents, and like *Ye*, causes systemic disease in rodent models of infection. Spread of *Yptb* occurs from peripheral sites in the intestine and Peyer's patches to mesenteric lymph nodes, and the systemic tissues of the reticuloendothelial system, the spleen, and liver. Interestingly, this is not a linear process and involves at least some fraction of bacteria that bypass the lymph nodes to go directly to the systemic tissues (Barnes et al. 2006). This may involve spread via the circulation either as free bacteria or potentially subsequent to uptake by intestinal phagocytic cells that traffic the bacteria to systemic sites, which has been found to occur in *Salmonella* infection (Vazquez-Torres et al. 1999). Thus, both *Ye* and *Yptb* have been used extensively as models of systemic *Yersinia* infection in mice. Below we discuss key aspects of the *Yersinia*–host interaction from the standpoint of how the innate immune system detects *Yersinia* infection, and the mechanisms by which *Yersinia* modulates these inflammatory responses.

1.2 Features of Yersinia*-Induced Cell Death*

The YopJ protein of *Yp* and *Yptb* (termed YopP in *Y. enterocolitica*) has both deubiquitinase and acyltransferase activities (Mittal et al. 2006; Mukherjee et al. 2006; Zhou et al. 2005), which potently blocks nuclear factor kappa B (NF-κB) and mitogen-activated protein kinase (MAPK) signaling (Palmer et al. 1998; Schesser et al. 1998). Blockade of NF-κB and MAPK signaling in the context of TLR engagement results in both inhibition of cytokine production and death of *Yersinia*-infected cells (Mills et al. 1997; Monack et al. 1997; Ruckdeschel et al. 1998). Among the sequenced strains of pathogenic *Yersiniae*, YopJ and YopP share 95–98 % identity across the full length of the protein sequence; however, key polymorphisms impact both YopJ/P enzymatic activity and translocation, which tunes the precise outcome of *Yersinia* infection (Brodsky and Medzhitov 2008; Ruckdeschel et al. 2001; Zauberman et al. 2006; Zheng et al. 2011). In particular, while YopJ is an important virulence factor that facilitates the systemic spread of *Yptb* to systemic sites following oral infection (Monack et al. 1998), elevated YopJ activity in the context of *Yptb* or *Yp* attenuates the infection and induces a more robust adaptive immune response (Brodsky and Medzhitov 2008; Zauberman et al. 2009).

Yersinia-infected cells have been observed to exhibit features of apoptosis, pyroptosis, or necrosis, depending on the state of the cells and the cell type involved (Bergsbaken and Cookson 2007; Erfurth et al. 2004; Grobner et al. 2006; Philip et al. 2014; Zheng et al. 2012). In particular, priming of cells in vitro via TLR stimulation results in protection from *Yersinia*-induced cell death relative to cells that encounter *Yersinia* prior to such stimulation (Bergsbaken and Cookson 2007; Brodsky et al. 2010). Furthermore, dendritic cells are significantly more resistant to YopJ-induced cell death than macrophages in vitro (Brodsky and Medzhitov 2008). A number of studies have revealed key players in cell death pathways during *Yersinia* infection, providing insight into mechanisms of *Yersinia*-induced cell death, but little is currently known about the precise nature and role of *Yersinia* death in vivo. *Yersinia* is thought to primarily replicate as an extracellular pathogen that evades phagocytosis by neutrophils and monocytic cells in lymphoid tissues. Cell death has therefore been viewed as a strategy for *Yersinia* to eliminate host phagocytes (Monack et al. 1998). However, several studies suggest that host cell death during *Yersinia* infection promotes anti-*Yersinia* immunity, although the downstream consequences of *Yersinia*-induced cell death on induction of immune responses in vivo are not entirely clear (Bergman et al. 2009; Brodsky and Medzhitov 2008; Zauberman et al. 2009). Interestingly, recent studies in *Y. pestis* models of pneumonic plague have revealed that the Plasminogen activator protease Pla targets Fas Ligand for degradation, which limits apoptosis in vivo and promotes pulmonary bacterial replication (Caulfield et al. 2014). If *Yersiniae* transit through an intracellular stage during in vivo infection (Grabenstein et al. 2004), excessive death of the infected cell due to the activity of YopJ may eliminate its intracellular replicative niche. Since cell death is a characteristic feature of infection by many pathogens, including *Yersinia, Yersinia* has served as a genetically tractable and powerful system to investigate the mechanisms and consequences of cell death in immune defense.

2 Mechanisms of *Yersinia*-Induced Cell Death

The original reports of *Yersinia*-induced death of macrophages identified a role for YopJ and described this death as apoptosis (Mills et al. 1997; Monack et al. 1997). Interestingly, in contrast to other forms of apoptosis, *Yersinia* infection also induced activation of caspase-1 (Bergsbaken and Cookson 2007; Lilo et al. 2008) via mechanisms that are distinct from other known activators of caspase-1 (Brodsky et al. 2010; Philip et al. 2014). In contrast to other pathogens, *Yersinia* infection appears to both activate and inhibit multiple pathways of cell death, depending on the availability of different host and bacterial inflammatory signals (Fig. 1). Below, we discuss the different pathways of cell death that are triggered during Yersinia infection, and the way in which they are induced.

2.1 YopJ-Induced Death via the Extrinsic Apoptosis Pathway

Macrophages stimulated with LPS in the presence of protein synthesis inhibitors or inhibitors of NF-κB signaling undergo cell death (Philip et al. 2014; Ruckdeschel et al. 2004; Zhang et al. 2005). Consistent with this, *Tlr4*$^{-/-}$ macrophages are resistant to YopJ-dependent apoptosis, as are cells deficient in the TLR3/4 adaptor TRIF, but not MyD88 (Haase et al. 2003; Philip et al. 2014; Ruckdeschel et al. 2004; Zhang et al. 2005). Importantly, infection of dendritic cells with *Yersinia* leads to the formation of a Fadd/caspase-8/Ripk1 complex and caspase-8 activation (Grobner et al. 2007). In macrophage-like cell lines, activation of caspase-8 is accompanied by *Yersinia*-induced cleavage of the pro-apoptotic protein Bid, leading to the classical hallmarks of apoptosis including cytochrome-c release, and caspase-3 and caspase-7 cleavage (Denecker et al. 2001). While early studies observed that treatment with broad-spectrum caspase inhibitors reduced the number of TUNEL$^+$ cells during *Yersinia* infection (Denecker et al. 2001; Monack et al. 1998), *Yersinia*-infected dendritic cells treated with a pan-caspase inhibitor undergo

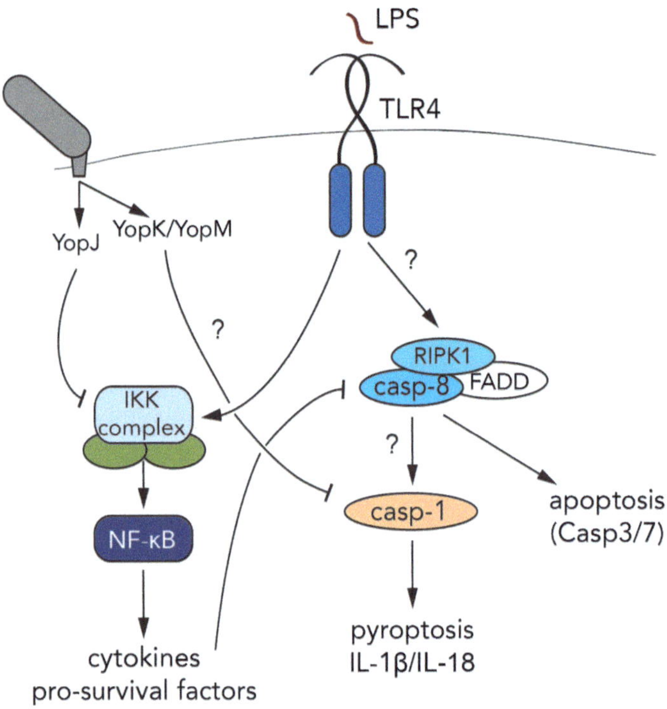

Fig. 1 Activation and inhibition of cell death signaling during *Yersinia* infection. TLR signaling induces upregulation of NF-κB- and MAPK-dependent gene expression, including proinflammatory mediators and pro-survival factors. These prosurvival factors limit induction of caspase-8-dependent apoptosis. Blockade of IKK and MKK function by YopJ de-represses caspase-8 activation, which triggers activation of multiple caspases, including caspase-1. This caspase-1 activation pathway is distinct from both canonical and non-canonical inflammasomes as it requires blockade of NF-κB and MAPK signaling by YopJ and does not engage other inflammasome sensors. *Yersinia* injection of YopK and YopM limits the activation of caspase-1 and pyroptosis in LPS-primed macrophages, thereby potentially allowing *Yersinia* to evade these responses. These responses are induced in murine macrophages by both *Y. pestis* and *Y. pseudotuberculosis* indicating that either murine cells sense *Y. pestis* hepta-acylated LPS enough to trigger death signaling or that alternative TLRs are also linked to these pathways

substantial cell death that exhibits morphological features of necrosis (Grobner et al. 2006).

While the mechanisms responsible for this alternative form of cell death were not clear initially, more recent studies uncovered a caspase-independent pathway of cell death driven by Receptor Interacting Protein Kinases 1 and 3 (Ripk1 and Ripk3), termed programmed necrosis (Degterev et al. 2008; He et al. 2011). Intriguingly, treatment of *Yersinia*-infected cells with caspase inhibitors induces cell death that is blocked either by Ripk3 deficiency or pharmacological inhibitors of Ripk1 or Ripk3, indicating that this cell death is programmed necrosis (Philip et al. 2014; Weng et al. 2014). Importantly however, Ripk3 deficiency or treatment with necrostatin has no impact on the levels of *Yersinia*-induced cell death on its own, suggesting that the primary mode of cell death that takes place in response to the activity of YopJ is caspase-8-mediated apoptosis (Philip et al. 2014).

Interestingly, YopJ-dependent apoptosis is also associated with caspase-1 activation (Brodsky et al. 2010; Lilo et al. 2008; Zheng et al. 2011), and the extent of YopJ-mediated NF-κB inhibition correlates with the degree of caspase-1 activation (Zheng et al. 2011) (Fig. 1). This is consistent with the finding that deletion of IKKβ in macrophages induces spontaneous inflammasome activation (Greten et al. 2007). Although Nlrp3 and the inflammasome adaptor Asc are required for YopJ-dependent secretion of IL-1β and IL-18 (Zheng et al. 2011), the mechanism by which YopJ activates caspase-1 is unclear, as caspase-1 processing and YopJ-dependent cell death still occur equivalently to wild-type cells in the absence of Asc, Nlrc4, or Nlrp3 (Brodsky et al. 2010; Philip et al. 2014). Intriguingly, during *Yersinia* infection, YopJ triggers a unique pathway of caspase-1 activation that requires activation of caspase-8 (Philip et al. 2014; Weng et al. 2014). Caspase-8 was also found to play a role in inflammasome priming in the context of canonical inflammasome stimuli (Gurung et al. 2014; Man et al. 2013). However, the caspase-8-dependent activation of caspase-1 during *Yersinia* infection occurs when priming itself is inhibited by the NF-κB-blocking activity of YopJ. Moreover, both the presence and enzymatic activity of caspase-8 were necessary for activation of caspase-1 by *Yersinia* infection (Philip et al. 2014; Weng et al. 2014).

Distinct inflammasome complexes with different functions have been identified and could potentially account for these observations. Caspase-1 could be recruited to an Nlrp3/Asc complex that regulates IL-1β and IL-18 production and to a separate caspase-8-containing complex that activates cell death. Indeed, during *Salmonella* infection, a complex containing catalytically active caspase-1, but not Asc triggers cell death but not cytokine secretion, while a distinct Asc-containing focus mediates caspase-1 processing and cytokine secretion (Broz et al. 2010).

2.2 Inflammasome Sensing of Yersinia *Infection and Its Evasion by YopK*

Nlrp12 was also recently found to induce inflammasome activation in response to *Y. pestis* infection, and both Nlrp3 and Nlrp12 contributed to host defense against *Yersinia*, presumably via induction of caspase-1-dependent IL-1β and IL-18 (Brodsky et al. 2010; Vladimer et al. 2013). Nlrp12 inflammasome activation occurs under conditions when *Yp* LPS is altered to provide increased stimulatory activity to TLR4, as *Y. pestis* normally evades TLR4 by producing a tetra-acylated LPS molecule when grown at 37 °C. YopJ may also activate an Nlrp12 inflammasome, although this remains to formally be demonstrated. Whether Nlrp12 also plays a cryptic role in other instances of TLR4 priming, or whether NLRP12 activation depends on the interaction between increased TLR stimulation and an activity of the *Yersinia* T3SS (potentially YopJ) remains to be determined. NLRP12 protein levels are upregulated by LPS, similarly to Nlrp3 raising the possibility that Nlrp12 activation in this context may be due to enhanced upregulation of Nlrp12 in response to elevated levels of *Yp* hexa-acylated LPS. Interestingly, in the absence of Nlrp12, there is no difference in levels of IL-1β secretion in response to parental *Yp* and *Yp* expressing hexa-acylated LPS (Vladimer et al. 2013), suggesting that NLRP12 responds preferentially to this LPS modification. Whether Nlrp12 also responds to other hexa-acylated forms of LPS, or whether the Nlrp12 inflammasome responds to a combination of signals, such as YopJ together with hexa-acylated LPS, remains to be established.

In addition to Nlrp12, the other primary NLR protein that responds to *Yersinia* infection is Nlrp3, which is triggered in response to the *Yersinia* T3SS in the absence of all known secreted effectors, particularly upon deletion of YopK (Brodsky et al. 2010). The T3SS of bacterial pathogens could be viewed as a 'Pattern of Pathogenesis,' analogous to pore-forming toxins of gram-positive pathogens that are essential to the virulence of the organism, but which also disrupt host membrane integrity and modulate signaling networks, resulting in increased host sensing (Vance et al. 2009). Interestingly, the pore-forming activity of the *Yersinia* T3SS is not required for its ability to induce inflammasome activation (Kwuan et al. 2013; Zwack et al. 2014). Rather, the ability to inject components of the T3SS translocon was particularly important to activate the T3SS-induced inflammasome. Notably, increased translocation of the T3SS pore-forming translocon proteins is responsible for inflammasome activation in response to *Yptb* infection (Zwack et al. 2014).

The precise mechanism by which the *Yersinia* T3SS triggers inflammasome activation remains unclear; however, a secreted effector protein, termed YopK, specifically prevents inflammasome activation in response to the *Yersinia* translocon proteins (Brodsky et al. 2010). YopK is a general regulator of *Yersinia* translocation, as *yopK* mutant *Yersinia* strains exhibit significantly greater level of translocation than their wild-type counterparts (Holmstrom et al. 1995, 1997). Interestingly, YopK-deficient *Yersinia* inject increased levels of the translocon proteins YopB and YopD into host cells, and limiting this translocation prevents inflammasome activation by the *Yersinia* T3SS (Zwack et al. 2014). How YopK prevents hyper-injection of translocon components into the cell remains unclear. While YopK interacts biochemically with translocon components (Brodsky et al. 2010; Thorslund et al. 2011), it is also translocated into the cytosol of infected cells and has been proposed to regulate translocation from within the cytosol (Dewoody et al. 2011). Interactions between YopK and the scaffolding protein Rack1 are important for the ability of YopK to regulate translocation, but whether these interactions are also important for YopK to limit inflammasome activation is not known (Thorslund et al. 2011).

In contrast to other gram-negative pathogens with T3SSs, such as *Salmonella* and *Shigella*, *Yersinia* does not trigger large amounts of NLRC4 inflammasome activation, and the contribution of Nlrc4 to anti-*Yersinia* inflammasome responses is only revealed in the absence of NLRP3 (Brodsky et al. 2010). This is likely due to the counter-regulation between the *Yersinia* T3SS and flagellin, which is downregulated at 37 °C (Minnich and Rohde 2007). Like the inner rod protein of the *Salmonella* pathogenicity island I T3SS, the *Yersinia* homolog YscI may also trigger the Naip/Nlrc4 inflammasome at low levels, as YscI possesses the conserved C-terminal sequence required to trigger this inflammasome (Miao et al. 2010). Interestingly, the *Yersinia* T3SS activates both the canonical Nlrp3 inflammasome, and the more recently described non-canonical caspase-11 inflammasome, which is activated in response to direct binding of cytosolic LPS (Casson et al. 2013). This non-canonical inflammasome requires priming by LPS and signaling through a TRIF-IFN-IFNAR feedback loop to upregulate caspase-11, which directly senses the presence of cytosolic LPS (Broz et al. 2012; Gurung et al. 2012; Kayagaki et al. 2011; Rathinam et al. 2012; Shi et al. 2014). It is possible that the canonical and non-canonical inflammasomes are triggered by distinct patterns of pathogenesis present within the cytosol during *Yersinia* infection (injection of translocon Yops vs. LPS). An alternative possibility, which remains to be tested, is that *Yersinia* lacking YopK trigger both canonical and non-canonical pathways via a common trigger independently of cytosolic LPS.

2.3 Evasion of Inflammasome by YopM

YopM is a T3SS effector that was originally found to play a role in the suppression of innate immune responses during *Yp* infection (Nemeth and Straley 1997). Interestingly, YopM interacted with two host kinases, Rsk1 and Prk2, and was found to play a role in systemic induction of the immuno-suppressive cytokine IL-10 during systemic mouse infection with *Yptb* (McPhee et al. 2010, 2012). YopM is a leucine-rich repeat (LRR)-containing effector protein, whose LRR comprises the majority of the protein length (Evdokimov et al. 2001). Surprisingly, YopM is highly polymorphic among the different pathogenic *Yersinia* species, and even within strains of the same species. The length of the LRR domain is highly variable, containing between 13 and 21 repeats in the different Yersinia strains, and is followed by an unstructured tail that is conserved in different YopM isoforms. Consistent with other LRR-containing proteins, the YopM LRR domain adopts an overall horseshoe-like structure, and this fold is predicted to serve as a binding platform for host cell proteins (Evdokimov et al. 2001). In the host cytosol, YopM binds to and activates two protein kinases, Rsk1 and Prk2 (Hentschke et al. 2010; McDonald et al. 2003). The C-terminal tail of YopM is required for both Rsk1 binding in macrophages and virulence of *Yersinia* in mice (McCoy et al. 2010; McPhee et al. 2010). However, it is currently not clear whether the interaction between YopM and Rsk1 or Prk2 is specifically responsible for the contribution of YopM to *Yersinia* virulence, or whether an alternative function of YopM is affected by ablation of the C terminus.

Notably, a particular YopM isoform from the YPIII strain of *Yp* and some *Yptb* strains binds to and inhibits caspase-1 in macrophages (LaRock and Cookson 2012). This inhibitory effect of YopM on inflammasome activation is revealed in LPS-primed macrophages, in the presence of YopK (Chung et al. 2014; LaRock and Cookson 2012). During infection with YopK-deficient *Yersinia*, both primed and unprimed macrophages undergo rapid inflammasome activation as a consequence of the activity of the T3SS (described above). These findings suggest an interaction between the macrophage activation state and the activity of various *Yersinia* effector proteins that determines the cellular response to *Yersinia* infection (Ruckdeschel and Richter 2002). Interestingly, recent findings indicate that in primed macrophages, YopJ, which normally induces inflammasome activation (as described above), actually functions together with YopM to prevent inflammasome activation (Ratner et al. 2016; Schoberle et al. 2016). While YopM can bind either to caspase-1 directly, thereby acting as a decoy substrate, or interact with the small GTPase IQGAP1, which also limits inflammasome activation (Chung et al. 2014; LaRock and Cookson 2012), the mechanism for how YopJ limits inflammasome activation in primed cells is not clear.

As might be expected, in naïve or unstimulated innate immune cells, YopJ limits inflammasome priming due to its ability to interfere with induction of both MyD88- and Trif-dependent responses (Rosadini et al. 2015), thereby likely interfering with upregulation of inflammasome pathway constituents in innate cells (Ratner et al. 2016; Rosadini et al. 2015). YopM interactions with caspase-1 prevent both recruitment of caspase-1 to inflammasomes and caspase-1 auto-cleavage (LaRock and Cookson 2012). Initial studies observed that translocon insertion by wild-type *Yptb* triggers caspase-1 in LPS-activated macrophages (Bergsbaken and Cookson 2007), but with delayed kinetics compared with a *yopM* mutant. Purified YopM bound to cleaved caspase-1 and inhibited caspase-1 activity in vitro. Binding to and inhibition of caspase-1 required Asp residue 271 located in the 10th LRR of YopM. This Asp residue is located in a consensus caspase-1 cleavage motif (YLTD); however, caspase-1 did not cleave YopM, and not all YopM isoforms have this motif. These data suggest that the YLTD motif in YopM acts as a pseudosubstrate inhibitor of caspase-1, similar to endogenous inhibitors such as Flightless-1 (LaRock and Cookson 2012).

Direct evidence that inhibition of caspase-1 by YopM is functionally linked to bacterial virulence was demonstrated by the findings that the virulence deficiency of YopM deletion is overcome by infection of caspase-1-deficient mice (LaRock and Cookson 2012; Chung et al. 2014). The above data suggest a mechanism for how YopM inhibits pro-inflammatory responses, and also raise a number of important questions. For example, how do YopM isoforms that lack the YLTD motif also inhibit caspase-1? Is YopM a bifunctional effector, with the LRR region acting as a caspase-1 inhibitor and the C-terminal tail working in conjunction with RSK1? It will also be interesting to determine if inhibition of caspase-1 is linked to the remarkable demonstration that purified *Y. enterocolitica* YopM can penetrate into cultured cells and inhibit transcription of pro-inflammatory cytokine genes (Ruter et al. 2014).

3 Interactions Between *Yersinia* and Cell Death Pathways In Vivo

3.1 Impact of Yersinia*-Induced Cell Death on Pathogenesis and Host Defense*

Early studies observed that macrophages and dendritic cells infected by *Yersinia* exhibit characteristics of apoptosis, specifically membrane blebbing, nuclear condensation, DNA fragmentation, and formation of large cytoplasmic vacuoles (Monack et al. 1997; Ruckdeschel et al. 1997). Apoptosis has classically been viewed as immunologically silent, but growing evidence suggests that during infection or in the context of tissue stress, apoptosis may in fact promote inflammatory responses (Green et al. 2009; Torchinsky et al. 2009). Furthermore, apoptotic cells can be phagocytosed, and their associated microbial antigens used to prime $CD8^+$ T-cell responses (Heath and Carbone 2001). Therefore, while cell death during *Yersinia* infection is thought to be apoptotic, it may not be immunologically silent. Below, we discuss the current state of understanding how *Yersinia*-induced cell death contributes to bacterial virulence or host defense in vivo.

3.2 Activation of Cell Death by Yersinia *Virulence Factors In Vivo*

A number of studies indicate that YopJ/P promotes *Yersinia* virulence in vivo. Oral infection with *Yptb* and *Ye* demonstrate that YopJ/P contributes to systemic disease and barrier dysfunction (Jung et al. 2012; Meinzer et al. 2012; Monack et al. 1998). YopJ is dispensable for colonization of the Peyer's patches (PPs) and mesenteric lymph nodes (mLNs), especially at higher infectious doses; however, YopJ-deficient *Yersinia* had significantly reduced levels of spleen colonization after oral infection (Monack et al. 1998). Spleens and mLNs from mice infected with YopJ-sufficient bacteria had a higher percentage of $Mac1^+$ $TUNEL^+$ and total $TUNEL^+$ cells compared to YopJ-deficient bacteria, consistent with the role of YopJ in apoptosis in vivo. Furthermore, in competitive index experiments, YopJ-deficient *Yersinia* showed colonization defects in PPs, mLNs, and spleen. YopJ-deficient *Yersinia* were not defective for splenic replication following intraperitoneal infection, indicating that YopJ primarily regulates dissemination from mucosal tissues, rather than replication at systemic sites (Monack et al. 1998). Consistently, YopJ-deficient *Yp* are still able to cause systemic infection in a rat model of bubonic plague, despite a defect in induction of apoptosis and cytokine inhibition (Lemaitre et al. 2006). These findings imply that apoptosis may be utilized by *Yersinia* to eliminate immune cells at mucosal surfaces resulting in barrier dysfunction and dissemination. An alternative possibility is that YopJ-mediated blockade of cytokine production during early stages of infection limits the ability of the host to control bacterial dissemination.

Paradoxically, ectopic expression of a hypercytotoxic YopP from *Ye* in *Yptb* results in its attenuation in oral mouse infection (Brodsky and Medzhitov 2008). While both *Yptb* and *Ye* cause cell death in cultured macrophages and infected tissues, infection with YopP-expressing *Ye* induced a significant increase in TUNEL$^+$ CD11b$^+$, CD11c$^+$, and B220$^+$ cells in mLNs relative to mice infected with the YopP-expressing strain (Brodsky and Medzhitov 2008). Similarly, *Yp* strains expressing YopP had higher cytotoxic potency than strains expressing YopJ, both in vitro and in tissues of infected mice; furthermore, expression of YopP in *Yp* also resulted in lower virulence following subcutaneous, but not intranasal or intravenous routes of infection (Zauberman et al. 2009). Interestingly, subcutaneous administration of *Yp* expressing YopP protected against infection with virulent *Y. pestis*, regardless of the route of challenge. These observations suggest that YopJ contributes to dissemination of *Yersinia* from barrier surfaces, but may be less important once bacteria have spread to systemic sites. Whether YopJ or additional immunosuppressive virulence mechanisms play a role in dampening the early inflammatory response to *Yersinia* infection in pneumonic plague (Lathem et al. 2007) also remains to be determined. Notably, pathogenesis of pneumonic plague is influenced by levels of apoptosis, as degradation of Fas Ligand by *Yp* Pla protease results in reduced apoptosis, increased bacterial burdens, and reduced levels of inflammatory cytokines in lung homogenates (Caulfield et al. 2014). This implies a common role for apoptosis in the induction of inflammatory responses in multiple organ systems during *Yersinia* infection.

Consistent with observations that YopJ promotes systemic dissemination following oral infection, YopJ contributes to gut barrier disruption (Jung et al. 2012; Meinzer et al. 2012). Specifically, YopJ can induce TLR2-dependent IL-1β secretion in PPs, which was associated with increased barrier permeability, suggesting that TLR2 signaling mediates YopJ-dependent gut disruption (Jung et al. 2012). Conversely, TLR2-deficient mice have been reported to be more susceptible to oral infection by *Yptb*, due to a loss of TLR2-dependent Reg3β expression in the gut epithelium (Dessein et al. 2009). Thus, the precise role of TLR2 in *Yersinia* infection remains to be further dissected. Notably, IL-1α production during intestinal infection is associated with pathological intestinal inflammation and increased dissemination of *Ye* (Dube et al. 2001), but the role of YopP or TLR2 in this context has not been examined.

In addition to cell death induced by the activity of a bacterial virulence factor, CD8$^+$ cytotoxic T cells also induce death of *Yersinia*-infected cells and are important for control of *Yersinia* infection, as demonstrated by the more severe disease that occurs in infected $\beta 2m^{-/-}$, anti-CD8α-treated, or perforin-deficient mice (Bergman et al. 2009). CD8$^+$ T-cell-mediated killing of bacteria-associated cells targeted them for phagocytosis by uninfected macrophages and could bypass the anti-phagocytic activity of *Yersinia* Yops (Bergman et al. 2009). Interestingly however, CD8$^+$ T cells were not responsible for increased resistance to YopP-expressing *Y. pestis*, suggesting that the increased cytotoxicity of these bacteria induces immune clearance through an alternative pathway (Zauberman et al. 2009). Indeed, the more cytotoxic YopP may bypass the requirement for CD8$^+$ T-cell-mediated killing due to the elevated cytotoxicity induced by the bacteria. These studies collectively suggest that regulation of cytotoxicity during *Yersinia* infection impacts virulence and that a balance between the cytokine-blocking and death-inducing functions of YopJ is required for optimal virulence. Specifically, absence of YopJ results in failure of *Yersinia* to suppress cytokine production or induce cell death and causes a defect in dissemination (Fig. 2). However, *Yptb* and

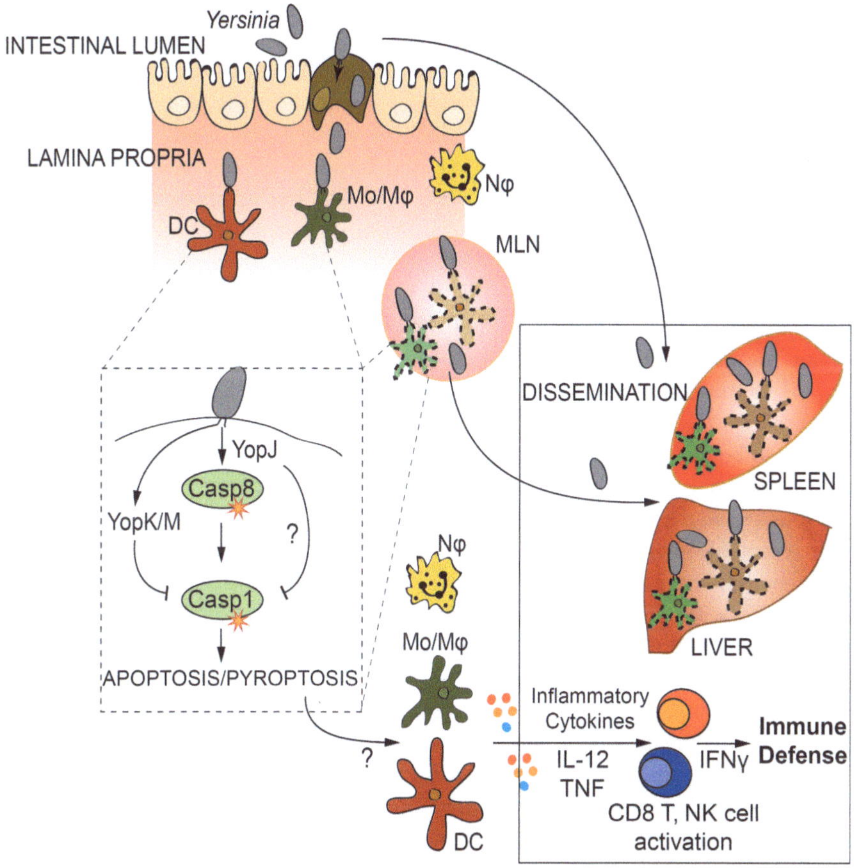

Fig. 2 Interplay of cell death signaling in virulence and host defense during intestinal *Yersinia* infection in vivo. Intestinal infection with *Yersinia* leads to dissemination of the bacteria to systemic tissues from a replicating pool in the intestine as well as spread from lamina propria to mesenteric lymph nodes (MLN). Infection of monocyte/macrophage (Mf/Mo) and DC populations and injection of YopJ results in activation of cell death pathways through caspase-8 and caspase-1. Depending on the activation state of these cells however, the combined activities of YopK, YopM, and YopJ may also limit the induction of these pathways, thus potentially limiting the activation of downstream inflammatory responses, for example, release of inflammatory mediators and T-cell responses that control anti-*Yersinia* immune defense

Y. pestis expressing YopP, which enables stronger inhibition of cytokine production and elevated levels of cell death, are also significantly attenuated in vivo. Thus, while the relative contributions of bacteria-induced and T-cell-induced cell death during *Yersinia* infection in vivo are not yet defined, activation of cell death in vivo either in response to YopJ activity, or as a consequence of T-cell-mediated cytotoxicity plays an important role in immune control of *Yersinia* infection.

3.3 In Vivo Interactions Between Yersinia *and Inflammasome Responses*

Given the interconnected nature of death signaling pathways, a current challenge remains defining the degree to which inflammasome activation and pyroptosis occur in vivo, and their precise contribution to host defense. Interestingly, in the absence of YopK, *Yptb* infection is significantly attenuated, and this attenuation is reversed in caspase-1/11-deficient mice, implying that YopK promotes virulence at least in part by enabling *Yersinia* to evade inflammasome-mediated responses (Brodsky et al. 2010). Similarly, YopM deficiency results in significant attenuation that is reversed in a *Casp1/11*$^{-/-}$ genetic background deficient, thus also implicating YopM in inflammasome evasion in vivo (LaRock and Cookson 2012). In both cases, whether the role of these virulence factors is to limit inflammasome-induced pyroptosis, or whether it is to limit the production of inflammasome-dependent cytokine responses, remains to be defined. Analysis of dual IL-1R/IL-18R-deficient animals would help address this, although other IL-1 family cytokines and inflammatory mediators are still released in the context of inflammasome activation and pyroptosis.

It is notable that *Yersinia* employs two distinct non-overlapping strategies to target the inflammasome. These strategies may function in a context- specific way in vivo, as the bacteria will encounter diverse cell types in distinct activation states during infection of different tissues over the course of the infection, whose responses will likely differ depending on their activation state and balance of injected effectors that they encounter. Interestingly, in primed bone marrow derived macrophages, YopJ functions together with YopM to *limit* inflammasome activation, which is very distinct from the function of YopJ in other contexts (Ratner et al. 2016; Schoberle et al. 2016). Furthermore, in vivo during intravenous infection, there was an additive effect of *yopJ* and *yopM* deficiency, providing further support for an in vivo role of YopJ activity in virulence. Again the specific nature of how YopJ activity promotes virulence in vivo—i.e., whether it is through blockade of cytokine production, inducing apoptosis, or preventing pyroptosis, remains to be determined.

During pneumonic plague infection in murine models, *Y. pestis* induces minimal inflammatory responses during the initial stages of infection within the first 24–36 h (Lathem et al. 2007). This may be due to the combination of both the non-stimulatory, tetra-acylated LPS that *Y. pestis* produces within mammalian hosts as well as the blockade of NF-κB and MAPK signaling by YopJ discussed above (Palmer et al. 1998; Rebeil et al. 2006; Ruckdeschel et al. 1997). Interestingly, as with *Yptb* intestinal infections, YopK is also an important *Yp* virulence factor that modulates cell death pathways in the context of pneumonic plague and may therefore also contribute to *Yersinia* immune evasion in this context (Peters et al. 2013). Notably, while caspase-1 activation does occur in the lung, accompanied by production of IL-1β and IL-18, *Y. pestis* also induces expression of the IL-1 receptor antagonist (IL-1RA), thereby limiting the ability of the host to respond to the presence of these cytokines (Pechous et al. 2016; Sivaraman et al. 2015). This pre-inflammatory phase may contribute to dissemination of the bacteria throughout the tissue and to systemic sites, resulting in a profound subsequent inflammatory response that is accompanied by neutrophilic infiltrate and immunopathologic tissue destruction (Lathem et al. 2007; Sivaraman et al. 2015). Whether YopJ-induced apoptosis of innate immune cells contributes to host defense or to bacterial dissemination in this context has not been teased apart, in large part because it has not been possible to separate blockade of cytokine production and induction of apoptosis during *Yersinia* infection (Philip and Brodsky 2012; Philip et al. 2014).

4 Concluding Remarks

Like all pathogens, *Yersinia* employs multiple virulence factors and strategies to balance the need to disable host defenses with the need to evade recognition by innate immune responses. The T3SS confers the ability to subvert host signaling pathways but also exposes *Yersinia* to detection by cytosolic sensing systems. Notably, *Yersinia* has solved this problem by means of multiple virulence factors that limit detection—in particular, two secreted effector proteins, YopK and YopM, prevent inflammasome activation by limiting the translocation of molecules that trigger inflammasome responses, and by direct blocking of caspase-1 activation, respectively. Whether other pathogens with T3SS utilize similar strategies to evade inflammasome activation remains undefined. While no sequence homologs of YopK exist, other pathogens also modulate inflammasome activation during infection, and understanding how pathogens interfere with inflammasome activation provides insight into inflammasome functioning (Shin and Brodsky 2015; Wynosky-Dolfi et al. 2014). Interestingly, YopJ, classically viewed as an activator of cell death responses, was recently found to act as a redundant factor with YopM in preventing inflammasome activation (Schoberle et al. 2016). How macrophage priming alters their responsiveness, and how macrophage priming switches the response to YopJ from activating to inhibiting inflammasome activation remains to be defined. It is possible that priming of macrophages upregulates a novel set of transcriptional responses that provide a target for YopJ activity that is absent from unprimed cells. Treatment of cells with LPS and pharmacological inhibitors of both MAPK and IKK results in robust cell death that in many ways phenocopies the activity of YopJ. Whether it is possible to similarly pharmacologically mimic the inflammasome inhibition function of YopJ that has been observed in primed macrophages remains to be determined.

The mechanisms and signaling pathways engaged by the host to conduct various cell death programs during *Yersinia* infection have been characterized extensively in vitro. Future studies will need to define how modulation of different cell death pathways impacts replication, dissemination, and host cytokine production in vivo. Addressing this question is likely to shed significant new light on our understanding of infection dynamics and may provide new targets for intervention in the context of many types of infections that modulate cell death pathways. An important question remains whether some commensal bacteria might also utilize T3SSs for purposes of long-term colonization, and consequently, whether such bacteria might modulate key host signaling components and pathways in order to limit activation of inflammasomes or other inflammatory cell death pathways. Indeed, a 'pathobiont' *E. coli* that expresses a T3SS was recently identified as being expanded in a genetically susceptible mouse strain lacking Nlrc4 and results in increased susceptibility to a mouse model of colitis (Ayres et al. 2012). How this strain avoids inducing overt disease in the absence of a colitis-inducing stimulus is unclear, but may involve limiting induction of inflammasome or other inflammatory death signaling pathways in a manner conceptually similar to those described here. Despite the nearly 20 years since the initial discovery that *Yersinia* induces cell death in infected macrophages, a great deal likely remains to be discovered. The availability of CRISPR-based approaches to rapidly make knockouts and point mutations in cellular signaling components that interact with bacterial virulence factors is likely to facilitate rapid progress in this area.

Acknowledgments We would like to thank Sunny Shin and members of the Brodsky and Shin labs for scientific discussion. Work in the Brodsky lab is supported by the NIH-NIAID and a BWF Investigator in the Pathogenesis of Infectious Disease Award.

References

Achtman M, Zurth K, Morelli G, Torrea G, Guiyoule A, Carniel E (1999) *Yersinia pestis*, the cause of plague, is a recently emerged clone of *Yersinia pseudotuberculosis*. Proc Natl Acad Sci USA 96(24):14043–14048

Ayres JS, Trinidad NJ, Vance RE (2012) Lethal inflammasome activation by a multidrug-resistant pathobiont upon antibiotic disruption of the microbiota. Nat Med 18(5):799–806. doi:10.1038/nm.2729

Balada-Llasat JM, Mecsas J (2006) Yersinia has a tropism for B and T cell zones of lymph nodes that is independent of the type III secretion system. PLoS Pathog 2(9):e86

Barnes PD, Bergman MA, Mecsas J, Isberg RR (2006) *Yersinia pseudotuberculosis* disseminates directly from a replicating bacterial pool in the intestine. J Exp Med 203(6):1591–1601

Ben Ari T, Gershunov A, Gage KL, Snall T, Ettestad P, Kausrud KL, Stenseth NC (2008) Human plague in the USA: the importance of regional and local climate. Biol Lett 4(6):737–740. doi:10.1098/rsbl.2008.0363

Bergman MA, Loomis WP, Mecsas J, Starnbach MN, Isberg RR (2009) CD8(+) T cells restrict *Yersinia pseudotuberculosis* infection: bypass of anti-phagocytosis by targeting antigen-presenting cells. PLoS Pathog 5(9):e1000573. doi:10.1371/journal.ppat.1000573

Bergsbaken T, Cookson BT (2007) Macrophage activation redirects *Yersinia*-infected host cell death from apoptosis to caspase-1-dependent pyroptosis. PLoS Pathog 3(11):e161

Bhaduri S, Wesley IV, Bush EJ (2005) Prevalence of pathogenic *Yersinia enterocolitica* strains in pigs in the United States. Appl Environ Microbiol 71(11):7117–7121. doi:10.1128/AEM.71.11.7117-7121.2005

Brodsky IE, Medzhitov R (2008) Reduced secretion of YopJ by Yersinia limits in vivo cell death but enhances bacterial virulence. PLoS Pathog 4(5):e1000067. doi:10.1371/journal.ppat.1000067

Brodsky IE, Palm NW, Sadanand S, Ryndak MB, Sutterwala FS, Flavell RA, Bliska JB, Medzhitov R (2010) A Yersinia effector protein promotes virulence by preventing inflammasome recognition of the type III secretion system. Cell Host Microbe 7(5):376–387

Broz P, Newton K, Lamkanfi M, Mariathasan S, Dixit VM, Monack DM (2010) Redundant roles for inflammasome receptors NLRP3 and NLRC4 in host defense against Salmonella. J Exp Med 207(8):1745–1755. doi:10.1084/jem.20100257

Broz P, Ruby T, Belhocine K, Bouley DM, Kayagaki N, Dixit VM, Monack DM (2012) Caspase-11 increases susceptibility to Salmonella infection in the absence of caspase-1. Nature 490(7419):288–291. doi:10.1038/nature11419

Campisi L, Cummings RJ, Blander JM (2014) Death-defining immune responses after apoptosis. Am J Transplant 14(7):1488–1498. doi:10.1111/ajt.12736

Casson CN, Copenhaver AM, Zwack EE, Nguyen HT, Strowig T, Javdan B, Bradley WP, Fung TC, Flavell RA, Brodsky IE, Shin S (2013) Caspase-11 activation in response to bacterial secretion systems that access the host cytosol. PLoS Pathog 9(6):e1003400. doi:10.1371/journal.ppat.1003400

Caulfield AJ, Walker ME, Gielda LM, Lathem WW (2014) The Pla protease of Yersinia pestis degrades fas ligand to manipulate host cell death and inflammation. Cell Host Microbe 15(4):424–434. doi:10.1016/j.chom.2014.03.005

Chain PS, Carniel E, Larimer FW, Lamerdin J, Stoutland PO, Regala WM, Georgescu AM, Vergez LM, Land ML, Motin VL, Brubaker RR, Fowler J, Hinnebusch J, Marceau M, Medigue C, Simonet M, Chenal-Francisque V, Souza B, Dacheux D, Elliott JM, Derbise A, Hauser LJ, Garcia E (2004) Insights into the evolution of *Yersinia pestis* through whole-genome comparison with *Yersinia pseudotuberculosis*. Proc Natl Acad Sci USA 101(38):13826–13831. doi:10.1073/pnas.0404012101

Chung LK, Philip NH, Schmidt VA, Koller A, Strowig T, Flavell RA, Brodsky IE, Bliska JB (2014) IQGAP1 is important for activation of caspase-1 in macrophages and is targeted by *Yersinia pestis* type III effector YopM. MBio 5(4):e01402–e01414. doi:10.1128/mBio.01402-14

Cornelis GR (2006) The type III secretion injectisome. Nat Rev Microbiol 4(11):811–825. doi:nrmicro1526 [pii] 10.1038/nrmicro1526

Degterev A, Hitomi J, Germscheid M, Ch'en IL, Korkina O, Teng X, Abbott D, Cuny GD, Yuan C, Wagner G, Hedrick SM, Gerber SA, Lugovskoy A, Yuan J (2008) Identification of RIP1 kinase as a specific cellular target of necrostatins. Nat Chem Biol 4(5):313–321. doi:10.1038/nchembio.83

Denecker G, Declercq W, Geuijen CA, Boland A, Benabdillah R, van Gurp M, Sory MP, Vandenabeele P, Cornelis GR (2001) *Yersinia enterocolitica* YopP-induced apoptosis of macrophages involves the apoptotic signaling cascade upstream of bid. J Biol Chem 276(23):19706–19714

Dessein R, Gironella M, Vignal C, Peyrin-Biroulet L, Sokol H, Secher T, Lacas-Gervais S, Gratadoux JJ, Lafont F, Dagorn JC, Ryffel B, Akira S, Langella P, Nunez G, Sirard JC, Iovanna J, Simonet M, Chamaillard M (2009) Toll-like receptor 2 is critical for induction of Reg3 beta expression and intestinal clearance of *Yersinia pseudotuberculosis*. Gut 58(6):771–776. doi:10.1136/gut.2008.168443

Dewoody R, Merritt PM, Houppert AS, Marketon MM (2011) YopK regulates the Yersinia pestis type III secretion system from within host cells. Mol Microbiol 79(6):1445–1461. doi:10.1111/j.1365-2958.2011.07534.x

Dube PH, Revell PA, Chaplin DD, Lorenz RG, Miller VL (2001) A role for IL-1 alpha in inducing pathologic inflammation during bacterial infection. Proc Natl Acad Sci USA 98(19):10880–10885. doi:10.1073/pnas.191214498

Durand EA, Maldonado-Arocho FJ, Castillo C, Walsh RL, Mecsas J (2010) The presence of professional phagocytes dictates the number of host cells targeted for Yop translocation during infection. Cell Microbiol 12(8):1064–1082. doi:10.1111/j.1462-5822.2010.01451.x

Erfurth SE, Grobner S, Kramer U, Gunst DS, Soldanova I, Schaller M, Autenrieth IB, Borgmann S (2004) *Yersinia enterocolitica* induces apoptosis and inhibits surface molecule expression and cytokine production in murine dendritic cells. Infect Immun 72(12):7045–7054

Evdokimov AG, Anderson DE, Routzahn KM, Waugh DS (2001) Unusual molecular architecture of the *Yersinia pestis* cytotoxin YopM: a leucine-rich repeat protein with the shortest repeating unit. J Mol Biol 312(4):807–821. doi:10.1006/jmbi.2001.4973

Grabenstein JP, Marceau M, Pujol C, Simonet M, Bliska JB (2004) The response regulator PhoP of *Yersinia pseudotuberculosis* is important for replication in macrophages and for virulence. Infect Immun 72(9):4973–4984

Green DR, Ferguson T, Zitvogel L, Kroemer G (2009) Immunogenic and tolerogenic cell death. Nat Rev Immunol 9(5):353–363. doi:10.1038/nri2545

Greten FR, Arkan MC, Bollrath J, Hsu LC, Goode J, Miething C, Goktuna SI, Neuenhahn M, Fierer J, Paxian S, Van Rooijen N, Xu Y, O'Cain T, Jaffee BB, Busch DH, Duyster J, Schmid RM, Eckmann L, Karin M (2007) NF-kappaB is a negative regulator of IL-1beta secretion as revealed by genetic and pharmacological inhibition of IKKbeta. Cell 130(5):918–931. doi:10.1016/j.cell.2007.07.009

Grobner S, Autenrieth SE, Soldanova I, Gunst DS, Schaller M, Bohn E, Muller S, Leverkus M, Wesselborg S, Autenrieth IB, Borgmann S (2006) Yersinia YopP-induced apoptotic cell death in murine dendritic cells is partially independent from action of caspases and exhibits necrosis-like features. Apoptosis 11(11):1959–1968. doi:10.1007/s10495-006-0189-3

Grobner S, Adkins I, Schulz S, Richter K, Borgmann S, Wesselborg S, Ruckdeschel K, Micheau O, Autenrieth IB (2007) Catalytically active Yersinia outer protein P induces cleavage of RIP and caspase-8 at the level of the DISC independently of death receptors in dendritic cells. Apoptosis 12(10):1813–1825. doi:10.1007/s10495-007-0100-x

Gurung P, Malireddi RK, Anand PK, Demon D, Walle LV, Liu Z, Vogel P, Lamkanfi M, Kanneganti TD (2012) Toll or interleukin-1 receptor (TIR) domain-containing adaptor inducing interferon-beta (TRIF)-mediated caspase-11 protease production integrates Toll-like receptor 4 (TLR4) protein- and Nlrp3 inflammasome-mediated host defense against enteropathogens. J Biol Chem 287(41):34474–34483. doi:10.1074/jbc.M112.401406

Gurung P, Anand PK, Malireddi RK, Vande Walle L, Van Opdenbosch N, Dillon CP, Weinlich R, Green DR, Lamkanfi M, Kanneganti TD (2014) FADD and Caspase-8 Mediate Priming and Activation of the Canonical and Noncanonical Nlrp3 Inflammasomes. J Immunol 192 (4):1835–1846. doi:10.4049/jimmunol.1302839

Haase R, Kirschning CJ, Sing A, Schrottner P, Fukase K, Kusumoto S, Wagner H, Heesemann J, Ruckdeschel K (2003) A dominant role of Toll-like receptor 4 in the signaling of apoptosis in bacteria-faced macrophages. J Immunol 171(8):4294–4303

He S, Liang Y, Shao F, Wang X (2011) Toll-like receptors activate programmed necrosis in macrophages through a receptor-interacting kinase-3-mediated pathway. Proc Natl Acad Sci USA 108(50):20054–20059. doi:10.1073/pnas.1116302108

Heath WR, Carbone FR (2001) Cross-presentation, dendritic cells, tolerance and immunity. Annu Rev Immunol 19:47–64. doi:10.1146/annurev.immunol.19.1.47

Hentschke M, Berneking L, Belmar Campos C, Buck F, Ruckdeschel K, Aepfelbacher M (2010) Yersinia virulence factor YopM induces sustained RSK activation by interfering with dephosphorylation. PLoS One 5(10). doi:10.1371/journal.pone.0013165

Hinnebusch BJ, Rosso ML, Schwan TG, Carniel E (2002) High-frequency conjugative transfer of antibiotic resistance genes to *Yersinia pestis* in the flea midgut. Mol Microbiol 46(2):349–354

Holmstrom A, Rosqvist R, Wolf-Watz H, Forsberg A (1995) Virulence plasmid-encoded YopK is essential for *Yersinia pseudotuberculosis* to cause systemic infection in mice. Infect Immun 63 (6):2269–2276

Holmstrom A, Petterson J, Rosqvist R, Hakansson S, Tafazoli F, Fallman M, Magnusson KE, Wolf-Watz H, Forsberg A (1997) YopK of *Yersinia pseudotuberculosis* controls translocation of Yop effectors across the eukaryotic cell membrane. Mol Microbiol 24(1):73–91

Jung C, Meinzer U, Montcuquet N, Thachil E, Chateau D, Thiebaut R, Roy M, Alnabhani Z, Berrebi D, Dussaillant M, Pedruzzi E, Thenet S, Cerf-Bensussan N, Hugot JP, Barreau F (2012) *Yersinia pseudotuberculosis* disrupts intestinal barrier integrity through hematopoietic TLR-2 signaling. J Clin Invest 122(6):2239–2251. doi:10.1172/JCI58147

Kayagaki N, Warming S, Lamkanfi M, Vande Walle L, Louie S, Dong J, Newton K, Qu Y, Liu J, Heldens S, Zhang J, Lee WP, Roose-Girma M, Dixit VM (2011) Non-canonical inflammasome activation targets caspase-11. Nature 479(7371):117–121. doi:10.1038/nature10558

Koberle M, Klein-Gunther A, Schutz M, Fritz M, Berchtold S, Tolosa E, Autenrieth IB, Bohn E (2009) *Yersinia enterocolitica* targets cells of the innate and adaptive immune system by injection of Yops in a mouse infection model. PLoS Pathog 5(8):e1000551. doi:10.1371/journal.ppat.1000551

Kwit N, Nelson C, Kugeler K, Petersen J, Plante L, Yaglom H, Kramer V, Schwartz B, House J, Colton L, Feldpausch A, Drenzek C, Baumbach J, DiMenna M, Fisher E, Debess E, Buttke D, Weinburke M, Percy C, Schriefer M, Gage K, Mead P (2015) Human Plague—United States, 2015. MMWR Morb Mortal Wkly Rep 64(33):918–919

Kwuan L, Adams W, Auerbuch V (2013) Impact of host membrane pore formation by the *Yersinia pseudotuberculosis* type III secretion system on the macrophage innate immune response. Infect Immun 81(3):905–914. doi:10.1128/IAI.01014-12

LaRock CN, Cookson BT (2012) The Yersinia virulence effector YopM binds caspase-1 to arrest inflammasome assembly and processing. Cell Host Microbe 12(6):799–805. doi:10.1016/j.chom.2012.10.020

Lathem WW, Price PA, Miller VL, Goldman WE (2007) A plasminogen-activating protease specifically controls the development of primary pneumonic plague. Science 315(5811): 509–513

Lemaitre N, Sebbane F, Long D, Hinnebusch BJ (2006) *Yersinia pestis* YopJ suppresses tumor necrosis factor alpha induction and contributes to apoptosis of immune cells in the lymph node but is not required for virulence in a rat model of bubonic plague. Infect Immun 74(9):5126–5131

Lilo S, Zheng Y, Bliska JB (2008) Caspase-1 activation in macrophages infected with *Yersinia pestis* KIM requires the type III secretion system effector YopJ. Infect Immun 76(9):3911–3923. doi:IAI.01695-07 [pii] 10.1128/IAI.01695-07

Maldonado-Arocho FJ, Green C, Fisher ML, Paczosa MK, Mecsas J (2013) Adhesins and host serum factors drive Yop translocation by yersinia into professional phagocytes during animal infection. PLoS Pathog 9(6):e1003415. doi:10.1371/journal.ppat.1003415

Man SM, Tourlomousis P, Hopkins L, Monie TP, Fitzgerald KA, Bryant CE (2013) Salmonella infection induces recruitment of Caspase-8 to the inflammasome to modulate IL-1beta production. J Immunol 191(10):5239–5246. doi:10.4049/jimmunol.1301581

Marketon MM, DePaolo RW, DeBord KL, Jabri B, Schneewind O (2005) Plague bacteria target immune cells during infection. Science 309(5741):1739–1741. doi:10.1126/science.1114580

McCoy MW, Marre ML, Lesser CF, Mecsas J (2010) The C-terminal tail of *Yersinia pseudotuberculosis* YopM is critical for interacting with RSK1 and for virulence. Infect Immun 78(6):2584–2598. doi:10.1128/IAI.00141-10

McDonald C, Vacratsis PO, Bliska JB, Dixon JE (2003) The yersinia virulence factor YopM forms a novel protein complex with two cellular kinases. J Biol Chem 278(20):18514–18523. doi:10.1074/jbc.M301226200

McPhee JB, Mena P, Bliska JB (2010) Delineation of regions of the Yersinia YopM protein required for interaction with the RSK1 and PRK2 host kinases and their requirement for

interleukin-10 production and virulence. Infect Immun 78(8):3529–3539. doi:10.1128/IAI.00269-10
McPhee JB, Mena P, Zhang Y, Bliska JB (2012) Interleukin-10 induction is an important virulence function of the *Yersinia pseudotuberculosis* type III effector YopM. Infect Immun 80(7):2519–2527. doi:10.1128/IAI.06364-11
Meinzer U, Barreau F, Esmiol-Welterlin S, Jung C, Villard C, Leger T, Ben-Mkaddem S, Berrebi D, Dussaillant M, Alnabhani Z, Roy M, Bonacorsi S, Wolf-Watz H, Perroy J, Ollendorff V, Hugot JP (2012) *Yersinia pseudotuberculosis* effector YopJ subverts the Nod2/RICK/TAK1 pathway and activates caspase-1 to induce intestinal barrier dysfunction. Cell Host Microbe 11(4):337–351. doi:10.1016/j.chom.2012.02.009
Miao EA, Mao DP, Yudkovsky N, Bonneau R, Lorang CG, Warren SE, Leaf IA, Aderem A (2010) Innate immune detection of the type III secretion apparatus through the NLRC4 inflammasome. Proc Natl Acad Sci U S A 107(7):3076–3080. doi:10.1073/pnas.0913087107
Mills SD, Boland A, Sory MP, van der Smissen P, Kerbourch C, Finlay BB, Cornelis GR (1997) *Yersinia enterocolitica* induces apoptosis in macrophages by a process requiring functional type III secretion and translocation mechanisms and involving YopP, presumably acting as an effector protein. Proc Natl Acad Sci USA 94(23):12638–12643
Minnich SA, Rohde HN (2007) A rationale for repression and/or loss of motility by pathogenic Yersinia in the mammalian host. Adv Exp Med Biol 603:298–310. doi:10.1007/978-0-387-72124-8_27
Mittal R, Peak-Chew SY, McMahon HT (2006) Acetylation of MEK2 and I kappa B kinase (IKK) activation loop residues by YopJ inhibits signaling. Proc Natl Acad Sci USA 103 (49):18574–18579
Monack DM, Mecsas J, Ghori N, Falkow S (1997) *Yersinia* signals macrophages to undergo apoptosis and YopJ is necessary for this cell death. Proc Natl Acad Sci USA 94(19):10385–10390
Monack DM, Mecsas J, Bouley D, Falkow S (1998) *Yersinia*-induced apoptosis in vivo aids in the establishment of a systemic infection of mice. J Exp Med 188(11):2127–2137
Mukherjee S, Keitany G, Li Y, Wang Y, Ball HL, Goldsmith EJ, Orth K (2006) *Yersinia* YopJ acetylates and inhibits kinase activation by blocking phosphorylation. Science 312 (5777):1211–1214
Munoz-Pinedo C (2012) Signaling pathways that regulate life and cell death: evolution of apoptosis in the context of self-defense. Adv Exp Med Biol 738:124–143. doi:10.1007/978-1-4614-1680-7_8
Nemeth J, Straley SC (1997) Effect of *Yersinia pestis* YopM on experimental plague. Infect Immun 65(3):924–930
Palmer LE, Hobbie S, Galan JE, Bliska JB (1998) YopJ of *Yersinia pseudotuberculosis* is required for the inhibition of macrophage TNF-alpha production and downregulation of the MAP kinases p38 and JNK. Mol Microbiol 27(5):953–965
Pechous RD, Sivaraman V, Stasulli NM, Goldman WE (2016) Pneumonic plague: the darker side of *Yersinia pestis*. Trends Microbiol 24(3):190–197. doi:10.1016/j.tim.2015.11.008
Peters KN, Dhariwala MO, Hughes Hanks JM, Brown CR, Anderson DM (2013) Early apoptosis of macrophages modulated by injection of *Yersinia pestis* YopK promotes progression of primary pneumonic plague. PLoS Pathog 9(4):e1003324. doi:10.1371/journal.ppat.1003324
Philip NH, Brodsky IE (2012) Cell death programs in Yersinia immunity and pathogenesis. Front Cell Infect Microbiol 2:149. doi:10.3389/fcimb.2012.00149
Philip NH, Dillon CP, Snyder AG, Fitzgerald P, Wynosky-Dolfi MA, Zwack EE, Hu B, Fitzgerald L, Mauldin EA, Copenhaver AM, Shin S, Wei L, Parker M, Zhang J, Oberst A, Green DR, Brodsky IE (2014) Caspase-8 mediates caspase-1 processing and innate immune defense in response to bacterial blockade of NF-kappaB and MAPK signaling. Proc Natl Acad Sci USA 111(20):7385–7390. doi:10.1073/pnas.1403252111
Rathinam VA, Vanaja SK, Waggoner L, Sokolovska A, Becker C, Stuart LM, Leong JM, Fitzgerald KA (2012) TRIF licenses caspase-11-dependent NLRP3 inflammasome activation by gram-negative bacteria. Cell. doi:10.1016/j.cell.2012.07.007

Ratner D, Orning MP, Starheim KK, Marty-Roix R, Proulx MK, Goguen JD, Lien E (2016) Manipulation of IL-1beta and IL-18 production by *Yersinia pestis* effectors YopJ and YopM and redundant impact on virulence. J Biol Chem. doi:10.1074/jbc.M115.697698

Rebeil R, Ernst RK, Jarrett CO, Adams KN, Miller SI, Hinnebusch BJ (2006) Characterization of late acyltransferase genes of *Yersinia pestis* and their role in temperature-dependent lipid A variation. J Bacteriol 188(4):1381–1388. doi:10.1128/JB.188.4.1381-1388.2006

Rosadini CV, Zanoni I, Odendall C, Green ER, Paczosa MK, Philip NH, Brodsky IE, Mecsas J, Kagan JC (2015) A single bacterial immune evasion strategy dismantles both MyD88 and TRIF signaling pathways downstream of TLR4. Cell Host Microbe 18(6):682–693. doi:10.1016/j.chom.2015.11.006

Rosenberg JC (1968) Doctors afield: Alexandre Yersin. N Engl J Med 278(5):261–263. doi:10.1056/NEJM196802012780507

Ruckdeschel K, Richter K (2002) Lipopolysaccharide desensitization of macrophages provides protection against *Yersinia enterocolitica*-induced apoptosis. Infect Immun 70(9):5259–5264

Ruckdeschel K, Machold J, Roggenkamp A, Schubert S, Pierre J, Zumbihl R, Liautard JP, Heesemann J, Rouot B (1997) *Yersinia enterocolitica* promotes deactivation of macrophage mitogen-activated protein kinases extracellular signal-regulated kinase-1/2, p38, and c-Jun NH2-terminal kinase. Correlation with its inhibitory effect on tumor necrosis factor-alpha production. J Biol Chem 272(25):15920–15927

Ruckdeschel K, Harb S, Roggenkamp A, Hornef M, Zumbihl R, Kohler S, Heesemann J, Rouot B (1998) *Yersinia enterocolitica* impairs activation of transcription factor NF-kappaB: involvement in the induction of programmed cell death and in the suppression of the macrophage tumor necrosis factor alpha production. J Exp Med 187(7):1069–1079

Ruckdeschel K, Richter K, Mannel O, Heesemann J (2001) Arginine-143 of *Yersinia enterocolitica* YopP crucially determines isotype-related NF-kappaB suppression and apoptosis induction in macrophages. Infect Immun 69(12):7652–7662

Ruckdeschel K, Pfaffinger G, Haase R, Sing A, Weighardt H, Hacker G, Holzmann B, Heesemann J (2004) Signaling of apoptosis through TLRs critically involves toll/IL-1 receptor domain-containing adapter inducing IFN-beta, but not MyD88, in bacteria-infected murine macrophages. J Immunol 173(5):3320–3328

Ruter C, Silva MR, Grabowski B, Lubos ML, Scharnert J, Poceva M, von Tils D, Flieger A, Heesemann J, Bliska JB, Schmidt MA (2014) Rabbit monoclonal antibodies directed at the T3SS effector protein YopM identify human pathogenic Yersinia isolates. Int J Med Microbiol 304(3–4):444–451. doi:10.1016/j.ijmm.2014.02.003

Schesser K, Spiik AK, Dukuzumuremyi JM, Neurath MF, Pettersson S, Wolf-Watz H (1998) The yopJ locus is required for *Yersinia*-mediated inhibition of NF-kappaB activation and cytokine expression: YopJ contains a eukaryotic SH2-like domain that is essential for its repressive activity. Mol Microbiol 28(6):1067–1079

Schoberle TJ, Chung LK, McPhee JB, Bogin B, Bliska JB (2016) Uncovering an important role for YopJ in the inhibition of Caspase-1 in activated macrophages and promoting *Yersinia pseudotuberculosis* virulence. Infect Immun. doi:10.1128/IAI.00843-15

Shi J, Zhao Y, Wang Y, Gao W, Ding J, Li P, Hu L, Shao F (2014) Inflammatory caspases are innate immune receptors for intracellular LPS. Nature 514(7521):187–192. doi:10.1038/nature13683

Shin S, Brodsky IE (2015) The inflammasome: learning from bacterial evasion strategies. Semin Immunol 27(2):102–110. doi:10.1016/j.smim.2015.03.006

Sivaraman V, Pechous RD, Stasulli NM, Eichelberger KR, Miao EA, Goldman WE (2015) *Yersinia pestis* activates both IL-1beta and IL-1 receptor antagonist to modulate lung inflammation during pneumonic plague. PLoS Pathog 11(3):e1004688. doi:10.1371/journal.ppat.1004688

Thorslund SE, Edgren T, Pettersson J, Nordfelth R, Sellin ME, Ivanova E, Francis MS, Isaksson EL, Wolf-Watz H, Fallman M (2011) The RACK1 signaling scaffold protein selectively interacts with *Yersinia pseudotuberculosis* virulence function. PLoS ONE 6(2):e16784. doi:10.1371/journal.pone.0016784

Torchinsky MB, Garaude J, Martin AP, Blander JM (2009) Innate immune recognition of infected apoptotic cells directs T(H)17 cell differentiation. Nature 458(7234):78–82. doi:10.1038/nature07781

Vance RE, Isberg RR, Portnoy DA (2009) Patterns of pathogenesis: discrimination of pathogenic and nonpathogenic microbes by the innate immune system. Cell Host Microbe 6(1):10–21. doi:10.1016/j.chom.2009.06.007

Vazquez-Torres A, Jones-Carson J, Baumler AJ, Falkow S, Valdivia R, Brown W, Le M, Berggren R, Parks WT, Fang FC (1999) Extraintestinal dissemination of Salmonella by CD18-expressing phagocytes. Nature 401(6755):804–808. doi:10.1038/44593

Viboud GI, Bliska JB (2005) Yersinia outer proteins: role in modulation of host cell signaling responses and pathogenesis. Annu Rev Microbiol 59:69–89. doi:10.1146/annurev.micro.59.030804.121320

Vladimer GI, Marty-Roix R, Ghosh S, Weng D, Lien E (2013) Inflammasomes and host defenses against bacterial infections. Curr Opin Microbiol 16(1):23–31. doi:10.1016/j.mib.2012.11.008

Weng D, Marty-Roix R, Ganesan S, Proulx MK, Vladimer GI, Kaiser WJ, Mocarski ES, Pouliot K, Chan FK, Kelliher MA, Harris PA, Bertin J, Gough PJ, Shayakhmetov DM, Goguen JD, Fitzgerald KA, Silverman N, Lien E (2014) Caspase-8 and RIP kinases regulate bacteria-induced innate immune responses and cell death. Proc Natl Acad Sci USA 111 (20):7391–7396. doi:10.1073/pnas.1403477111

Wesley IV, Bhaduri S, Bush E (2008) Prevalence of *Yersinia enterocolitica* in market weight hogs in the United States. J Food Prot 71(6):1162–1168

Wren BW (2003) The yersiniae–a model genus to study the rapid evolution of bacterial pathogens. Nat Rev Microbiol 1(1):55–64

Wynosky-Dolfi MA, Snyder AG, Philip NH, Doonan PJ, Poffenberger MC, Avizonis D, Zwack EE, Riblett AM, Hu B, Strowig T, Flavell RA, Jones RG, Freedman BD, Brodsky IE (2014) Oxidative metabolism enables Salmonella evasion of the NLRP3 inflammasome. J Exp Med 211(4):653–668. doi:10.1084/jem.20130627

Zauberman A, Cohen S, Mamroud E, Flashner Y, Tidhar A, Ber R, Elhanany E, Shafferman A, Velan B (2006) Interaction of *Yersinia pestis* with macrophages: limitations in YopJ-dependent apoptosis. Infect Immun 74(6):3239–3250

Zauberman A, Tidhar A, Levy Y, Bar-Haim E, Halperin G, Flashner Y, Cohen S, Shafferman A, Mamroud E (2009) *Yersinia pestis* endowed with increased cytotoxicity is avirulent in a bubonic plague model and induces rapid protection against pneumonic plague. PLoS ONE 4 (6):e5938. doi:10.1371/journal.pone.0005938

Zhang Y, Ting AT, Marcu KB, Bliska JB (2005) Inhibition of MAPK and NF-kappa B pathways is necessary for rapid apoptosis in macrophages infected with *Yersinia*. J Immunol 174 (12):7939–7949

Zheng Y, Lilo S, Brodsky IE, Zhang Y, Medzhitov R, Marcu KB, Bliska JB (2011) A Yersinia effector with enhanced inhibitory activity on the NF-kappaB pathway activates the NLRP3/ASC/caspase-1 inflammasome in macrophages. PLoS Pathog 7(4):e1002026. doi:10.1371/journal.ppat.1002026

Zheng Y, Lilo S, Mena P, Bliska JB (2012) YopJ-induced caspase-1 activation in Yersinia-infected macrophages: independent of apoptosis, linked to necrosis, dispensable for innate host defense. PLoS ONE 7(4):e36019. doi:10.1371/journal.pone.0036019

Zhou H, Monack DM, Kayagaki N, Wertz I, Yin J, Wolf B, Dixit VM (2005) *Yersinia* virulence factor YopJ acts as a deubiquitinase to inhibit NF-kappa B activation. J Exp Med 202 (10):1327–1332

Zimbler DL, Schroeder JA, Eddy JL, Lathem WW (2015) Early emergence of Yersinia pestis as a severe respiratory pathogen. Nat Commun 6:7487. doi:10.1038/ncomms8487

Zwack EE, Snyder AG, Wynosky-Dolfi MA, Ruthel G, Philip NH, Marketon MM, Francis MS, Bliska JB, Brodsky IE (2014) Inflammasome activation in response to the Yersinia Type III secretion system requires hyperinjection of translocon proteins YopB and YopD. MBio 6(1). doi:10.1128/mBio.02095-14

The Orchestra and Its Maestro: *Shigella's* Fine-Tuning of the Inflammasome Platforms

Anna-Karin Hermansson, Ida Paciello and Maria Lina Bernardini

Abstract *Shigella* spp. are the causative agents of bacillary dysentery, leading to extensive mortality and morbidity worldwide. These facultative intracellular bacteria invade the epithelium of the colon and the rectum, inducing a severe inflammatory response from which the symptoms of the disease originate. *Shigella* are human pathogens able to manipulate and subvert the innate immune system surveillance. *Shigella* dampens inflammasome activation in epithelial cells. In infected macrophages, inflammasome activation and IL-1β and IL-18 release lead to massive neutrophil recruitment and greatly contribute to inflammation. Here, we describe how *Shigella* hijacks and finely tunes inflammasome activation in the different cell populations involved in pathogenesis: epithelial cells, macrophages, neutrophils, DCs, and B and T lymphocytes. *Shigella* emerges as a "sly" pathogen that switches on/off the inflammasome mechanisms in order to optimize the interaction with the host and establish a successful infection.

Contents

A.-K. Hermansson · I. Paciello · M.L. Bernardini (✉)
Dipartimento di Biologia e Biotecnologie "C. Darwin", Sapienza—Università di Roma, Piazzale Aldo Moro, 5, 00185 Rome, Italy
e-mail: marialina.bernardini@uniroma1.it

A.-K. Hermansson
e-mail: anna-karin.hermansson@uniroma1.it

I. Paciello
e-mail: ida.paciello@uniroma1.it

M.L. Bernardini
Dipartimento di Biologia e Biotecnologie "C. Darwin", Istituto Pasteur Italia—Fondazione Cenci Bolognetti, Sapienza—Università di Roma, Piazzale Aldo Moro, 5, 00185 Rome, Italy

S. Backert (ed.), *Inflammasome Signaling and Bacterial Infections*, Current Topics in Microbiology and Immunology 397,
DOI 10.1007/978-3-319-41171-2_5

1 Introduction

Shigella is the causative agent of shigellosis, or bacillary dysentery, and the source of extensive mortality and morbidity worldwide. *Shigella* spp. belong to the *Enterobacteriacea* family and are closely related to *Escherichia coli*. There are four species in the *Shigella* genus: *S. dysenteriae*, *S. flexneri*, *S. boydii,* and *S. sonnei*. *S. dysenteriae* is the only species capable of producing Shiga toxin, but all strains can cause shigellosis, or bacillary dysentery. The percentage of cases attributed to each strain and serotype varies largely between countries, but *S. flexneri* strains are the main cause of endemic shigellosis in low-income countries and thus also of worldwide mortality (Kotloff et al. 2013), whereas *S. sonnei* is the prevailing strain in middle- and high-income countries (Kotloff et al. 1999). A shift towards *S. sonnei* has been noticed, also in countries such as Bangladesh and Pakistan, probably due to improvements in nutrition, sanitation, and socio-economic status (Ud-Din et al. 2013; Zafar et al. 2009; Khalil et al. 1998). *Shigella* spreads through the faecal–oral route, mainly via contaminated food or water. Patients with shigellosis typically present with frequent passage of small liquid stools with visible blood. Other common symptoms are abdominal cramps, tenesmus, fever, and anorexia. It is estimated that 69 % of episodes and 61 % of deaths occur in children (Kotloff et al. 1999).

Man is the only natural host for *Shigella,* and mice are naturally resistant to oral ingestion of *Shigella*. No animal model perfectly recapitulates the disease *Shigella* induces in humans. The majority of in vivo studies of *Shigella* pathogenesis exploit the murine pulmonary model of shigellosis, in which intranasal administration of *Shigella* leads to an inflammatory reaction in lungs, paralleling the severe inflammation of dysenteric patients' intestinal mucosa (Voino-Yasenetsky and Voino-Yasenetskaya 1962). *Shigella* has an extremely low infectious dose: ingestion of 10–100 bacteria can be sufficient to cause disease (DuPont et al. 1989), something that is probably in part due to *Shigella*'*s* high resistance to acidic environments (Yang et al. 2015; Gorden and Small 1993), such as in the stomach. Low pH, bile salts, and temperature are some of the factors that induce the expression of the genes of the type III secretion system (T3SS), *Shigella*'*s* main virulence weapon (Yang et al. 2015). The T3SS is encoded by a pathogenicity island present on a large virulence plasmid common to all *Shigella* species. *Shigella* lacking the virulence plasmid are non-invasive and avirulent (Sansonetti et al. 1981, 1982, 1983). Once *Shigella* reaches the colon and rectum, it invades the epithelium. Studies on monolayers of polarized intestinal Caco2 cells showed that *S. flexneri* could only to invade cells from the basolateral side, but not from the apical side, leading to the hypothesis that *Shigella* invades the intestinal epithelium from the basolateral side after entry through microfold cells (M cells) scattered on the follicle-associated epithelium (Wassef et al. 1989). The M cell has a pocket shape on the basolateral side, where macrophages reside, to which transcytosed bacteria are delivered. After uptake by macrophages, *Shigella* induces a rapid inflammatory cell death, which is associated with caspase-1 activation (Zychlinsky et al. 1992; Hilbi et al. 1997). Following the model proposed by Sansonetti (2001a), once *Shigella* has induced cell death in the macrophage, it escapes from the dying macrophage and infects the epithelium from the basolateral side. A more recent study from the group of Sansonetti showed that *Shigella* targets the colonic crypts, both in the human ex vivo model where colonic biopsies are infected and in the guinea pig intrarectal model, and penetrates colonocytes at the crypt mouth from the apical side (Arena et al. 2015). It is possible that both mechanisms, entry through M cells and invasion through the basolateral side and apical invasion of colonocytes, occur contemporaneously.

Shigella can inject a variety of effector substances through the T3SS, of which some induce actin rearrangements, leading to the uptake of bacteria by the epithelial cell (Carayol and Tran Van Nhieu 2013). Also mediated through the T3SS, *Shigella* then escapes from the endocytic vacuole into the cytoplasm where it multiplies (Hale et al. 1979; Ménard et al. 1993; Bârzu et al. 1997). *Shigella* does not express flagella, but inside the cell it moves by inducing actin polymerization at one pole, providing a propulsive force (Egile et al. 1999; Bernardini et al. 1989). Cell to cell spread occurs by the creation of pseudopodia that are engulfed by neighbouring cells via a clathrin-dependent endocytic pathway, mainly at tricellular junctions (Fukumatsu et al. 2012). Infected epithelial cells secrete IL-8, a chemokine that recruits large numbers of neutrophils (Philpott et al. 2000; Sansonetti et al. 1999; Niebuhr et al. 2002; Pédron et al. 2003; Singer and Sansonetti 2004). Neutrophils are highly inflammatory and contribute to the tissue destruction seen in shigellosis, but they are also essential for the resolution of infection (Weinrauch et al. 2002; Sansonetti et al. 1999).

Mucosal destruction, particularly in the epithelial areas that overlay the lymphoid follicles, is a feature of natural (Mathan and Mathan 1991) and experimental shigellosis (Sansonetti et al. 1991). The colonic and rectal mucosae of patients with acute shigellosis show abundant signs of inflammation. Necrotic erosions of the surface epithelium with abscesses rich in infiltrating neutrophils, extending from small ulcers on the epithelium, have been observed. There is generally a massive infiltration of neutrophils, macrophages, lymphocytes and natural killer cells (Raqib et al. 1995a, b; Mathan and Mathan 1986). Cell death and activation of caspase-1 are also frequent: in one study, 43 % of T cells were TUNEL-positive, while only 5.4 % of neutrophils and 10 % of macrophages showed TUNEL positivity, indicating that neutrophils are more resistant to *Shigella*-induced cell death than T cells and macrophages (Raqib et al. 2002a). Extensive expression of IL-1β and IL-18 was also seen (Raqib et al. 2002a). Caspase-1 was strongly increased in inflamed mucosa, and activation was seen in neutrophils, macrophages, and T cells. Caspase-3 activation was also observed in the lamina propria and lymphoid aggregates as well as Fas receptor and Fas-L activation, granzyme A, and perforin production, indicating that several cell death pathways are active: pyroptosis, apoptosis, and $CD8^+$-induced cytotoxicity (Raqib et al. 2002a). Cell infiltration and cytokine production often persist during the convalescent stage, weeks after resolution of symptoms (Raqib et al. 1995a). Activation of caspase-3, but not caspase-1, was seen in the epithelium (Bagchi et al. 2010). There are thus several signs of inflammasome activation in infected mucosa: caspase-1 activation, IL-1β and IL-18 secretion, and cell death.

Through in vitro and in vivo studies, it has become clear that the effect of *Shigella* is cell type specific and that inflammasome activation plays its specific role in the interaction with each cell type. Here, we will review this important aspect of *Shigella*–host cell interactions.

2 Epithelial Cells

Several studies in vitro have shown that epithelial cells are the niche where intracellular *Shigella* proliferate (Sansonetti et al. 1986), in contrast to macrophages that succumb to rapid death upon contact with these bacteria. In vitro infected epithelial cells can survive during six to seven hours of infection in spite of the presence of around one hundred bacteria filling their cytoplasm (Sansonetti et al. 1986; Mantis et al. 1996; Cersini et al. 1998). Using these observations as a starting point, various groups have addressed the question about the molecular mechanisms allowing epithelial cells to survive the bacterial assault rather than the process leading them to death. The destruction of infected cells plays a pivotal role in eradicating pathogens, limiting bacterial proliferation, and emitting alarm signals. However, bacterial pathogens employ various mechanisms to delay the death of infected cells, and *Shigella* seems to belong to this group of "sly" pathogens (Pearson et al. 2013).

Shigella enters epithelial cells through the effectors secreted by the T3SS (Clerc et al. 1986; Parsot 2009). Effectors such as the Ipa proteins induce cytoskeletal rearrangement in epithelial cells promoting bacterial endocytosis. Once inside the cells, the IpaB and IpaC proteins help the bacterium to escape from the vacuole (Ménard et al. 1993; Bârzu et al. 1997). This process allows *Shigella* to proliferate within the cytoplasm of the infected epithelial cells. At this point, a series of bacterial effectors are finely upregulated following induction of secretion under intracellular growth conditions (Parsot 2009). These proteins (most of them called Osp) are considered the second wave of T3SS effectors. The characterized functions of some of the Osp proteins are consistent with the hypothesis that *Shigella* exploits various and powerful immune evasion strategies to lower the inflammatory reaction and to subvert the innate immune responses, such as programmed cell death. Epithelial cells infected with *Shigella* are reported to undergo death at late times of infection due to apoptosis or necrosis (Bergounioux et al. 2012; Carneiro et al. 2009; Clark and Maurelli 2007; Tattoli et al. 2008; Lembo-Fazio et al. 2011). Clark and Maurelli (2007) reported that at early times of infection, *Shigella* was able to protect epithelial cells from the pro-apoptotic activity of staurosporine. Cells showed caspase-9 activation and cytochrome c release, but not caspase-3 activation. Protection was mediated by a T3SS late secreted effector. Faherty et al. (2010) followed up this study through a microarray experimental approach, revealing that *Shigella*-mediated protection from staurosporine activity was associated with the expression of genes encoding pro-survival proteins interfering with the intrinsic and extrinsic pathways of apoptosis. Carneiro et al. (2009) demonstrated that *Shigella* invasion in non-myeloid cells induced necrotic cell death through a pathway dependent on Bnip3 and cyclophilin D, two molecules implicated in the host oxidative stress responses. In 2012, Bergounioux et al. described that *Shigella* invasion of epithelial cells rapidly induced a genotoxic response, calpain protease activation, and necrotic cell death. In 2011, our group showed that *Shigella* invasion of epithelial cells triggered the up-regulation of both pro-apoptotic and anti-apoptotic pathways, resulting in the induction of mitochondria-dependent apoptosis and cell necrosis at late times of infection (Lembo-Fazio et al. 2011). The majority of these studies also highlighted the activation of inflammatory pathways along with that of pro-survival pathways mediated by key transcriptional regulators such as JUN and NF-κB.

Recently, Nlrc4/Naip inflammasome activation in intestinal epithelial cells (IEC) was shown to be involved in the protection against mucosal infection with the facultative intracellular pathogen *Salmonella enterica* serovar Typhimurium (*S.* Typhimurium) and the extracellular murine pathogen *Citrobacter rodentium* (Sellin et al. 2014; Liu et al. 2012; Nordlander et al. 2014). Both pathogens activate the inflammasome through recognition of proteins belonging to the T3SS. However, neither IL-1α nor IL-1β were involved in this process where the induction of pyroptosis plays the main role. T3SS is a key virulence factor of *Shigella,* and this pathogen lyses the membrane surrounding the phagocytic vacuole: both features could favour the activation of the Nlrc4 inflammasome in IEC. Dupont et al. (2009) observed an inflammasome platform through confocal microscopy, including NLRC4, NLRP3, and ASC, in HeLa cells infected with *Shigella* at early times of infection. At late times, caspase-1 activation was also remarked together with pyroptotic death. This was the first report where activation of caspase-1 was associated with *Shigella* infection of epithelial cells, though further investigations as to whether the observed inflammasome platforms were functional were not made. In 2013, two reports (Hagar et al. 2013; Kayagaki et al. 2013) described the crucial role of caspase-11 in mice and of the putative human functional orthologues caspase-4 and caspase-5, in eliciting a non-canonical inflammasome able to act independently of caspase-1 in response to intracellular LPS. Then, Shi et al. (2014) characterized the molecular mechanisms underlying the activation of the non-canonical inflammasome according to which intracellular LPS directly binds with high affinity and specificity to the CARD domains of human caspase-4/caspase-5 and murine caspase-11. In order to gain access to the cytosol of epithelial cells and to deliver LPS that can bind to caspase-4, Gram-negative pathogens must escape from the vacuole. Thus, *Shigella* could potentially activate this process. A study from our group found that expression of caspase-4 was strongly upregulated in *Shigella*-infected HeLa cells (Lembo-Fazio et al. 2011). Kobayashi et al. (2013) demonstrated that a late T3SS secreted effector, OspC3, was able to bind to caspase-4 in *Shigella*-infected epithelial cells. OspC3 inhibits the assembly of the active tetramer form of caspase-4, consisting of two p19 and two p10 subunits, by binding to the p19 subunit. This immune evasion strategy allows *Shigella* to escape from caspse-4 surveillance. In accordance with the immuno-evasive role of OspC3, a *Shigella ΔospC3* mutant augmented the pyroptotic rate of infected epithelial cells and increased IL-18 release. In other words, in epithelial cells, *Shigella* seems to block inflammasome activation, mainly sustained by caspase-4, thus prolonging the lifespan of the epithelial cells. Whether and to what extent caspase-1 is involved in this process is still unclear.

In 2013, we showed that intracellular *Shigella* modify the lipid A structure of LPS, passing from a hexa-acylated form which prevails in laboratory medium-grown bacteria, to tetra- and tri-acylated forms typical of intracellular *Shigella* (Paciello et al. 2013). The activation of caspase-11 and caspase-4 is dependent on the number of acyl chains of lipid A, in parallel to what has already been demonstrated for the activation of the TLR4 complex (Shi et al. 2014; Akashi et al. 2001; Hagar et al. 2013), and the tetra-acylated forms of lipid A are poorly able to activate both the TLR4 complex and caspase-4. Therefore, changing the number of acyl chains could be considered another immune evasion mechanism of *Shigella,* employed to avoid caspase-4 activation.

These new insights concerning the strategies of *Shigella* to reduce/abrogate caspase-4 LPS binding depict a complex scenario where delayed induction of cell death through apoptosis and necrosis or also pyroptosis, triggered by DAMPs following *Shigella* infection, might be beneficial for both the host and the bacteria. Epithelial cells act as sentinels of the innate immune system, and through delayed death, they might gain time to produce the alarm signals necessary to activate downstream responses. Intracellular *Shigella* activate NOD1 by releasing peptidoglycan fragments. This process initiates the inflammatory reaction by NF-κB and API-1 leading to IL-8 production and neutrophil transmigration in infected tissues (Girardin et al. 2001; Nigro et al. 2008). Delayed death might also be a strategy to achieve enough time for replication, indicated by the fact that *Shigella* activates various pro-survival pathways.

Recent studies in vivo on *S.* Typhimurium and the roles of Nlrc4-mediated inflammasome and caspase-4/caspase-11 activation in epithelial cells (Sellin et al. 2014; Knodler et al. 2010, 2014) suggest that pyroptotic cell death is instrumental for the removal of infected IECs from the mucosa, dislodging them into the intestinal lumen. This could be a highly important mechanism for the host to lower the bacterial burden in the intestinal mucosa and eventually eradicate the pathogen. *S.* Typhimurium seems to be able to activate the inflammasome in epithelial cells, while the related pathogen *Shigella* exploits evasion strategies to escape from inflammasome activation. This difference could be explained by the observation that epithelial cells are the replicative niche of *Shigella*, which is not the case for *Salmonella*. Considering that the infectious dose provoking shigellosis is very low (10–100 bacteria), *Shigella* cannot risk to be eliminated along with the infected epithelial cells at an early stage of infection, while at a later stage, it might be advantageous, leading to a spread of bacteria in the environment. This model would explain the redundant immune evasion strategies applied by *Shigella* to reduce the inflammasome activation in epithelial cells.

3 Macrophages

Zychlinsky et al. provided the first reports of the phenomena accompanying pyroptosis in 1992 and 1994. They found that fully invasive *Shigella* (*S. flexneri*) were able to induce death of infected macrophages (Zychlinsky et al. 1992), accompanied by the release of IL-1α and IL-1β (Zychlinsky et al. 1994a). At the time, cell death was classified as apoptosis due to nuclear changes. The same type of death was induced by clinical isolates of *S. flexneri* and *S. sonnei*, consolidating the relevance of the observations (Guichon and Zychlinsky 1997). The T3SS secreted protein IpaB was found to be essential in the process (Zychlinsky et al. 1994b). IpaB seems to have a dual role. The first role is to form a translocon in the host cell membrane in concert with another protein secreted by the T3SS, IpaC, in order to promote the transport of T3SS secreted effectors across the cell membrane (Blocker et al. 1999). The second role is lysis of the membrane of the phagocytic vacuole, allowing bacteria to proliferate within the cytosol (High et al. 1992; De Geyter et al. 2000). This function is also necessary to trigger macrophage death (Zychlinsky et al. 1994b). *Shigella*-induced cell death and IL-1β release in murine and human macrophages were found to depend on caspase-1, initially named interleukin-1 converting enzyme (ICE) (Chen et al. 1996; Hilbi et al. 1997; Alnemri et al. 1996), while caspase-3 and caspase-11 were not involved (Hilbi et al. 1998). Caspase-1 activation was seen as a further confirmation of apoptotic cell death, since all caspases were thought to be involved in apoptosis. Purified IpaB was reported to bind directly to caspase-1, provoking cell death (Hilbi et al. 1997, 1998; Chen et al. 1996). These pioneering studies were the first to associate the virulence of a pathogen to the death of macrophages. However, at that time, controversy arose concerning the type of death induced in *Shigella*-infected macrophages. One group found that infected human macrophages died by oncosis and that cell death and IL-1β release were two independent events, since a pan-caspase inhibitor blocked IL-1β release but not cytotoxicity (Fernandez-Prada et al. 1997). Likewise, in the monocyte cell line U937, *Shigella* was reported to induce oncosis or apoptosis, depending on the stage of differentiation and on how the cells were differentiated (Nonaka et al. 1999).

In these same years, several relevant aspects of shigellosis were described concerning the pathogenesis of *Shigella*. Macrophage death fuelling the inflammation induced by bacteria in infected tissues played a major role in this model, and much attention was paid to how inflammation contributes to bacterial invasion (Sansonetti 2001b).

Macrophage death induced by *Shigella* was clearly inflammatory, something that was also supported by studies in dysenteric patients (Raqib et al. 2002a). This did not match the general view of apoptosis, considered a silent death. Clarifications were sought through animal models. The importance of inflammasome activation in shigellosis was demonstrated in the mouse intranasal model by Sansonetti et al. (2000). *Caspase 1*$^{-/-}$ mice were unable to control bacterial load and contain the inflammatory reaction induced by *Shigella* and eventually died. The pattern of infection and inflammation in *IL-18*$^{-/-}$ defective animals paralleled that observed with *caspase 1*$^{-/-}$. On the contrary, the foci of bacterial infection in lungs of *IL-1β*$^{-/-}$ mice were restricted compared to the wild-type animals. The authors suggested that IL-1β was involved only in the first phases of infection. In the rabbit ileal loop model of infection, blocking IL-1β and IL-1α activity through administration of IL-1 receptor antagonist (IL-1ra) (Sansonetti et al. 1995) resulted in decreased inflammation and lower bacterial loads. These two studies indicate that IL-1β is detrimental in shigellosis. However, it is also possible that the fine-tuning of IL-1β secretion, such as the timing of secretion, can determine whether it will be beneficial or harmful. Further studies would be needed to clarify this.

In 2002, the team of Tschopp described the assembly of a multimeric complex involved in caspase-1 activation and secretion of IL-1β and called it "inflammasome" (Martinon et al. 2002). Since then, many studies on *Shigella*-infected macrophages have sought to clarify the molecular mechanisms leading to inflammasome activation and modulation. Binding of the T3SS secreted protein IpaB to cholesterol, allowing insertion of the translocon on the host cell membranes had been observed, (Lafont et al. 2002; Hayward et al. 2005), as well as the capacity of IpaB to destabilize model membrane (De Geyter et al. 2000). Contact between IpaB and cholesterol was also found to be a prerequisite for caspase-1-mediated cytotoxicity in macrophages (Schroeder and Hilbi 2007). Purified IpaB was described to spontaneously oligomerize and insert into plasma membranes of target cells, allowing potassium influx into the vacuole compartment and membrane destabilization (Senerovic et al. 2012). The authors proposed a model where IpaB-mediated membrane damages would lead to release of lysosome or phagolysosome content into the cytosol, triggering inflammasome activation and cell death. They found that Nlrc4, caspase-1, and ASC were necessary for IL-1β release induced by *Shigella* or purified IpaB, while Nlrp3 was not. Suzuki and collaborators (Suzuki et al. 2007) also reported that Nlrc4 (Ipaf) was responsible for pyroptotic death of *Shigella*-infected macrophages, while Nlrp3 was needed neither for IL-1β secretion nor cell death. ASC was found to be involved in the processing of IL-1β but not in the pyroptotic death. These results were consistent with work published by different authors, showing that macrophages from *Asc*$^{-/-}$ mice were fully competent to undergo pyroptosis but unable to secrete mature IL-1β in response to various known inflammasome-activating stimuli (Mariathasan et al. 2004; Broz et al. 2010). Two different types of inflammasomes were suggested to form upon activation of CARD-containing cytosolic receptors, one independent of ASC, called the "death complex", and one involving ASC, the "ASC focus", necessary for IL-1 secretion (Broz et al. 2010). Whether IpaB plays a direct role by binding to caspase-1, as originally described, or acts indirectly by promoting vacuolar membrane destabilization and favouring T3SS recognition by NLRC4 (see later) is still not fully elucidated.

Bacterial flagellin was the first recognized NLRC4 activator, and the flagellated bacteria *Salmonella* and *Legionella* are both able to induce Nlrc4 inflammasome formation (Franchi et al. 2006; Amer et al. 2006). *Shigella* does not express *fliC*, the gene encoding the flagellar filament structural protein, and consequently lacks flagella (Suzuki et al. 2007). Instead, T3SS rod and needle proteins were later discovered to activate NLRC4, and another subfamily of NLRs, the NAIPs, were seen to act as adaptors, conferring ligand specificity. Mice express seven Naip homologous (one to seven), while humans only have one (NAIP). In mice, Naip6 and 5 recognize flagellin (Kofoed and Vance 2011; Zhao et al. 2011), Naip2 responds to T3SS rod protein, e.g. *Shigella* MxiI (Zhao et al. 2011; Kofoed and Vance 2011; Suzuki et al. 2014a; Miao et al. 2010b), and Naip1 can bind directly to T3SS needle protein, e.g. *Shigella* MxiH. Human NAIP recognizes T3SS needle protein (Yang et al. 2013), and one isoform of NAIP also recognizes flagellin (Kortmann et al. 2015).

In 2014, another study (Suzuki et al. 2014b) indicated that *Shigella* actively promotes inflammasome activation in macrophages. GLMN (glomulin/flagellar-associated protein 68), a Cullin-RING-E3 ligase inhibitor, is a negative regulator of the NLRP3/NLRC4 inflammasomes. The late T3SS effector IpaH 7.8, a member of the IpaH family of *Shigella*-secreted proteins (Rohde et al. 2007), targets GLMN for ubiquitination and proteasome degradation, thus releasing the break on NLRP3/NLRC4 and facilitating inflammasome activation. In this study, IL-1β secretion was found to depend on Nlrc4, with some contribution from Nlrp3. In the mouse intranasal in vivo model, GLMN degradation induced by wild-type *Shigella* resulted in stronger inflammation as well as increased bacterial load compared to an IpaH7.8 mutant.

Other reports imply that *Shigella* might also exert a dampening effect on macrophage inflammasome activation. Our group demonstrated that the presence of purified hypo-acylated LPS, derived from *Shigella* growing in epithelial cells, induced reduced secretion of IL-1β upon activation of the Nlrc4 inflammasome in macrophages compared to LPS from extracellular *Shigella* (Paciello et al. 2013). This scenario is likely to correspond to the phase in infection where *Shigella* exits epithelial cells, encountering other immune cells. Another interesting link between *Shigella*-infected epithelial cells and macrophage inflammasome activation might be derived from the observation that infected epithelial cells release ATP, a process that *Shigella* dampens through secretion of the effector protein IpgD (Puhar et al. 2013). ATP is a well-known activator of the NLRP3 inflammasome, and inhibiting ATP release might thus be another way of preventing inflammasome activation.

The involvement of the other inflammatory caspases (human caspase-4/caspase-5 and mouse caspase-11) in macrophage death remains obscure. An early report (Hilbi et al. 1998) indicated that cell death was independent of caspase-11. However, Suzuki et al. (2005) described a caspase-1/TLR-4 independent cell death pathway in *Shigella*-infected macrophages and found that cytosolic lipid A could induce this cell death. Since caspase-4/caspase-5/caspase-11 was later described as cytosolic sensors for LPS (Shi et al. 2014), their involvement might be hypothesized.

The contrasting results sometimes indicate one type of cell death, sometimes another, historically surely depended on ignorance regarding the existence of the inflammasome and different methodologies used to detect cell death. However, other parameters, such as timing and multiplicity of infection and variations inherent in cell lines and cell differentiation methods, are also likely to be important for the type of cell death induced by *Shigella*.

It has been suggested that macrophages undergo pyroptosis in order to eliminate a possible replicative niche and expose intracellular bacteria to cells with a higher bactericidal capacity, such as neutrophils (Miao et al. 2011), and aforesaid mechanism has also been shown for *S.* Typhimurium (Miao et al. 2010a).

It is difficult to figure out how *Shigella* might benefit from pyroptosis. We may presume that pyroptotic cell death could be a way of escaping from macrophages and to prevent them from killing bacteria. The results of Suzuki et al. (2014b), indicating that Shigella actively promotes inflammasome activation by secretion of the effector protein IpaH7.8 as described previously, speak in favour of this. However, together with the data of LPS modulation found by our group, it becomes evident that *Shigella* can finely tune inflammasome activation depending on the different phases of infection.

4 Neutrophils

Neutrophils play an essential role in shigellosis. Colonic biopsies from patients with acute shigellosis caused by *S. dysenteriae 1* or *S. flexneri* show oedema and neutrophil infiltration, especially in the surface epithelium and the lamina propria and to a lesser extent also in the crypt epithelium. In one-third of the patients, this picture persisted also one month after infection. (Islam et al. 1997a). Neutrophils in mucosa of infected patients were more resistant to *Shigella*-induced cell death compared to macrophages and T cells, but caspase-1 activation has been detected in all three cell types (Raqib et al. 2002a), suggesting potential inflammasome activation also in neutrophils.

In contrast to epithelial cells, macrophages, and dendritic cells (DCs), where *Shigella* can escape from the phagosome into the cytosol, rabbit neutrophils were demonstrated to be capable of containing *Shigella* in the phagosome. A role for bactericidal/permeability increasing protein was shown (Mandic-Mulec et al. 1997). Another study demonstrated neutrophil elastase (NE) as the key factor allowing containment of *Shigella* to the phagosome in both human and murine neutrophils. After NE inhibition, bacteria were capable of escaping to the cytosol. Inhibition of NE resulted in a doubling of intracellular bacteria and of cytotoxicity as measured by LDH release after infection with high MOIs (Weinrauch et al. 2002). LDH release is seen both upon pyroptotic and necrotic cell death, and the type of cell death was not classified further. It would be highly interesting to investigate whether incapacity to contain the bacteria in the phagosome leads to caspase-1 activation.

Another report also described LDH release from *Shigella*-infected neutrophils in vitro. Cell death was classified as necrosis, due to release of LDH and absence of typical DNA laddering. DNA fragmentation in pyroptosis is different from that in apoptosis, and laddering is not always seen and thus cannot completely rule out a pyroptotic cell death. However, the hypothesis that infected neutrophils undergo pyroptosis is questionable. Miao et al. (2010a) reported that mouse neutrophils express caspase-1, but not Nlrc4, suggesting that they were unable to induce pyroptosis in response to flagellin or T3SS proteins delivered by pathogens into the cytoplasm. Neutrophils could instead efficiently phagocyte and destroy bacteria escaping from infected macrophages killed by pyroptosis.

Whether neutrophils express NLRs or not is controversial, and contradictory results have been described. Another study found that human and mouse neutrophils express mRNA for various NLRs, including NLRP3 and NLRC4, in quantities similar to macrophages (Chen et al. 2014), where inflammasome activation is well characterized, and an increase of expression of these NLRs was seen upon LPS stimulation. Mouse neutrophils were also seen to secrete IL-1β in response to *S.* Typhimurium, in levels similar to those induced in macrophages. This was abrogated in *caspase-1*$^{-/-}$, *caspase-11*$^{-/-}$, *Asc*$^{-/-}$, and *Nlrc4*$^{-/-}$ cells, indicating that secretion was dependent on Nlrc4 inflammasome formation. These results suggest that neutrophils are competent for inflammasome activation. However, IL-1β secretion was not followed by cell death (Chen et al. 2014). In a mouse model of *Staphylococcus aureus* infection, neutrophils turned out to be the main source of IL-1β, and this was dependent on TLR2, NOD2, FPR1, and the ASC/Nlrp3 (Cho et al. 2012). Inflammasome activation in neutrophils thus seems to be associated more with IL-1β secretion than with pyroptosis. It is unclear whether neutrophils contribute to IL-1β secretion in *Shigella* infection. If this is the case, Il-1β recruiting neutrophils, which also secrete IL-1β, would lead to a vicious circle, fuelling inflammation.

The importance of neutrophils in clearing *Shigella* infection, and the association of neutrophil depletion with death of the host, also speaks against pyroptosis as a mechanism used by this cell population. Neutrophils have been demonstrated to contribute to tissue damage but also to the resolution of the infection in various models. IL-8 secreted by infected epithelial cells upon NOD1 activation recruits neutrophils, and injection of a neutralizing anti-IL-8 antibody before infection in the rabbit-ligated ileal loop model resulted in inhibition of neutrophil recruitment and epithelial damage. However, the price was high: a threefold increase in the villus tissue and a doubling of bacteria in the blood (Sansonetti et al. 1999). Infection with wild-type *Shigella* in a model of human intestinal xenographs on SCID mice resulted in the production of IL-1β and IL-8 from the xenographs and tissue damage with leucocyte infiltrates, where neutrophils were the dominating cell type (Zhang et al. 2001). Depletion of neutrophils with monoclonal antibodies prior to infection did not have a major influence on tissue damage, but was associated with a more than 20-fold increase in intracellular bacteria (Zhang et al. 2001). In zebrafish larvae infected with *Shigella*, neutrophils were capable of killing *Shigella,* while macrophages were not. Macrophage death was seen both at sublethal and at lethal doses of infection, while neutrophil death was seen only with the lethal dose, indicating that overwhelming numbers of bacteria lead to detrimental neutrophil depletion (Mostowy et al. 2013).

Pyroptosis in macrophages might be a strategy to eliminate a hypothetical replicative niche for intracellular bacteria and expose them to the more potent bactericidal neutrophils. IL-1β release from macrophages might also be involved in the recruitment of neutrophils, as has been shown in various other conditions (Shaftel et al. 2007; Presicce et al. 2015; Miller et al. 2007).

These data altogether show the high complexity of the inflammatory response against *Shigella,* where individual components can be beneficial or detrimental, depending on which other components are activated contemporaneously.

5 Dendritic Cells

DCs can produce a wide range of cytokines in response to various stimuli and are important orchestrators of innate immunity. They have a high capacity to capture antigen and present them to lymphocytes and are thus essential in the initiation of an antigen-specific response. The longevity of a protective adaptive immune response after natural *Shigella* infection is not well known. Repeated exposure to *Shigella* infection is considered necessary to induce a protective response (Raqib et al. 2002a, b; Nothelfer et al. 2014; Salgado-Pabon et al. 2013). This is supported by the different kinetics seen in the induction of adaptive responses in children and adults in endemic areas. Adaptive immune responses induced in children are lower in their magnitude and take longer time to develop compared to adults, indicating previous exposure in adults (Raqib et al. 2002b). Since DCs are key cells in the induction of adaptive immune responses, the fate of infected DCs in *Shigella* infection could be important to explain the poor adaptive immune response.

Contrasting results concerning the cellular fate of infected human monocyte-derived dendritic cells (MoDCs) have been reported. One group found signs of apoptosis as seen by AnnexinV^{+}/PI^{-} after infection with *Shigella* (Kim et al. 2008). Other groups have reported membrane permeability, evaluated by LDH release around 60–90 min after addition of bacteria (Edgeworth et al. 2002; Osorio et al. 2007), something we have also found (Hermansson et al. unpublished results). *S. sonnei* induced lower levels of LDH release compared to *S. flexneri* and *S. dysenteriae* (Osorio et al. 2007). LDH release is not seen in apoptotic cell death, but in necrotic or pyroptotic cell death. Edgeworth et al. (2002) reported that treatment of cells with YVAD, a caspase-1-specific inhibitor, resulted in reduced cytotoxicity, while treatment with z-VAD, a pan-caspase inhibitor, almost completely abolished cell death. The concept of pyroptosis is not mentioned in the paper, as it had just recently been described (Cookson and Brennan 2001) at the time these results were published, and the existence of the inflammasome was yet unknown. However, the partial dependence of cell death on caspase-1 indicates inflammasome activation, while dependence on other caspases could point to a contribution of apoptosis or involvement of other inflammatory caspases such as caspase-4.

Infection of MoDCs with *S. flexneri* and *S. sonnei* also resulted in release of IL-1β (Edgeworth et al. 2002; Osorio et al. 2007), another sign of inflammasome activation, strengthening the hypothesis of pyroptotic cell death. IL-1β release is also likely to be involved in the direction of the type of adaptive response that will be induced (see further Sect. 6). MoDCs seem less prone to IL-18 release upon *Shigella* infection compared to monocyte-derived macrophages (Edgeworth et al. 2002).

Escape from the phagocytic vacuole into the cytosol is a mechanism which could lead to inflammasome activation, since bacteria in the cytoplasm can be recognized by cytosolic receptors needed for the assembly of the inflammasome complex. Escape of *S. flexneri* was described by Kim et al. (2008), even though in that case it was reported to be correlated to apoptosis.

In the mouse DC line DC2.4, *Shigella* infection resulted in a robust activation of caspase-1, which was completely abrogated after silencing of Nlrc4. Caspase-1 activation was also reduced after silencing of Naip1, although to a lesser extent compared to Nlrc4 silencing (Yang et al. 2013). Naip1 was shown to recognize T3SS needle protein from various bacteria, amongst which MxiH from *Shigella,* and work as an adaptor between the bacterial molecule and Nlrc4. Both BMDCs and DC2.4 cells were found to have higher transcript levels of *Naip1* than BMDMs, and delivery of MxiH to the cytoplasm of DC2.4 cells induced caspase-1 activation, which was stronger than that seen in BMDMs. MxiH-induced caspase-1 activation was also dependent on Nlrc4 and Naip1 (Yang et al. 2013). Conclusively, despite contrasting results concerning type of cell death, inflammasome activation in DCs is likely.

6 Lymphocytes

T and B lymphocytes form the adaptive immune response. There are several indications that IgG antibodies against the O-antigen of LPS are important in protective immunity after natural *Shigella* infection (Robbins et al. 1992; Black et al. 1987; Coster et al. 1999; Ferreccio et al. 1991; Cohen et al. 1991). Since the LPS O-antigen differs between serotypes, this also confers that immunity is serotype specific. In an endemic area, adults are likely to have encountered *Shigella* repeatedly. In a study from Bangladesh, where *Shigella* is endemic, strong and rapid increase in *Shigella*-specific faecal sIgA was seen in infected adults. An increase in IgA and IgG plasma cells in the rectal mucosa of *Shigella*-infected adults also points to a role for humoral adaptive immunity in shigellosis (Islam et al. 1997b). *Shigella* has been shown to interact with B cells in ex vivo studies. When human colon biopsies were infected with *Shigella*, bacteria came in contact with

B cells and occasionally also invaded them (Nothelfer et al. 2014). Further in vitro experiments showed that B cells were quite resistant to *Shigella* invasion: despite high MOIs used, few cells were infected. Cell death was seen both in infected and in not infected B cells, but it was less extensive compared to what is seen in vitro in macrophages or DCs, and it was classified as apoptosis due to AnnexinV$^+$/PI$^-$ staining (Nothelfer et al. 2014). However, after footpad injection of bacteria in mice, a high percentage of B cell death in the draining lymph nodes was found (Nothelfer et al. 2014), indicating that B cells might be targeted by *Shigella* and that they are sensitive to *Shigella*-induced death.

Shigella flexneri did not prime antigen-specific CD8$^+$ T cells in the mouse pulmonary infection model (Jehl et al. 2011). The primary type of T cell-induced response was Th17 (CD4$^+$ IL-17A and IL-22 producing cells). After repeated infection, CD4$^+$ INFγ producing cells were also seen (Sellge et al. 2010).

IL-17 has an important role in infection with various extracellular bacteria but also some intracellular, especially at epithelial and mucosal surfaces (Jin and Dong 2013). It can induce neutrophil expansion and survival, as well as recruitment of both neutrophils and other immune cells such as lymphocytes, DCs, and monocytes. It also has a strong upregulatory effect on antimicrobial peptide production (Onishi and Gaffen 2010). Treating mice that had once been infected with *Shigella*, and thus had the possibility to mount a protective adaptive response, with anti-IL-17A mAbs during renewed infection resulted in increased weight loss and higher bacterial loads, indicating that Th17 cells are important in protective acquired immunity against *Shigella* (Sellge et al. 2010). The capacity of IL-1β to induce a Th17 response (Acosta-Rodriguez et al. 2007; Lasigliè et al. 2011) has been demonstrated in various conditions and is likely to provide a link between inflammasome activation by *Shigella* and T cells.

An in vitro study showed that *Shigella* was able to invade activated CD4$^+$ T cells, but not unactivated T cells (Ottoson et al. 2001). Using a MOI of 10 resulted in a percentage of dead cells that was slightly lower than the percentage of invaded cells. Whether invasion corresponded to cell death was not studied further, and the type of cell death was not characterized. However, *Shigella* infection resulted in impaired T cell migration (Ottoson et al. 2001).

Extensive death of CD3$^+$ T cells in the rectal mucosa of infected patients has been seen, based on TUNEL positivity (Raqib et al. 2002a). They were therefore classified as apoptotic, though at that time the process of pyroptosis had not yet been described and distinguished from apoptosis. DNA nicking is seen also in pyroptosis, and pyroptotic cells can be TUNEL-positive. Caspase-1 activity in the same cells was also described, indicating that T cells might die from pyroptosis in response to *Shigella*. This is not unlikely, and it has recently been demonstrated that the death of HIV-infected CD4$^+$ T cells is a result of caspase-1-dependent pyroptosis (Doitsh et al. 2014). Death of T cells might contribute to difficulties in mounting an efficient protective secondary response, as well as rendering patients more susceptible to other infections during the convalescence period.

Also in mice infected with *S. dysenteriae,* cell death was seen in T cells, both in $CD4^+$ and in $CD8^+$ T cells. This was determined through AnnexinV positivity and labelled as apoptosis, but Annexin V alone cannot exclude other types of cell death since also membrane permeability will result in Annexin V positivity (Bagchi et al. 2010).

7 *Shigella* and the Inflammasome—Who's Holding the Conducting Baton?

The picture emerging from these data shows how *Shigella* can tune up and down inflammasome activation, pro-death, and pro-survival pathways, depending on the cell population (Fig. 1). In epithelial cells, *Shigella* induces mainly pro-survival (Carneiro et al. 2009; Niebuhr et al. 2002; Pendaries et al. 2006; Puhar et al. 2013; Kobayashi et al. 2013; Paciello et al. 2013). *Shigella* can move from an infected cell to the adjacent one without leaving the monolayer. In contrast to other facultative pathogens, the death of the infected epithelial cell is therefore not needed in order to allow *Shigella* to reinfect another cell. In macrophages, vacuolar escape instead leads to rapid induction of pyroptotic cell death. The Nlrc4-dependent inflammasome seems to be the main platform activated (Senerovic et al. 2012; Suzuki et al. 2014a, b). Eventually, a Nlrp3 platform can also contribute to IL-1β production (Suzuki et al. 2014b). A competition between defence mechanisms has also been observed. In *Shigella*-infected macrophages, activation of caspase-1 and Nlrc4 inhibited autophagy (Suzuki et al. 2007). The inverse relationship was seen in *Legionella penumophilia*-infected macrophages, and it was proposed that low-grade infection might be resolved by autophagy, but when the autophagic machinery is overwhelmed, pyroptosis mechanisms will set in Byrne et al. (2013). Most likely, various defence mechanisms are activated contemporaneously in *Shigella*-infected cells, such as autophagy, pyroptosis and other types of cell death, and depending on factors such as number of bacteria and time after infection, one specific mechanism takes overhand. This might be true also for different kinds of inflammasomes, where constitutive inflammasomes might be activated at early stages and inducible ones at later stages (Franchi et al. 2012; Bauernfeind et al. 2009; O'Connor et al. 2003; Schauvliege et al. 2002; Lin et al. 2000).

Whether *Shigella* induces pyroptosis in DCs and T cell is not fully clarified. Both cell types are competent for inflammasome-dependent cell death, and both cell types die after *Shigella* infection (Doitsh et al. 2014; Edgeworth et al. 2002; Raqib et al. 2002a). Either way, both cell types are highly important in the formation of an adaptive immune response. Man is the only natural host for *Shigella*, and it has been suggested that interfering with the induction of an adaptive response might allow *Shigella* to infect the same individual several times. Thus, promotion of pyroptosis in infected cells might be a way for *Shigella* to take advantage of the host defence.

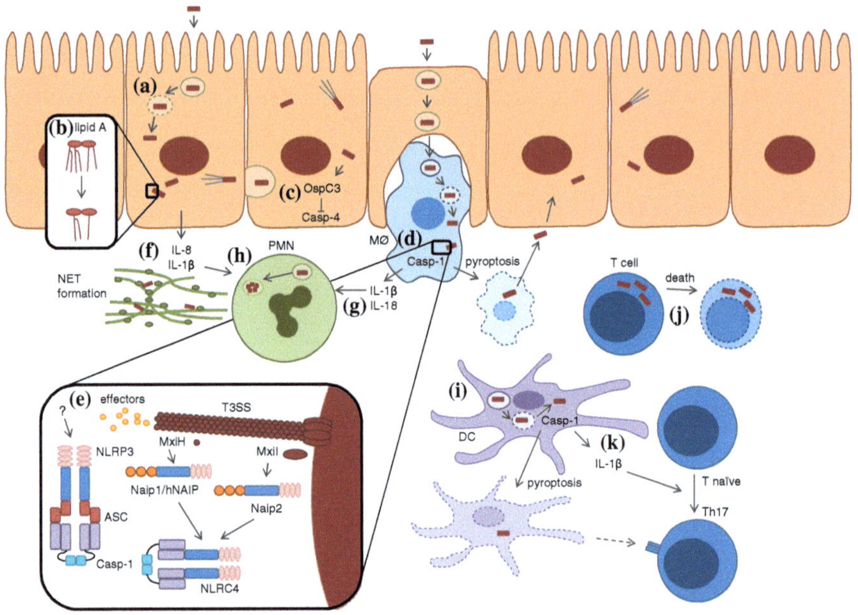

Fig. 1 Various cell populations affected by *Shigella* infection and inflammasome activation. *Shigella* invades the mucosa of the colon and the rectum and escapes from the endocytic vacuole into the cytosol (**a**). In order to avoid detection and diminish inflammation, it modifies its lipid A (**b**) and inhibits activation of capsase-4 by secreting the effector protein OspC3 (**c**). Bacterial escape from the phagocytic vacuole in macrophages leads to caspase-1 activation and pyroptotic cell death (**d**) by recognition of T3SS rod and needle protein, inducing the NLRC4 inflammasome and possibly other components that activate the NLRP3 inflammasome (**e**). Secretion of IL-8 and IL-1β from infected colonocytes and macrophages recruits neutrophils (**f**, **g**), which are capable of killing *Shigella* through phagocytosis and formation of NETs (**h**). Infected DCs and T cells also die (**i**, **j**), leading to poor induction of adaptive immunity. Cytokines released also affect the type of adaptive immunity induced (**k**)

Neutrophils are one of the few cell types that can resist *Shigella*-induced cell death, most likely due to their capacity to prevent bacterial escape into the cytosol (Miao et al. 2011).

The cell type-specific effect of *Shigella* might depend on the capacity of bacteria to distinguish different types of cells and activate appropriate effector proteins. It could also be the result of different “hard wiring” of specific cell types, where epithelial cells might be more resistant to death pathways and easier to manipulate towards survival compared to macrophages, who might be predisposed to respond with pyroptosis. The discovery that bacterial populations are not homogenous is highly interesting. The heterogeneity of *S.* Typhimurium infecting epithelial cells was recently discovered, with a major part of bacteria replicating within membrane-bound vesicles and a minor part replicating in the cytosol (Knodler et al. 2010). The two populations were transcriptionally distinct, and the cytosolic bacteria were suggested to serve as a reservoir for dissemination. A similar mechanism could hypothetically result in various *Shigella* populations adapted to invade different cell types.

Increasing knowledge of the role of IL-1β and IL-18 in shigellosis would be welcome. IL-1β is a highly potent pro-inflammatory cytokine. By secretion of IL-1β, inflammasome activation has consequences on a number of other cells, for example through leucocyte recruitment and direction of adaptive responses. A protective role of IL-1β has been shown in some infections (Cai et al. 2012; Van't Wout et al. 1988), but it can also be deleterious (Ceballos-Olvera et al. 2011), and the same is true for IL-18 (Faure et al. 2014; Ceballos-Olvera et al. 2011). One study in mice and one in rabbits indicate that IL-1β plays a positive role for *Shigella* rather than for the host (Sansonetti et al. 1995, 2000).

A relationship between IL-18 (and IL-23) secretion and IL-17 release was recently shown in *Pseudomonas aeruginosa* infection (Faure et al. 2014). Activation of Nlrc4 by the *P. aeruginosa* T3SS leading to IL-18 secretion promoted excessive neutrophil recruitment, which was eventually deleterious. IL-18 secretion also dampened the beneficial IL-17-mediated antimicrobial host response. Whether IL-18 plays a similar role in colonic mucosa during natural shigellosis is unknown. It is noteworthy that IL-17 has been reported to promote the expression of defensins in colonocytes (Ishigame et al. 2009). Defensins play a crucial role in controlling *Shigella* proliferation in mice intestine (Shim et al. 2010). *Shigella* downregulates the expression of antimicrobial peptides via the T3SS effectors (Sperandio et al. 2008). A closer evaluation of the axis inflammasome-IL-17-antimicrobial peptides during shigellosis would be interesting. One study in mice indicated a protective role for IL-18 (Sansonetti et al. 2000), but confirming studies are needed, preferably also in other models, given that *Shigella* is a human pathogen, and extrapolation from animal studies must be done with some caution.

In conclusion, if we let the epithelial cells be the string section, adaptive immunity the brass and IL-1β percussions, we may conclude that the maestro *Shigella* conducts its orchestra well.

References

Acosta-Rodriguez EV, Napolitani G, Lanzavecchia A, Sallusto F (2007) Interleukins 1β and 6 but not transforming growth factor-β are essential for the differentiation of interleukin 17-producing human T helper cells. Nat Immunol 8(9):942–949

Akashi S, Nagai Y, Ogata H, Oikawa M, Fukase K, Kusumoto S, Kawasaki K, Nishijima M, Hayashi S, Kimoto M, Miyake K (2001) Human MD-2 confers on mouse Toll-like receptor 4 species-specific lipopolysaccharide recognition. Int Immunol 13(12):1595–1599

Alnemri ES, Livingston DJ, Nicholson DW, Salvesen G, Thornberry NA, Wong WW, Yuan J (1996) Human ICE/CED-3 protease nomenclature. Cell 87(2):171

Amer A, Franchi L, Kanneganti TD, Body-Malapel M, Özören N, Brady G, Meshinchi S, Jagirdar R, Gewirtz A, Akira S, Núñez G (2006) Regulation of *Legionella* phagosome maturation and infection through flagellin and host ipaf. J Biol Chem 281(46):35217–35223

Arena ET, Campbell-Valois FX, Tinevez JY, Nigro G, Sachse M, Moya-Nilges M, Nothelfer K, Marteyn B, Shorte SL, Sansonetti PJ (2015) Bioimage analysis of *Shigella* infection reveals targeting of colonic crypts. Proc Natl Acad Sci USA 112(25):E3282–E3290

Bagchi AK, Sinha AK, Adhikari R, Maiti P, Mukherjee J, Panda A, Saha DR (2010) Selective deletion of CD8(+) cells upregulated by caspases-1 via IL-18 in mice immunized with major outer membrane protein of *Shigella dysenteriae 1* following infection. J Clin Immunol 30 (3):408–418

Bârzu S, Benjelloun-Touimi Z, Phalipon A, Sansonetti P, Parsot C (1997) Functional analysis of the *Shigella flexneri* IpaC invasin by insertional mutagenesis. Infect Immun 65(5):1599–1605

Bauernfeind F, Horvath G, Stutz A, Alnemri ES, MacDonald K, Speert D, Fernandes-Alnemri T, Wu J, Monks BG, Fitzgerald KA, Hornung V, Latz E (2009) NF-κB activating pattern recognition and cytokine receptors license NLRP3 inflammasome activation by regulating NLRP3 expression. J Immunol 183(2):787–791

Bergounioux J, Elisee R, Prunier AL, Donnadieu F, Sperandio B, Sansonetti P, Arbibe L (2012) Calpain activation by the *Shigella flexneri* effector VirA regulates key steps in the formation and life of the bacterium's epithelial niche. Cell Host Microbe 11(3):240–252

Bernardini ML, Mounier J, d'Hauteville H, Coquis-Rondon M, Sansonetti PJ (1989) Identification of icsA, a plasmid locus of *Shigella flexneri* that governs bacterial intra- and intercellular spread through interaction with F-actin. Proc Natl Acad Sci USA 86(10):3867–3871

Black RE, Levine MM, Clements ML, Losonsky G, Herrington D, Berman S, Formal SB (1987) Prevention of shigellosis by a *Salmonella typhi-Shigella sonnei* bivalent vaccine. J Infect Dis 155(6):1260–1265

Blocker A, Gounon P, Larquet E, Niebuhr K, Cabiaux V, Parsot C, Sansonetti P (1999) The tripartite type III secreton of *Shigella flexneri* inserts IpaB and IpaC into host membranes. J Cell Biol 147(3):683–693

Broz P, von Moltke J, Jones JW, Vance RE, Monack DM (2010) Differential requirement for Caspase-1 autoproteolysis in pathogen-induced cell death and cytokine processing. Cell Host Microbe 8(6):471–483

Byrne BG, Dubuisson JF, Joshi AD, Persson JJ, Swanson MS (2013) Inflammasome components coordinate autophagy and pyroptosis as macrophage responses to infection. MBio 4(1): e00620-12

Cai S, Batra S, Wakamatsu N, Pacher P, Jeyaseelan S (2012) NLRC4 inflammasome-mediated production of IL-1β modulates mucosal immunity in the lung against gram-negative bacterial infection. J Immunol 188(11):5623–5635

Carayol N, Tran Van Nhieu G (2013) The inside story of *Shigella* invasion of intestinal epithelial cells. Cold Spring Harb Perspect Med 3(10):a016717

Carneiro LA, Travassos LH, Soares F, Tattoli I, Magalhaes JG, Bozza MT, Plotkowski MC, Sansonetti PJ, Molkentin JD, Philpott DJ, Girardin SE (2009) *Shigella* induces mitochondrial dysfunction and cell death in nonmyleoid cells. Cell Host Microbe 5(2):123–136

Ceballos-Olvera I, Sahoo M, Miller MA, Del Barrio L, Re F (2011) Inflammasome-dependent pyroptosis and IL-18 protect against *Burkholderia pseudomallei* lung infection while IL-1β is deleterious. PLoS Pathog 7(12):e1002452

Cersini A, Salvia AM, Bernardini ML (1998) Intracellular multiplication and virulence of *Shigella flexneri* auxotrophic mutants. Infect Immun 66(2):549–557

Chen Y, Smith MR, Thirumalai K, Zychlinsky A (1996) A bacterial invasin induces macrophage apoptosis by binding directly to ICE. EMBO J 15(15):3853–3860

Chen KW, Groß CJ, Sotomayor FV, Stacey KJ, Tschopp J, Sweet MJ, Schroder K (2014) The neutrophil NLRC4 inflammasome selectively promotes IL-1β maturation without pyroptosis during acute *Salmonella* challenge. Cell Rep 8(2):570–582

Cho JS, Guo Y, Ramos RI, Hebroni F, Plaisier SB, Xuan C, Granick JL, Matsushima H, Takashima A, Iwakura Y, Cheung AL, Cheng G, Lee DJ, Simon SI, Miller LS (2012) Neutrophil-derived IL-1β is sufficient for abscess formation in immunity against *Staphylococcus aureus* in mice. PLoS Pathog 8(11):e1003047

Clark CS, Maurelli AT (2007) *Shigella flexneri* inhibits staurosporine-induced apoptosis in epithelial cells. Infect Immun 75(5):2531–2539

Clerc P, Ryter A, Mounier J, Sansonetti PJ (1986) Plasmid-mediated intracellular multiplication of *Shigella flexneri*. Ann Inst Pasteur Microbiol 137A(3):315–320

Cohen D, Slepon R, Green MS (1991) Sociodemographic factors associated with serum anti-*Shigella* lipopolysaccharide antibodies and shigellosis. Int J Epidemiol 20(2):546–550

Cookson BT, Brennan MA (2001) Pro-inflammatory programmed cell death. Trends Microbiol 9 (3):113–114

Coster TS, Hoge CW, VanDeVerg LL, Hartman AB, Oaks EV, Venkatesan MM, Cohen D, Robin G, Fontaine-Thompson A, Sansonetti PJ, Hale TL (1999) Vaccination against shigellosis with attenuated *Shigella flexneri 2a* strain SC602. Infect Immun 67(7):3437–3443

De Geyter C, Wattiez R, Sansonetti P, Falmagne P, Ruysschaert JM, Parsot C, Cabiaux V (2000) Characterization of the interaction of IpaB and IpaD, proteins required for entry of *Shigella flexneri* into epithelial cells, with a lipid membrane. Eur J Biochem 267(18):5769–5776

Doitsh G, Galloway NL, Geng X, Yang Z, Monroe KM, Zepeda O, Hunt PW, Hatano H, Sowinski S, Muñoz-Arias I, Greene WC (2014) Cell death by pyroptosis drives CD4 T-cell depletion in HIV-1 infection. Nature 505(7484):509–514

DuPont HL, Levine MM, Hornick RB, Formal SB (1989) Inoculum size in shigellosis and implications for expected mode of transmission. J Infect Dis 159(6):1126–1128

Dupont N, Lacas-Gervais S, Bertout J, Paz I, Freche B, Van Nhieu GT, van der Goot FG, Sansonetti PJ, Lafont F (2009) *Shigella* phagocytic vacuolar membrane remnants participate in the cellular response to pathogen invasion and are regulated by autophagy. Cell Host Microbe 6(2):137–149

Edgeworth JD, Spencer J, Phalipon A, Griffin GE, Sansonetti PJ (2002) Cytotoxicity and interleukin-1β processing following *Shigella flexneri* infection of human monocyte-derived dendritic cells. Eur J Immunol 32(5):1464–1471

Egile C, Loisel TP, Laurent V, Li R, Pantaloni D, Sansonetti PJ, Carlier MF (1999) Activation of the CDC42 effector N-WASP by the *Shigella flexneri* IcsA protein promotes actin nucleation by Arp2/3 complex and bacterial actin-based motility. J Cell Biol 146(6):1319–1332

Faherty CS, Merrell DS, Semino-Mora C, Dubois A, Ramaswamy AV, Maurelli AT (2010) Microarray analysis of *Shigella flexneri*-infected epithelial cells identifies host factors important for apoptosis inhibition. BMC Genom 11:272

Faure E, Mear JB, Faure K, Normand S, Couturier-Maillard A, Grandjean T, Balloy V, Ryffel B, Dessein R, Chignard M, Uyttenhove C, Guery B, Gosset P, Chamaillard M, Kipnis E (2014) *Pseudomonas aeruginosa* type-3 secretion system dampens host defense by exploiting the NLRC4-coupled inflammasome. Am J Respir Crit Care Med 189(7):799–811

Fernandez-Prada CM, Hoover DL, Tall BD, Venkatesan MM (1997) Human monocyte-derived macrophages infected with virulent *Shigella flexneri in vitro* undergo a rapid cytolytic event similar to oncosis but not apoptosis. Infect Immun 65(4):1486–1496

Ferreccio C, Prado V, Ojeda A, Cayyazo M, Abrego P, Guers L, Levine MM (1991) Epidemiologic patterns of acute diarrhea and endemic *Shigella* infections in children in a poor periurban setting in Santiago. Chile. Am J Epidemiol 134(6):614–627

Franchi L, Amer A, Body-Malapel M, Kanneganti TD, Ozoren N, Jagirdar R, Inohara N, Vandenabeele P, Bertin J, Coyle A, Grant EP, Nunez G (2006) Cytosolic flagellin requires Ipaf for activation of caspase-1 and interleukin 1β in *Salmonella*-infected macrophages. Nat Immunol 7(6):576–582

Franchi L, Kamada N, Nakamura Y, Burberry A, Kuffa P, Suzuki S, Shaw MH, Kim YG, Núñez G (2012) NLRC4-driven interleukin-1β production discriminates between pathogenic and commensal bacteria and promotes host intestinal defense. Nat Immunol 13(5):449–456

Fukumatsu M, Ogawa M, Arakawa S, Suzuki M, Nakayama K, Shimizu S, Kim M, Mimuro H, Sasakawa C (2012) *Shigella* targets epithelial tricellular junctions and uses a noncanonical clathrin-dependent endocytic pathway to spread between cells. Cell Host Microbe 11(4): 325–336

Girardin SE, Tournebize R, Mavris M, Page AL, Li X, Stark GR, Bertin J, DiStefano PS, Yaniv M, Sansonetti PJ, Philpott DJ (2001) CARD4/Nod1 mediates NF-κB and JNK activation by invasive *Shigella flexneri*. EMBO Rep 2(8):736–742

Gorden J, Small PL (1993) Acid resistance in enteric bacteria. Infect Immun 61(1):364–367

Guichon A, Zychlinsky A (1997) Clinical isolates of *Shigella* species induce apoptosis in macrophages. J Infect Dis 175(2):470–473

Hagar JA, Powell DA, Aachoui Y, Ernst RK, Miao EA (2013) Cytoplasmic LPS activates caspase-11: implications in TLR4-independent endotoxic shock. Science 341(6151):1250–1253

Hale TL, Morris RE, Bonventre PF (1979) *Shigella* infection of henle intestinal epithelial cells: role of the host cell. Infect Immun 24(3):887–894

Hayward RD, Cain RJ, McGhie EJ, Phillips N, Garner MJ, Koronakis V (2005) Cholesterol binding by the bacterial type III translocon is essential for virulence effector delivery into mammalian cells. Mol Microbiol 56(3):590–603

High N, Mounier J, Prévost MC, Sansonetti PJ (1992) IpaB of *Shigella flexneri* causes entry into epithelial cells and escape from the phagocytic vacuole. EMBO J 11(5):1991–1999

Hilbi H, Chen Y, Thirumalai K, Zychlinsky A (1997) The interleukin 1β-converting enzyme, caspase 1, is activated during *Shigella flexneri*-induced apoptosis in human monocyte-derived macrophages. Infect Immun 65(12):5165–5170

Hilbi H, Moss JE, Hersh D, Chen Y, Arondel J, Banerjee S, Flavell RA, Yuan J, Sansonetti PJ, Zychlinsky A (1998) *Shigella*-induced apoptosis is dependent on caspase-1 which binds to IpaB. J Biol Chem 273(49):32895–32900

Ishigame H, Kakuta S, Nagai T, Kadoki M, Nambu A, Komiyama Y, Fujikado N, Tanahashi Y, Akitsu A, Kotaki H, Sudo K, Nakae S, Sasakawa C, Iwakura Y (2009) Differential roles of interleukin-17A and −17F in host defense against mucoepithelial bacterial infection and allergic responses. Immunity 30(1):108–119

Islam D, Veress B, Bardhan PK, Lindberg AA, Christensson B (1997a) In situ characterization of inflammatory responses in the rectal mucosae of patients with shigellosis. Infect Immun 65 (2):739–749

Islam D, Veress B, Bardhan PK, Lindberg AA, Christensson B (1997b) Quantitative assessment of IgG and IgA subclass producing cells in rectal mucosa during shigellosis. J Clin Pathol 50 (6):513–520

Jehl SP, Doling AM, Giddings KS, Phalipon A, Sansonetti PJ, Goldberg MB, Starnbach MN (2011) Antigen-specific CD8(+) T cells fail to respond to *Shigella flexneri*. Infect Immun 79 (5):2021–2030

Jin W, Dong C (2013) IL-17 cytokines in immunity and inflammation. Emerg Microbes Infect 2 (9):e60

Kayagaki N, Wong MT, Stowe IB, Ramani SR, Gonzalez LC, Akashi-Takamura S, Miyake K, Zhang J, Lee WP, Muszynski A, Forsberg LS, Carlson RW, Dixit VM (2013) Noncanonical inflammasome activation by intracellular LPS independent of TLR4. Science 341(6151): 1246–1249

Khalil K, Khan SR, Mazhar K, Kaijser B, Lindblom GB (1998) Occurrence and susceptibility to antibiotics of *Shigella* species in stools of hospitalized children with bloody diarrhea in Pakistan. Am J Trop Med Hyg 58(6):800–803

Kim DW, Chu H, Joo DH, Jang MS, Choi JH, Park SM, Choi YJ, Han SH, Yun CH (2008) OspF directly attenuates the activity of extracellular signal-regulated kinase during invasion by *Shigella flexneri* in human dendritic cells. Mol Immunol 45(11):3295–3301

Knodler LA, Vallance BA, Celli J, Winfree S, Hansen B, Montero M, Steele-Mortimer O (2010) Dissemination of invasive *Salmonella* via bacterial-induced extrusion of mucosal epithelia. Proc Natl Acad Sci USA 107(41):17733–17738

Knodler LA, Crowley SM, Sham HP, Yang H, Wrande M, Ma C, Ernst RK, Steele-Mortimer O, Celli J, Vallance BA (2014) Noncanonical inflammasome activation of caspase-4/caspase-11 mediates epithelial defenses against enteric bacterial pathogens. Cell Host Microbe 16(2): 249–256

Kobayashi T, Ogawa M, Sanada T, Mimuro H, Kim M, Ashida H, Akakura R, Yoshida M, Kawalec M, Reichhart JM, Mizushima T, Sasakawa C (2013) The *Shigella* OspC3 effector inhibits caspase-4, antagonizes inflammatory cell death, and promotes epithelial infection. Cell Host Microbe 13(5):570–583

Kofoed EM, Vance RE (2011) Innate immune recognition of bacterial ligands by NAIPs determines inflammasome specificity. Nature 477(7366):592–595

Kortmann J, Brubaker SW, Monack DM (2015) Cutting edge: inflammasome activation in primary human macrophages is dependent on flagellin. J Immunol Author Choice 195(3):815–819

Kotloff KL, Winickoff JP, Ivanoff B, Clemens JD, Swerdlow DL, Sansonetti PJ, Adak GK, Levine MM (1999) Global burden of *Shigella* infections: implications for vaccine development and implementation of control strategies. Bull World Health Organ 77(8):651–666

Kotloff KL, Nataro JP, Blackwelder WC, Nasrin D, Farag TH, Panchalingam S, Wu Y, Sow SO, Sur D, Breiman RF, Faruque AS, Zaidi AK, Saha D, Alonso PL, Tamboura B, Sanogo D, Onwuchekwa U, Manna B, Ramamurthy T, Kanungo S, Ochieng JB, Omore R, Oundo JO, Hossain A, Das SK, Ahmed S, Qureshi S, Quadri F, Adegbola RA, Antonio M, Hossain MJ, Akinsola A, Mandomando I, Nhampossa T, Acácio S, Biswas K, O'Reilly CE, Mintz ED, Berkeley LY, Muhsen K, Sommerfelt H, Robins-Browne RM, Levine MM (2013) Burden and aetiology of diarrhoeal disease in infants and young children in developing countries (the global enteric multicenter study, GEMS): a prospective, case-control study. Lancet 382 (9888):209–222

Lafont F, Tran Van Nhieu G, Hanada K, Sansonetti P, van der Goot FG (2002) Initial steps of *Shigella* infection depend on the cholesterol/sphingolipid raft-mediated CD44-IpaB interaction. EMBO J 21(17):4449–4457

Lasiglié D, Traggiai E, Federici S, Alessio M, Buoncompagni A, Accogli A, Chiesa S, Penco F, Martini A, Gattorno M (2011) Role of IL-1β in the development of human T(H)17 cells: lesson from NLPR3 mutated patients. PLoS ONE 6(5):e20014

Lembo-Fazio L, Nigro G, Noël G, Rossi G, Chiara F, Tsilingiri K, Rescigno M, Rasola A, Bernardini ML (2011) Gadd45α activity is the principal effector of *Shigella* mitochondria-dependent epithelial cell death *in vitro* and *ex vivo*. Cell Death Dis 2:e122

Lin XY, Choi MSK, Porter AG (2000) Expression analysis of the human caspase-1 subfamily reveals specific regulation of the CASP5 gene by lipopolysaccharide and interferon-γ. J Biol Chem 275(51):39920–39926

Liu Z, Zaki MH, Vogel P, Gurung P, Finlay BB, Deng W, Lamkanfi M, Kanneganti TD (2012) Role of inflammasomes in host defense against *Citrobacter rodentium* infection. J Biol Chem 287(20):16955–16964

Mandic-Mulec I, Weiss J, Zychlinsky A (1997) *Shigella flexneri* is trapped in polymorphonuclear leukocyte vacuoles and efficiently killed. Infect Immun 65(1):110–115

Mantis N, Prévost MC, Sansonetti P (1996) Analysis of epithelial cell stress response during infection by *Shigella flexneri*. Infect Immun 64(7):2474–2482

Mariathasan S, Newton K, Monack DM, Vucic D, French DM, Lee WP, Roose-Girma M, Erickson S, Dixit VM (2004) Differential activation of the inflammasome by caspase-1 adaptors ASC and Ipaf. Nature 430(6996):213–218

Martinon F, Burns K, Tschopp J (2002) The inflammasome: a molecular platform triggering activation of inflammatory caspases and processing of proIL-β. Mol Cell 10(2):417–426

Mathan MM, Mathan VI (1986) Ultrastructural pathology of the rectal mucosa in *Shigella* dysentery. Am J Pathol 123(1):25–38

Mathan MM, Mathan VI (1991) Morphology of rectal mucosa of patients with shigellosis. Rev Infect Dis 13(Suppl 4):S314–S318

Ménard R, Sansonetti PJ, Parsot C (1993) Nonpolar mutagenesis of the ipa genes defines IpaB, IpaC, and IpaD as effectors of *Shigella flexneri* entry into epithelial cells. J Bacteriol 175 (18):5899–5906

Miao EA, Leaf IA, Treuting PM, Mao DP, Dors M, Sarkar A, Warren SE, Wewers MD, Aderem A (2010a) Caspase-1-induced pyroptosis is an innate immune effector mechanism against intracellular bacteria. Nat Immunol 11(12):1136–1142

Miao EA, Mao DP, Yudkovsky N, Bonneau R, Lorang CG, Warren SE, Leaf IA, Aderem A (2010b) Innate immune detection of the type III secretion apparatus through the NLRC4 inflammasome. Proc Natl Acad Sci USA 107(7):3076–3080

Miao EA, Rajan JV, Aderem A (2011) Caspase-1-induced pyroptotic cell death. Immunol Rev 243 (1):206–214

Miller LS, Pietras EM, Uricchio LH, Hirano K, Rao S, Lin H, O'Connell RM, Iwakura Y, Cheung AL, Cheng G, Modlin RL (2007) Inflammasome-mediated production of IL-1β is required for neutrophil recruitment against *Staphylococcus aureus in vivo*. J Immunol 179 (10):6933–6942

Mostowy S, Boucontet L, Mazon Moya MJ, Sirianni A, Boudinot P, Hollinshead M, Cossart P, Herbomel P, Levraud JP, Colucci-Guyon E (2013) The zebrafish as a new model for the *in vivo* study of *Shigella flexneri* interaction with phagocytes and bacterial autophagy. PLoS Pathog 9 (9):e1003588

Niebuhr K, Giuriato S, Pedron T, Philpott DJ, Gaits F, Sable J, Sheetz MP, Parsot C, Sansonetti PJ, Payrastre B (2002) Conversion of PtdIns(4, 5)P(2) into PtdIns(5)P by *the S. flexneri* effector IpgD reorganizes host cell morphology. EMBO J 21(19):5069–5078

Nigro G, Fazio LL, Martino MC, Rossi G, Tattoli I, Liparoti V, De Castro C, Molinaro A, Philpott DJ, Bernardini ML (2008) Muramylpeptide shedding modulates cell sensing of *Shigella flexneri*. Cell Microbiol 10(3):682–695

Nonaka T, Kuwae A, Sasakawa C, Imajoh-Ohmi S (1999) *Shigella flexneri* YSH6000 induces two types of cell death, apoptosis and oncosis, in the differentiated human monoblastic cell line U937. FEMS Microbiol Lett 174(1):89–95

Nordlander S, Pott J, Maloy KJ (2014) NLRC4 expression in intestinal epithelial cells mediates protection against an enteric pathogen. Mucosal Immunol 7(4):775–785

Nothelfer K, Arena ET, Pinaud L, Neunlist M, Mozeleski B, Belotserkovsky I, Parsot C, Dinadayala P, Burger-Kentischer A, Raqib R, Sansonetti PJ, Phalipon A (2014) B lymphocytes undergo TLR2-dependent apoptosis upon *Shigella* infection. J Exp Med 211(6):1215–1229

O'Connor W Jr, Harton JA, Zhu X, Linhoff MW, Ting JP (2003) Cutting edge: CIAS1/cryopyrin/PYPAF1/NALP3/CATERPILLER 1.1 is an inducible inflammatory mediator with NF-κB suppressive properties. J Immunol 171(12):6329–6333

Onishi RM, Gaffen SL (2010) Interleukin-17 and its target genes: mechanisms of interleukin-17 function in disease. Immunology 129(3):311–321

Osorio M, Bray MD, Walker RI (2007) Vaccine potential for inactivated shigellae. Vaccine 25 (9):1581–1592

Ottoson NC, Pribila JT, Chan ASH, Shimizu Y (2001) Cutting edge: T cell migration regulated by CXCR4 chemokine receptor signaling to ZAP-70 tyrosine kinase. J Immunol 167(4):1857–1861

Paciello I, Silipo A, Lembo-Fazio L, Curcurù L, Zumsteg A, Noël G, Ciancarella V, Sturiale L, Molinaro A, Bernardini ML (2013) Intracellular *Shigella* remodels its LPS to dampen the innate immune recognition and evade inflammasome activation. Proc Natl Acad Sci USA 110 (46):E4345–E4354

Parsot C (2009) *Shigella* type III secretion effectors: how, where, when, for what purposes? Curr Opin Microbiol 12(1):110–116

Pearson JS, Giogha C, Ong SY, Kennedy CL, Kelly M, Robinson KS, Lung TW, Mansell A, Riedmaier P, Oates CV, Zaid A, Mühlen S, Crepin VF, Marches O, Ang CS, Williamson NA, O'Reilly LA, Bankovacki A, Nachbur U, Infusini G, Webb AI, Silke J, Strasser A, Frankel G, Hartland EL (2013) A type III effector antagonizes death receptor signalling during bacterial gut infection. Nature 501(7466):247–251

Pédron T, Thibault C, Sansonetti PJ (2003) The invasive phenotype of *Shigella flexneri* directs a distinct gene expression pattern in the human intestinal epithelial cell line Caco-2. J Biol Chem 278(36):33878–33886

Pendaries C, Tronchère H, Arbibe L, Mounier J, Gozani O, Cantley L, Fry MJ, Gaits-Iacovoni F, Sansonetti PJ, Payrastre B (2006) PtdIns5P activates the host cell PI3-kinase/Akt pathway during *Shigella flexneri* infection. EMBO J 25(5):1024–1034

Philpott DJ, Yamaoka S, Israël A, Sansonetti PJ (2000) Invasive *Shigella flexneri* activates NF-κB through a lipopolysaccharide-dependent innate intracellular response and leads to IL-8 expression in epithelial cells. J Immunol 165(2):903–914

Presicce P, Senthamaraikannan P, Alvarez M, Rueda CM, Cappelletti M, Miller LA, Jobe AH, Chougnet CA, Kallapur SG (2015) Neutrophil recruitment and activation in decidua with intra-amniotic IL-1β in the preterm rhesus macaque. Biol Reprod 92(2):56

Puhar A, Tronchère H, Payrastre B, Nhieu GT, Sansonetti PJ (2013) A *Shigella* effector dampens inflammation by regulating epithelial release of danger signal ATP through production of the lipid mediator PtdIns5P. Immunity 39(6):1121–1131

Raqib R, Lindberg AA, Wretlind B, Bardhan PK, Andersson U, Andersson J (1995a) Persistence of local cytokine production in shigellosis in acute and convalescent stages. Infect Immun 63 (1):289–296

Raqib R, Wretlind B, Andersson J, Lindberg AA (1995b) Cytokine secretion in acute shigellosis is correlated to disease activity and directed more to stool than to plasma. J Infect Dis 171 (2):376–384

Raqib R, Ekberg C, Sharkar P, Bardhan PK, Zychlinsky A, Sansonetti PJ, Andersson J (2002a) Apoptosis in acute shigellosis is associated with increased production of Fas/Fas ligand, perforin, caspase-1, and caspase-3 but reduced production of Bcl-2 and interleukin-2. Infect Immun 70(6):3199–3207

Raqib R, Qadri F, SarkEr P, Mia SM, Sansonnetti PJ, Albert MJ, Andersson J (2002b) Delayed and reduced adaptive humoral immune responses in children with shigellosis compared with in adults. Scand J Immunol 55(4):414–423

Robbins JB, Chu C, Schneerson R (1992) Hypothesis for vaccine development: protective immunity to enteric diseases caused by nontyphoidal salmonellae and shigellae may be conferred by serum IgG antibodies to the O-specific polysaccharide of their lipopolysaccharides. Clin Infect Dis 15(2):346–361

Rohde JR, Breitkreutz A, Chenal A, Sansonetti PJ, Parsot C (2007) Type III secretion effectors of the IpaH family are E3 ubiquitin ligases. Cell Host Microbe 1(1):77–83

Salgado-Pabon W, Celli S, Arena ET, Nothelfer K, Roux P, Sellge G, Frigimelica E, Bousso P, Sansonetti PJ, Phalipon A (2013) *Shigella* impairs T lymphocyte dynamics in vivo. Proc Natl Acad Sci USA 110(12):4458–4463

Sansonetti PJ (2001a) III. Shigellosis: from symptoms to molecular pathogenesis. Am J Physiol Gastrointest Liver Physiol 280(3):G319–G323

Sansonetti PJ (2001b) Rupture, invasion and inflammatory destruction of the intestinal barrier by *Shigella*, making sense of prokaryote-eukaryote cross-talks. FEMS Microbiol Rev 25(1):3–14

Sansonetti PJ, Kopecko DJ, Formal SB (1981) *Shigella sonnei* plasmids: evidence that a large plasmid is necessary for virulence. Infect Immun 34(1):75–83

Sansonetti PJ, Kopecko DJ, Formal SB (1982) Involvement of a plasmid in the invasive ability of *Shigella* flexneri. Infect Immun 35(3):852–860

Sansonetti PJ, Hale TL, Dammin GJ, Kapfer C, Collins HH Jr, Formal SB (1983) Alterations in the pathogenicity *of Escherichia coli* K-12 after transfer of plasmid and chromosomal genes from *Shigella flexneri*. Infect Immun 39(3):1392–1402

Sansonetti PJ, Ryter A, Clerc P, Maurelli AT, Mounier J (1986) Multiplication of *Shigella flexneri* within HeLa cells: lysis of the phagocytic vacuole and plasmid-mediated contact hemolysis. Infect Immun 51(2):461–469

Sansonetti PJ, Arondel J, Fontaine A, d'Hauteville H, Bernardini ML (1991) OmpB (osmo-regulation) and icsA (cell-to-cell spread) mutants of *Shigella flexneri*: vaccine candidates and probes to study the pathogenesis of shigellosis. Vaccine 9(6):416–422

Sansonetti PJ, Arondel J, Cavaillon JM, Huerre M (1995) Role of interleukin-1 in the pathogenesis of experimental shigellosis. J Clin Invest 96(2):884–892

Sansonetti PJ, Arondel J, Huerre M, Harada A, Matsushima K (1999) Interleukin-8 controls bacterial transepithelial translocation at the cost of epithelial destruction in experimental shigellosis. Infect Immun 67(3):1471–1480

Sansonetti PJ, Phalipon A, Arondel J, Thirumalai K, Banerjee S, Akira S, Takeda K, Zychlinsky A (2000) Caspase-1 activation of IL-1β and IL-18 are essential for *Shigella flexneri*-induced inflammation. Immunity 12(5):581–590

Schauvliege R, Vanrobaeys J, Schotte P, Beyaert R (2002) Caspase-11 gene expression in response to lipopolysaccharide and interferon-γ requires nuclear factor-κB and signal transducer and activator of transcription (STAT) 1. J Biol Chem 277(44):41624–41630

Schroeder GN, Hilbi H (2007) Cholesterol is required to trigger caspase-1 activation and macrophage apoptosis after phagosomal escape of *Shigella*. Cell Microbiol 9(1):265–278

Sellge G, Magalhaes JG, Konradt C, Fritz JH, Salgado-Pabon W, Eberl G, Bandeira A, Di Santo JP, Sansonetti PJ, Phalipon A (2010) Th17 cells are the dominant T cell subtype primed by *Shigella flexneri* mediating protective immunity. J Immunol 184(4):2076–2085

Sellin ME, Müller AA, Felmy B, Dolowschiak T, Diard M, Tardivel A, Maslowski KM, Hardt WD (2014) Epithelium-intrinsic NAIP/NLRC4 inflammasome drives infected enterocyte expulsion to restrict *Salmonella* replication in the intestinal mucosa. Cell Host Microbe 16 (2):237–248

Senerovic L, Tsunoda SP, Goosmann C, Brinkmann V, Zychlinsky A, Meissner F, Kolbe M (2012) Spontaneous formation of IpaB ion channels in host cell membranes reveals how *Shigella* induces pyroptosis in macrophages. Cell Death Dis 3:e384

Shaftel SS, Carlson T, Olschowka JA, Kyrkanides S, Matousek SB, O'Banion MK (2007) Chronic interleukin-1β expression in mouse brain leads to leukocyte infiltration and neutrophil-independent blood-brain barrier permeability without overt neurodegeneration. J Neurosci 27(35):9301–9309

Shi J, Zhao Y, Wang Y, Gao W, Ding J, Li P, Hu L, Shao F (2014) Inflammatory caspases are innate immune receptors for intracellular LPS. Nature 514(7521):187–192

Shim DH, Ryu S, Kweon MN (2010) Defensins play a crucial role in protecting mice against oral *Shigella flexneri* infection. Biochem Biophys Res Commun 401(4):554–560

Singer M, Sansonetti PJ (2004) IL-8 is a key chemokine regulating neutrophil recruitment in a new mouse model of *Shigella*-induced colitis. J Immunol 173(6):4197–4206

Sperandio B, Regnault B, Guo J, Zhang Z, Stanley SL Jr, Sansonetti PJ, Pedron T (2008) Virulent *Shigella* flexneri subverts the host innate immune response through manipulation of antimicrobial peptide gene expression. J Exp Med 205(5):1121–1132

Suzuki T, Nakanishi K, Tsutsui H, Iwai H, Akira S, Inohara N, Chamaillard M, Nuñez G, Sasakawa C (2005) A novel caspase-1/toll-like receptor 4-independent pathway of cell death induced by cytosolic Shigella in infected macrophages. J Biol Chem 280(14):14042–14050

Suzuki T, Franchi L, Toma C, Ashida H, Ogawa M, Yoshikawa Y, Mimuro H, Inohara N, Sasakawa C, Nuñez G (2007) Differential regulation of caspase-1 activation, pyroptosis, and autophagy via Ipaf and ASC in *Shigella*-infected macrophages. PLoS Pathog 3(8):e111

Suzuki S, Franchi L, He Y, Muñoz-Planillo R, Mimuro H, Suzuki T, Sasakawa C, Núñez G (2014a) *Shigella* type III secretion protein MxiI is recognized by Naip2 to induce Nlrc4 inflammasome activation independently of Pkcδ. PLoS Pathog 10(2):e1003926

Suzuki S, Mimuro H, Kim M, Ogawa M, Ashida H, Toyotome T, Franchi L, Suzuki M, Sanada T, Suzuki T, Tsutsui H, Núñez G, Sasakawa C (2014b) *Shigella* IpaH7.8 E3 ubiquitin ligase targets glomulin and activates inflammasomes to demolish macrophages. Proc Natl Acad Sci USA 111(40):E4254–E4263

Tattoli I, Carneiro LA, Jéhanno M, Magalhaes JG, Shu Y, Philpott DJ, Arnoult D, Girardin SE (2008) NLRX1 is a mitochondrial NOD-like receptor that amplifies NF-κB and JNK pathways by inducing reactive oxygen species production. EMBO Rep 9(3):293–300

Ud-Din AI, Wahid SU, Latif HA, Shahnaij M, Akter M, Azmi IJ, Hasan TN, Ahmed D, Hossain MA, Faruque AS, Faruque SM, Talukder KA (2013) Changing trends in the prevalence of Shigella species: emergence of multi-drug resistant *Shigella sonnei* biotype g in Bangladesh. PLoS ONE 8(12):e82601

Van't Wout JW, Van der Meer JW, Barza M, Dinarello CA (1988) Protection of neutropenic mice from lethal *Candida albicans* infection by recombinant interleukin 1. Eur J Immunol 18 (7):1143–1146

Voino-Yasenetsky MV, Voino-Yasenetskaya MK (1962) Experimental pneumonia caused by bacteria of the *Shigella* group. Acta Morphol Acad Sci Hung 11:439–454

Wassef JS, Keren DF, Mailloux JL (1989) Role of M cells in initial antigen uptake and in ulcer formation in the rabbit intestinal loop model of shigellosis. Infect Immun 57(3):858–863

Weinrauch Y, Drujan D, Shapiro SD, Weiss J, Zychlinsky A (2002) Neutrophil elastase targets virulence factors of enterobacteria. Nature 417(6884):91–94

Yang J, Zhao Y, Shi J, Shao F (2013) Human NAIP and mouse NAIP1 recognize bacterial type III secretion needle protein for inflammasome activation. Proc Natl Acad Sci USA 110 (35):14408–14413

Yang G, Wang L, Wang Y, Li P, Zhu J, Qiu S, Hao R, Wu Z, Li W, Song H (2015) hfq regulates acid tolerance and virulence by responding to acid stress in *Shigella flexneri*. Res Microbiol 166(6):476–485

Zafar A, Hasan R, Nizami SQ, von Seidlein L, Soofi S, Ahsan T, Chandio S, Habib A, Bhutto N, Siddiqui FJ, Rizvi A, Clemens JD, Bhutta ZA (2009) Frequency of isolation of various subtypes and antimicrobial resistance of *Shigella* from urban slums of Karachi. Pakistan. Int J Infect Dis 13(6):668–672

Zhang Z, Jin L, Champion G, Seydel KB, Stanley SL (2001) *Shigella* infection in a SCID mouse-human intestinal xenograft model: role for neutrophils in containing bacterial dissemination in human intestine. Infect Immun 69(5):3240–3247

Zhao Y, Yang J, Shi J, Gong YN, Lu Q, Xu H, Liu L, Shao F (2011) The NLRC4 inflammasome receptors for bacterial flagellin and type III secretion apparatus. Nature 477(7366):596–600

Zychlinsky A, Prevost MC, Sansonetti PJ (1992) *Shigella flexneri* induces apoptosis in infected macrophages. Nature 358(6382):167–169

Zychlinsky A, Fitting C, Cavaillon JM, Sansonetti PJ (1994a) Interleukin 1 is released by murine macrophages during apoptosis induced by *Shigella flexneri*. J Clin Invest 94(3):1328–1332

Zychlinsky A, Kenny B, Ménard R, Prévost MC, Holland IB, Sansonetti PJ (1994b) IpaB mediates macrophage apoptosis induced by *Shigella flexneri*. Mol Microbiol 11(4):619–627

Inflammasome Activation by *Helicobacter pylori* and Its Implications for Persistence and Immunity

Suneesh Kumar Pachathundikandi, Anne Müller and Steffen Backert

Abstract Infection with the Gram-negative pathogen *Helicobacter pylori* is the most prevalent chronic bacterial infection affecting about 50 % of the human world population and is the main risk factor for gastric cancer development. The pro-inflammatory cytokine IL-1β plays a crucial role in the development of gastric tumors, and polymorphisms in the IL-1 gene cluster resulting in increased IL-1β production have been associated with increased risk for gastric cancer. Recently, *Helicobacter pylori* was postulated to activate the inflammasome in human and mouse immune cells, and the molecular mechanisms and the bacterial virulence factors activating the inflammasome were elucidated in cell culture as well as animal models. It appears that *H. pylori*-induced IL-1β secretion is mediated by activation of toll-like receptor 2 (TLR-2), Nod-like receptor family member NLRP3 and caspase-1. The *cag* pathogenicity island-encoded type IV secretion system, lipopolysaccharide, vacuolating cytotoxin, and urease B subunit appear to play a role in inflammasome activation. In addition, recent results indicate that the TLR-2 → NLRP3 → caspase-1 → IL-18 axis is critical to *H. pylori*-specific immune regulation conferring protection against allergen-induced asthma and inflammatory bowel disease in murine models. The present chapter will review the proposed mechanisms of NLRP3 inflammasome activation during *H. pylori* infection and discuss the recent progress in this important research field.

S.K. Pachathundikandi · S. Backert (✉)
Department of Biology, Division of Microbiology, Friedrich Alexander University Erlangen-Nuremberg, Staudtstrasse 5, 91085 Erlangen, Germany
e-mail: steffen.backert@fau.de

A. Müller
Institute of Molecular Cancer Research, University of Zürich, Winterthurerstr. 190, 8057 Zurich, Switzerland

S. Backert (ed.), *Inflammasome Signaling and Bacterial Infections*,
Current Topics in Microbiology and Immunology 397,
DOI 10.1007/978-3-319-41171-2_6

Contents

1 Introduction

1.1 H. pylori *Is a Prime Example of Chronic Infections and Gastric Disease Development*

Helicobacter pylori is a Gram-negative pathogen colonizing the stomach mucosa of more than half of the human world population. The bacterium co-exists with its human host for at least the last 100,000 years and has even been useful for analyzing human migrations (Moodley and Linz 2009). This long history of co-existence has allowed co-evolution and adaptation of *H. pylori* to its host. *H. pylori* uses multiple colonization- and pathogenicity-associated factors to persist in a highly specialized niche, the stomach antrum (Salama et al. 2013). As it is exposed to the low pH of the gastric juice at the luminal side of the gastric mucosa, *H. pylori* expresses the urease enzyme to hydrolyze urea, which releases ammonia and locally neutralizes the pH in the proximity of the bacteria. The pathogen is also known to produce toxins and effector proteins exerting various damaging effects on host cells and gastric tissue. This includes the vacuolating toxin A (VacA) and the cytotoxin-associated gene pathogenicity island (*cag*PAI) encoding a type IV secretion system (T4SS) for delivery of the CagA protein into target cells. In addition, *H. pylori* uses a series of outer membrane proteins including BabA/B, SabA, HopQ, OipA, and others for adhering to the host cell surface. Dozens of other pathogenicity factors help the bacteria to colonize the host for almost the entire life span—if not eradicated by antibiotics (Amieva and El-Omar 2008; Atherton and Blaser 2009; Pachathundikandi et al. 2013). All individuals harboring *H. pylori* develop asymptomatic gastritis by activation of multiple chemokine mediators, including transcription factor NF-κB (nuclear factor of kappa light polypeptide gene enhancer in B-Cells), the master regulator of inflammation

(Backert and Naumann 2010). This inflammatory scenario can further progress into various pathologies such as peptic ulcers and gastric tumorigenesis (Polk and Peek 2010). Inflammation is normally beneficial for clearing infections and restoring homeostasis. However, chronic inflammation may increase the risk for tumorigenesis through the production of various cytokines, chemokines, growth factors, which give an overall advantage for host cell survival signaling, genome instability, angiogenesis, metabolic derangements, and subversion of anti-tumor immunity (Hanahan and Weinberg 2011).

1.2 Host Genetic Polymorphisms Are Crucial for Gastric Pathology

The clinical outcome of *H. pylori* infections depends on three main parameters—bacterial genotype, host genetic susceptibility, and environmental factors. Host gene polymorphisms have been studied in various human populations in order to provide a link with specific *H. pylori*-associated pathologies. The pro-inflammatory cytokine IL-1β is one of the major cytokines reported to be involved in the gastric tumorigenesis. In addition, polymorphisms in *TNF* (tumor necrosis factor-α), *IL1RN* (interleukin-1 receptor antagonist), *IL8* (interleukin-8), *IL10,* and *TLR4* (toll-like receptor-4) genes were reported to contribute to *H. pylori* infection-induced gastric tumorigenesis (Savage et al. 2004; Shanks and El-Omar 2009; Kim et al. 2012). Genetic mutations in gene of morphogens like TGF-β superfamily members such as bone morphogenetic protein-6 (BMP-6) and growth differentiation factor-15 (GDF-15) were also implicated in the gastric cancer (Katoh 2007). The transcription factor RUNX-3 (runt-related transcription factor-3) was reported to control genetic and epigenetic changes, representing another crucial element in gastric epithelial cell transformation and cancer development (Kitajima et al. 2008; Lim et al. 2011). The increased production of above cytokines, chemokines, morphogens, and signaling complexes in gastric epithelial stem cells might be giving a favorable combination of molecular switches leads to gastric cancer. Associations with polymorphisms in various cytokine and related genes such as *IL6, TLR1, 2, 9, 10* (Toll-like receptors), *NLRP3, NLRP12, NLRX1* (NLR family receptors) *CARD8* (caspase recruitment domain family, member-8), *CASP1* (caspase-1), and *HLAs* (human leukocyte antigen) were also reported in *H. pylori*-related pathologies (Azuma et al. 1998; Hong-mei et al. 2011; Mayerle et al. 2013; Castaño-Rodríguez et al. 2014; Ravishankar Ram et al. 2015). About 10–20 % of infected people have a lifetime risk of developing ulcer disease and a 1–2 % risk of developing distal gastric cancer (Kusters et al. 2006). *H. pylori*-associated gastric carcinoma represents a prime example of inflammation-induced malignancy and highlights the importance of host genetics in disease development. In this chapter, we review *H. pylori*-mediated inflammasome activation and its role in modulating and directing host specific immunity and gastric disease. An overall model for the involved signaling pathways is shown in Fig. 1.

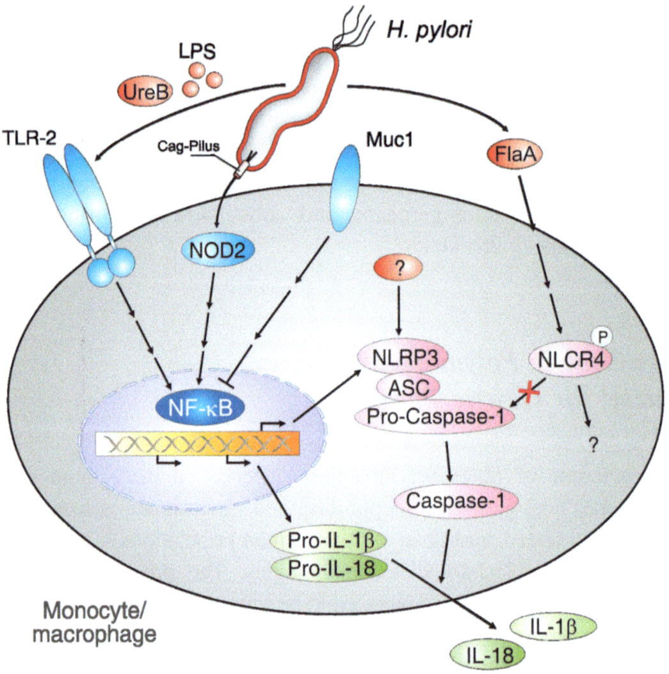

Fig. 1 Signaling pathways of inflammasome activation by *Helicobacter pylori. H. pylori* stimulate inflammasome formation in human and mouse DCs, macrophages, or monocytes. Host cell receptors including TLR-2, NOD2, and Muc1, which are targeted by various indicated or yet unknown bacterial factors, appear to play a role in upstream signaling events leading to transcription factor NF-κB activation and production of NLRP3 and 1L-1β and 1L-18 pro-forms. Engagement of NLRP3 and ASC induces the cleavage of autoproteolytic pro-caspase-1 and the subsequent processing and release of mature interleukin-1β (1L-1β) and 1L-18 cytokines. In addition, *H. pylori* flagellin-A was shown to activate the phosphorylation of NLCR4 by a yet unknown mechanism, but this signaling does not lead to caspase-1 activation and must have other yet unknown consequences for the host cells. For more details please see text

2 Importance of IL-1β in Gastric Cancer Development

In 1994, the World Health Organization (WHO) classified *H. pylori* as a type-I carcinogen due to its epidemiologic involvement in gastric cancer and mucosa-associated lymphoid tissue (MALT) lymphoma. Worldwide, approximately 700,000 people succumb to gastric cancer every year and the 5-year survival rate in the USA is <15 % (Polk and Peek 2010). Early studies have shown that enhanced production of IL-1β is related to specific polymorphisms in the *IL-1* gene cluster, which are prominent and directly related to gastric cancer development (El-Omar et al. 2000). IL-1β is a well-known pro-inflammatory cytokine and pyrogen having important roles in different immunologic responses such as leukocyte recruitment, production of other mediators of inflammation, and related immunopathologies. Moreover, IL-1β is implicated in epigenetic changes and suppression of gastric acid secretion, which ultimately leads to gastric atrophy and metaplasia, a prerequisite for gastric cancer development (El-Omar et al. 2000; Apte et al. 2006). To support this hypothesis, transgenic overexpression of human IL-1β in mice was shown to develop stepwise pathologic progression of gastric dysplasia and cancer (Tu et al. 2008). However, not every *H. pylori*-infected individual develops associated pathologies such as peptic ulcer or gastric cancer. This attracted further studies on host genetic polymorphisms to explain the increased production of IL-1β. As a potential hazardous cytokine at high concentration, the actions of IL-1β are endogenously controlled by an anti-inflammatory molecule called interleukin-1 receptor antagonist (IL-1RA), which competes with IL-1 cytokines for IL-1R receptor binding. The genetic polymorphisms such as −511 $C > T$, −31 $T > C$, and +3954 $C > T$ genotypes of *IL1B* and *IL1RN*, respectively, are correlated with gastric cancer appearance upon *H. pylori* infection (El-Omar et al. 2000, 2003). In addition, various other studies have linked the −511 $C > T$ and −31 $T > C$ polymorphisms in IL-1β to duodenal ulcers (Lee et al. 2003; Chakravorty et al. 2006). Moreover, a very recent comprehensive meta-analysis of published data correlated *IL1B* −31 $T > C$ polymorphism with increased gastric cancer risk during *H. pylori* infection (Ying et al. 2016). IL-1β induced the expression of cAMP response element-binding protein (CREB) and its target gene product CCAAT/enhancer-binding protein beta (C/EBPβ) transcription factor through ERK1/2 kinase signaling, which ultimately increased the proliferation of gastric carcinoma cells in vitro and in vivo (Resende et al. 2016). This demonstrates the direct influence of secreted IL-1β on gastric carcinoma proliferation. Several initial studies have reported significantly elevated levels of IL-1β in gastric biopsies from infected patients, but no significant increase in the related cytokine IL-18 (Noach et al. 1994; Basso et al. 1996; Jung et al. 1997). These studies mostly analyzed the total IL-1β levels in mucosal lysates or during ex vivo culture of infected mucosa. In addition, studies on IL-18 production have shown that *H. pylori* induced an increased IL-18 mRNA expression in the gastric mucosa of infected patients, but with no significant increase in caspase-1 activation and IL-18 secretion (Tomita et al. 2001). Another study has found that IL-18 mRNA and protein expression in *H. pylori*-infected epithelial cells and monocytes correlated and was dependent on the *cag*PAI status and OipA protein expression (Yamauchi et al. 2008). Koch et al. (2016) reported that *H. pylori*-induced IL-1β protein expression started as early as 60 min post-infection and increased steadily, which was also dependent on the T4SS apparatus contact in mouse bone marrow-derived macrophages (BMDMs). Taken together, these studies indicate that there is substantial genetic predisposition and cross talk of *H. pylori* with the host immune system, which is associated with chronicity of infection and gastric disease development.

3 Inflammasome Responses During *H. pylori* Infection

3.1 Inflammasome Activation by H. pylori

Many reports in the last decade confirmed that 'inflammasome'-mediated caspase-1 activation is required for inactive pro-IL-1β and pro-IL-18 cleavage and secretion of the corresponding active cytokines (Martinon et al. 2002; von Moltke et al. 2013). In most cases, inflammasome-mediated caspase-1 activation was also found to promote a specific type of cell death, called pyroptosis. It was identified that cleaved N-terminal domain of gasdermin-D (GSDMD) by inflammatory caspases, including caspase-1 and caspase-4/5/11, is responsible for pyroptosis induction during inflammasome activation (Shi et al. 2015). Studies in the last few years identified that *H. pylori* can also activate the inflammasome. The spatial and temporal production of pro-inflammatory cytokines such as IL-1β and IL-18 is crucial for an effective immune response and control of extra- and intracellular bacterial pathogens. NLR members such as NLRC4, NLRP1, NLRP3, and HIN20 domain containing protein AIM2 (absent in melanoma-2) have been shown to induce multimeric inflammasome complex formation to activate pro-caspase-1. In addition, NLRs like NLRP6, NLRP7, and NLRP12 have also been recently implicated in inflammasome formation (Broz and Monack 2013). Another recent study by Hitzler et al. (2012) reported the caspase-1-mediated processing of pro-IL-1β and pro-IL-18 into active forms using murine bone marrow-derived dendritic cells (BMDCs) as well as mouse models of *H. pylori* infection. This study demonstrated that many aspects of host immunity against *H. pylori* are influenced by the inflammasome, including caspase-1, IL-1β, and IL-18 activation. Mice infected with *H. pylori* strain SS1 exhibited elevated numbers of cells with active caspase-1 in mesenteric lymph nodes, where the priming of *H. pylori*-specific T cells occurs. Infected BMDCs from *caspase-1*$^{-/-}$ mice were not able to process pro-IL-1β and pro-IL-18. Interestingly, *caspase-1*$^{-/-}$ mice were able to control the infection better than wild-type animals, but with pronounced immunopathology (Hitzler et al. 2012). This was correlated with higher levels of gastric IL-17. Further studies with *Il1r*$^{-/-}$ and *Il18*$^{-/-}$ mice revealed the differential functions of these cytokines in mounting immunity against *H. pylori*. While IL-1β was required for protective immunity against *H. pylori*, IL-18 controlled the corresponding immunopathology (Hitzler et al. 2012). In another comprehensive mouse study, induction of pro-IL-1β by *H. pylori* has been attributed to TLR-2/NOD2-mediated signaling in a *cag*PAI-dependent manner (Kim et al. 2013). This study has also shown that caspase-1 activation by *H. pylori* is dependent on TLR-2 and NOD2 in isolated BMDCs. NLRP3 expression was significantly reduced in infected *Nod2*$^{-/-}$, *Tlr2*$^{-/-}$, and double deficient *Nod2*$^{-/-}$/*Tlr2*$^{-/-}$ dendritic cells (DCs). Pretreatment with bacterial lipopolysaccharide (LPS) was able to rescue this defective ability of DCs from the above knockout mice to secrete IL-1β in response to *H. pylori* infection. This report emphasizes that TLR-2-NOD2 signaling primes the expression of pro-IL-1β and NLRP3 proteins. Moreover, *H. pylori* colonization appears to be restricted by IL-1β signaling (Kim et al. 2013). Another study reported that *H. pylori*-induced NLRP3 inflammasome activation in BMDCs involved K^+ efflux, phagocytosis, and reactive oxygen species (ROS) production (Semper et al. 2014). They have also showed that inflammasome-mediated caspase-1 activation was not detected in BMDCs after treatment with heat-killed *H. pylori* and strains lacking a functional T4SS or VacA. *H. pylori* infection of *Nlrp3*$^{-/-}$ mice exhibited higher colonization and low inflammation, which can be attributed to reduced IL-17 and IFN-γ in the gastric mucosa and lower infiltration of $CD45^+$ and $CD4^+$ cells (Semper et al. 2014).

3.2 Regulation of Inflammasome Activation upon H. pylori *Infection*

The above studies highlighted that inflammasome component expression, activation of caspase-1, and expression of the cytokines IL-1β and IL-18 are important determinants of *H. pylori* infection in mice. Even though inflammasome activation and secretion of IL-1β and IL-18 plays an important role in the development of adaptive immunity, a tight control of the whole process is necessary to maintain normal homeostasis. In general, the uncontrolled activation of the inflammasome by bacterial pathogens or congenital problems has been linked to severe immunopathologies. The gene encoding the mucin-1 (*Muc1*) carries various polymorphisms, which have been implicated in the susceptibility to *H. pylori*-associated gastric diseases and gastric cancer development. In fact, Muc-1 expression plays an important role in preventing the adhesion of *H. pylori* to the gastric epithelium (Vinall et al. 2002; Saeki et al. 2011). A very recent report also showed that *Muc1* deficiency in infected mice promotes severe gastric pathologies (Ng et al. 2015). The authors further demonstrated that *Muc1*$^{-/-}$ mice exhibited high mortality rates after *H. pylori* infection, which was correlated with high numbers of mononuclear cell infiltration and severe atrophic gastritis. IL-1β was the only pro-inflammatory cytokine differentially elevated in mice after two months of infection and that correlated with atrophic gastritis symptoms. Muc-1 expression in mice protected against precancerous lesions and dysplasia. Interestingly, wild-type mice exhibited significantly more colonized bacteria than *Muc1*$^{-/-}$ mice during long-term infection (Ng et al. 2015). This phenotype in infected *Muc1*$^{-/-}$ mice was reminiscent of stomach-specific IL-1β overexpressing transgenic mice, which also developed gastritis, dysplasia, and cancer, even in the absence of infection (Tu et al. 2008). The quantification of mononuclear infiltrates at the site of infection in *Muc1*$^{-/-}$ mice revealed higher numbers of macrophages, B cells, neutrophils, and DCs but no significant difference in T cells as compared to wild-type cells. Moreover, Muc-1 expressed in immune cells regulated the severity of gastric disease. *Muc1*$^{-/-}$ macrophages, but not DCs produced significantly higher amounts of IL-1β and IL-18 when treated with *H. pylori* lysate. Infected *Muc1*$^{-/-}$/*caspase-1*$^{-/-}$ double-deficient mice did not reveal significant atrophic gastritis, which emphasizes that inflammasome activation induced the severe pathology in *Muc1*$^{-/-}$-infected mice (Tu et al. 2008). *H. pylori* lysate and *E. coli* LPS-treated *Muc1*$^{-/-}$ mice significantly increased NLRP3 expression by an increased activation of TLR signaling and activation of transcription factor NF-κB p65 phosphorylation. Infected *Muc1*$^{-/-}$/*Nlrp3*$^{-/-}$ double-deficient mice exhibited a significantly reduced production of active IL-1β in comparison to *Muc1*$^{-/-}$ mice, which implicates the role of Muc-1 in NLRP3 inflammasome regulation. In addition, Muc-1 deficiency reduced the expression of trefoil factor-2 (TFF2) through methylation of the *Tff2* gene, presumably via IL-1β, similar as observed in transgenic mice overexpressing IL-1β. TFF2 is important for maintaining the gastric epithelial integrity because *H. pylori* infection of *Tff2*$^{-/-}$ mice led to dysplasia (Fox et al. 2007; Peterson et al. 2010). In conclusion, *H. pylori*-mediated inflammasome activation and IL-1β secretion are tightly regulated in mice via Muc-1 and prevented the formation of precancerous lesions, dysplasia and gastric cancer.

3.3 Inflammasome Activation-Mediated Host-Specific Immunity in H. pylori *Infection*

A very recent study by Koch et al. (2015) supported previous findings of caspase-1 activation and IL-1β secretion in infected mouse BMDCs, which was dependent on TLR-2, NLRP3, and also ASC. A transposon mutant library screening identified multiple *H. pylori* genes involved in LPS biosynthesis and the urease locus controlled IL-1β secretion during infection of mice. Interestingly, it was demonstrated that NLRP3 transcription was induced by the *H. pylori* urease B (UreB) subunit and IL-1β transcription was dependent on LPS. In contrast, various NLRs such as NLRC4, NLRP6, and AIM2 were dispensable for inflammasome activation by *H. pylori*. It was also shown that TLR-2 is required for IL-1β secretion through activation of the inflammasome. Various other investigated factors such as TLR-4, TLR-5, TLR-9, NOD2, IL-1R, and P2RX7 were also dispensable for inflammasome-mediated active IL-1β production and secretion. However, the TLR adaptor protein MyD88, but not TRIF, was necessary for *H. pylori* LPS-mediated induction of pro-IL-1β transcription. Interestingly, infected *Tlr2*$^{-/-}$ and *Tlr4*$^{-/-}$ mice were also able to induce pro-IL-1β transcription when infected with bacteria and this emphasizes an interesting aspect of *H. pylori* LPS recognition through either TLR-2 or TLR-4, which is complimentary to other published work (Yokota et al. 2007; Cullen et al. 2011). *Nlrp3*$^{-/-}$ and *Tlr2*$^{-/-}$ mice recapitulated the phenotype of *caspase-1*$^{-/-}$ mice, i.e., more efficient infection control associated with stronger T-cell responses. These data led to the conclusion that *H. pylori* activates the inflammasome in a TLR-2 and NLRP3-dependent manner and that *H. pylori* benefits from inflammasome activation, which ensures persistent infection (Koch et al. 2015). Moreover, it was shown that the *cag*PAI and γ-glutamyl-transpeptidase (GGT) are not directly involved in inflammasome-mediated caspase-1 activation and subsequent IL-1β secretion (Koch et al. 2015). Further studies have identified UreB as a potential direct ligand of TLR-2 for inducing NLRP3 expression and

LPS for inducing pro-IL-1β expression. Therefore, these two important bacterial factors are working at the transcriptional and post-translational level to control IL-1β secretion. Recombinant UreB rescued the phenotype of *ureB* mutants, while heat treatment impaired this activity. In addition, it was remarkably shown that *H. pylori* urease is necessary for the protection against experimental asthma in mice, which is a hallmark of *H. pylori* infection.

Moreover, Engler et al. (2015) reported a benefit of *H. pylori*-induced activation of the NLRP3 inflammasome in alleviating inflammatory bowel disease (IBD) in mice. In fact, IBD is on the rise in human populations worldwide and this has been attributed to environmental factors, diet, antibiotic use, and changes in the intestinal microbiome. This study therefore opens up a new strategy for controlling IBD using *H. pylori* or its components. *H. pylori* infection or oral administration of *H. pylori* extracts significantly reduced the symptoms of dextran sodium sulfate (DSS) and T-cell transfer models of the IBD. However, infection was not beneficial on controlling other experimental autoimmune diseases, including autoimmune encephalomyelitis or type I diabetes (Engler et al. 2015). The authors have used histopathological and endoscopic methods to assess the symptoms associated with IBD during *H. pylori* infection or treatment with bacterial extracts. It was reported that mice treated with live *H. pylori* or bacterial extracts exhibited significantly alleviated symptoms of IBD when compared to positive controls. The *H. pylori*-mediated effect on IBD was also confirmed in a T-cell transfer-mediated model of colitis. Moreover, the authors have detected increased intestinal *Muc2* gene expression and formation of thick mucus as a protective layer in the *H. pylori*-treated group. It was correlated with the intestine-specific transcription factor CDX2 and transforming growth factor-β (TGF-β) signaling by *H. pylori*. Further analysis in $Nlrp3^{-/-}$, $Il18^{-/-}$, $Il18r^{-/-}$, and $Myd88^{-/-}$ mice have attributed the beneficial effect of infection to inflammasome activation and formation of active IL-18 by *H. pylori* extracts. It was previously shown that IL-18-mediated production of regulatory T cells (Tregs) induces a tolerogenic phenotype in *H. pylori*-infected mice (Oertli et al. 2012). The above series of studies therefore confirm and extend earlier experimental and epidemiological data that linked *H. pylori* infection to protection against IBD (Luther et al. 2010; Sonnenberg and Genta 2012). In a large epidemiological survey investigating a total of 65,515 patients, of which 1061 served as cases with IBD and 64,451 as controls, the histological presence of *H. pylori* was correlated with the histological presence of Crohn's disease, ulcerative colitis, and indeterminate colitis: this study confirmed an inverse association between *H. pylori* and IBD (odds ratio ~0.5) (Sonnenberg and Genta 2012). The general trend was further confirmed by a large meta-analysis comparing 23 individual studies, which examined *H. pylori* seropositivity relative to IBD diagnosis. Overall, within a total study population of almost 6000 individuals, 27.1 % of IBD patients revealed evidence of infection with *H. pylori* compared to 40.9 % of patients in the non-infected control group, suggesting a protective benefit of *H. pylori* infection against the development of IBD (Luther et al. 2010). Experimental data obtained in *Salmonella*-induced colitis and acute DSS-induced colitis confirmed a protective effect of *H. pylori* in both models, which could be linked to the bacteria's DNA (Higgins et al. 2011; Luther et al. 2011). The combined results suggest that *H. pylori* infection protects against chronic inflammatory conditions of the lower bowel as well as allergic disease manifestations, at least some of which can be attributed to NLRP3 inflammasome activation and the production and secretion of protective IL-18.

3.4 Inflammasome-Mediated IL-1β Secretion in Cultured Human and Mouse Cells

In one of the earlier studies, Basak et al. (2005) reported that purified *H. pylori* LPS induced the transcription of pro-IL-1β in THP1 macrophages, which was regulated by the transcription factors NF-κB and C/EBPβ. These authors demonstrated that LPS recognition by TLR-4 induced a PI3-kinase → Rac1 → p21 signaling cascade leading to NF-κB activation and PI3-kinase → Akt → p38 signaling activated C/EBPβ. It was also shown that Rac1 → PAK1 signaling is necessary for caspase-1-mediated processing of pro-IL-1β to the active form. *H. pylori*-infected peripheral blood mononuclear cells (PBMCs) derived monocytes, DCs, M1 macrophages significantly secreted IL-1β and other pro-inflammatory cytokines. However, M2 macrophages infection produced significantly less cytokines (Fehlings et al. 2012). Semper et al. (2014) found that *H. pylori* infection of PBMCs and isolated monocytes lead to IL-1β secretion in a T4SS-dependent manner without a role for CagA. However, this response was not VacA-dependent as seen in infected BMDCs. In another study, *H. pylori* infection of isolated human neutrophils increased the expression of NLRP3 and ASC in a T4SS-independent manner (Pérez-Figueroa et al. 2016). These authors also ruled out the involvement of TLR-2 and TLR-4 in IL-1β secretion from neutrophils. It was shown that isolated neutrophils express inherently active caspase-1 and this activity was further increased during *H. pylori* infection. Altogether, this study concluded that NLRP3 inflammasome activation and IL-1β secretion in human neutrophils during *H. pylori* infection is independent of the T4SS, and the host receptors TLR-2 and TLR-4 (Pérez-Figueroa et al. 2016). A very recent study showed that *H. pylori*-infected PMA-differentiated THP1 cells induced the secretion of IL-1β in a caspase-1 and NLRP3-dependent mechanism, which required ROS, K^+ efflux, and Ca^{2+} signaling (Kameoka et al. 2016). Taken together, there are only a very few studies with human cell lines for inflammasome activation in *H. pylori* infection. Thus, more studies are necessary in the future which will shed new light on this process. Another interesting study with purified *H. pylori* flagellin along with *Salmonella typhimurium* and *Yersinia enterocolitica* flagellins was conducted to understand the effect on NLRC4 phosphorylation and activation to induce IL-1β secretion. Each of the flagellins was transferred into target cells and once reached in cytosol induced NLRC4 phosphorylation (Matusiak et al. 2015). However, unlike the activities of flagellins from Salmonella and Yersinia, *H. pylori* flagellin was unable to induce inflammasome activation and IL-1β production. In addition, *H. pylori* gene HP0940, encoding a serine/threonine kinase, is known to induce cytokine secretion from gastric epithelial cells (Kim et al. 2010). In concurrence, mouse RAW264.7 cells treated with the purified HP0940 protein were reported to activate caspase-1 and IL-1β secretion in a dose- and time-dependent manner and also reduced the cell viability through apoptosis (Tenguria et al. 2014). These studies indicate that *H. pylori*-induced inflammasome activation is controlled by multiple bacterial factors in a cell-type specific fashion.

4 Concluding Remarks

Taken together, current data provide evidence that NLRP3-mediated inflammasome activation might be an early event in *H. pylori* infection, responsible for eliciting the strong adaptive immune response which is required to control the infection. Recent studies provided confidence that activation of the inflammasome by *H. pylori* is also surprisingly related to the control of chronic inflammatory and allergic diseases. It is also evident that stringent regulatory mechanisms exist to control the activation of inflammasome formation during *H. pylori* infection. Because of the different mechanisms utilized by the bacterium to evade clearance by the host immune system, the infection persists and the levels of IL-1β remain high. Furthermore, polymorphisms inducing elevated levels of IL-1β in the gastric mucosa have been associated with increased risk of gastric cancer development. It has to be noted that the risk was restricted to the non-cardia subsite, suggesting that the distribution of the inflammation is important in terms of gastric cancer risk. Altogether, these events pave the way for the development of gastric atrophy, which is considered as the initial step of gastric carcinogenesis, which can under conditions of continuous inflammation eventually lead to the development of gastric cancer. Moreover, little is currently known about the vast differences in the risk of gastric cancer development among human populations, which is likely to be influenced by human genetic predisposition, population ecology, and other parameters. In summary, many of the peculiarities that set *H. pylori* apart from other Gram-negative enteropathogens remain underexplored and deserve further work in future.

Acknowledgments This work was supported by grants of the German Science Foundation (project B10 in CRC-796 and A4 in CRC-1181) to SB.

References

Amieva MR, El-Omar EM (2008) Host-bacterial interactions in *Helicobacter pylori* infection. Gastroenterology 134:306–323

Apte RN, Dotan S, Elkabets M, White MR, Reich E, Carmi Y, Song X, Dvozkin T, Krelin Y, Voronov E (2006) The involvement of 1L-1 in tumorigenesis, tumor invasiveness, metastasis and tumor-host interactions. Cancer Metastasis Rev 25(3):387–408

Atherton JC, Blaser MJ (2009) Coadaptation of *Helicobacter pylori* and humans: ancient history, modern implications. J Clin Invest 119:2475–2487

Azuma T, Ito S, Sato F, Yamazaki Y, Miyaji H, Ito Y, Suto H, Kuriyama M, Kato T, Kohli Y (1998) The role of the HLA-DQA1 gene in resistance to atrophic gastritis and gastric adenocarcinoma induced by *Helicobacter pylori* infection. Cancer 82:1013–1018

Backert S, Naumann M (2010) What a disorder: proinflammatory signaling pathways induced by *Helicobacter pylori*. Trends Microbiol 18(11):479–486

Basak C, Pathak SK, Bhattacharyya A, Mandal D, Pathak S, Kundu M (2005) NF-kappaB- and C/EBPbeta-driven interleukin-1beta gene expression and PAK1-mediated caspase-1 activation play essential roles in interleukin-1beta release from *Helicobacter pylori* lipopolysaccharide-stimulated macrophages. J Biol Chem 280(6):4279–4288

Basso D, Scrigner M, Toma A, Navaglia F, Di Mario F, Rugge M, Plebani M (1996) *Helicobacter pylori* infection enhances mucosal interleukin-1 beta, interleukin-6, and the soluble receptor of interleukin-2. Int J Clin Lab Res 26(3):207–210

Broz P, Monack DM (2013) Newly described pattern recognition receptors team up against Intracellular pathogens. Nat Rev Immunol 13:551–565

Castaño-Rodríguez N, Kaakoush NO, Goh KL, Fock KM, Mitchell HM (2014) The NOD-like receptor signaling pathway in *Helicobacter pylori* infection and related gastric cancer: a case-control study and gene expression analyses. PLoS ONE 9(6):e98899

Chakravorty M, Ghosh A, Choudhury A, Santra A, Hembrum J, Roychoudhury S (2006) Interaction between 1L-1B gene promoter polymorphisms in determining susceptibility to *Helicobacter pylori* associated duodenal ulcer. Hum Mutat 27:411–419

Cullen TW, Giles DK, Wolf LN, Ecobichon C, Boneca IG, Trent MS (2011) *Helicobacter pylori* versus the host: remodeling of the bacterial outer membrane is required for survival in the gastric mucosa. PLoS Pathog 7(12):e1002454

El-Omar EM, Carrington M, Chow WH et al (2000) Interleukin-1 polymorphisms associated with increased risk of gastric cancer. Nature 404:398–402

El-Omar EM, Rabkin CS, Gammon MD, Vaughan TL, Risch HA, Schoenberg JB, Stanford JL, Mayne ST, Goedert J, Blot WJ, Fraumeni JF, Chow WH (2003) Increased risk of noncardia gastric cancer associated with proinflammatory cytokine gene polymorphisms. Gastroenterology 124:1193–1201

Engler DB, Leonardi I, Hartung ML, Kyburz A, Spath S, Becher B, Rogler G, Müller A (2015) *Helicobacter pylori*-specific protection against inflammatory bowel disease requires the NLRP3 inflammasome and 1L-18. Inflamm Bowel Dis 21:854–861

Fehlings M, Drobbe L, Moos V, Renner Viveros P, Hagen J, Beigier-Bompadre M, Pang E, Belogolova E, Churin Y, Schneider T, Meyer TF, Aebischer T, Ignatius R (2012) Comparative analysis of the interaction of *Helicobacter pylori* with human dendritic cells, macrophages, and monocytes. Infect Immun 80(8):2724–2734

Fox JG, Rogers AB, Whary MT et al (2007) Accelerated progression of gastritis to dysplasia in the pyloric antrum of TFF2$^{-/-}$ C57BL6 × Sv129 *Helicobacter pylori*-infected mice. Am J Pathol 171:1520–1528

Hanahan D, Weinberg RA (2011) Hallmarks of cancer: the next generation. Cell 144:646–674

Higgins PD, Johnson LA, Luther J, Zhang M, Sauder KL, Blanco LP, Kao JY (2011) Prior *Helicobacter pylori* infection ameliorates Salmonella typhimurium-induced colitis: mucosal crosstalk between stomach and distal intestine. Inflamm Bowel Dis 17(6):1398–1408

Hitzler I, Sayi A, Kohler E, Engler DB, Koch KN, Hardt WD, Müller A (2012) Caspase-1 has both proinflammatory and regulatory properties in *Helicobacter* infections, which are differentially mediated by its substrates IL-1β and IL-18. J Immunol 188:3594–3602

Hong-mei Z, Kai-feng P, Yang Z, Lian Z, Jun-ling M, Tong Z, Hui-juan S, Wen-qing L, Wei-cheng Y (2011) The correlation between polymorphisms of Toll-like receptor 2 and Toll-like receptor 9 and susceptibility to gastric cancer. Chin J Prev Med 45(07):588–592

Jung HC, Kim JM, Song IS, Kim CY (1997) *Helicobacter pylori* induces an array of pro-inflammatory cytokines in human gastric epithelial cells: quantification of mRNA for interleukin-8, -1 alpha/beta, granulocyte-macrophage colony-stimulating factor, monocyte chemoattractant protein-1 and tumour necrosis factor-alpha. J Gastroenterol Hepatol 12:473–480

Kameoka S, Kameyama T, Hayashi T, Sato S, Ohnishi N, Hayashi T, Murata-Kamiya N, Higashi H, Hatakeyama M, Takaoka A (2016) *Helicobacter pylori* induces IL-1β protein through the inflammasome activation in differentiated macrophagic cells. Biomed Res 37(1):21–27

Katoh M (2007) Dysregulation of stem cell signaling network due to germline mutation, SNP, *Helicobacter pylori* infection, epigenetic change and genetic alteration in gastric cancer. Cancer Biol Ther 6(6):832–839

Kim DJ, Park KS, Kim JH, Yang SH, Yoon JY, Han BG, Kim HS, Lee SJ, Jang JY, Kim KH, Kim MJ, Song JS, Kim HJ, Park CM, Lee SK, Lee BI, Suh SW (2010) *Helicobacter pylori* proinflammatory protein up-regulates NF-kappaB as a cell-translocating Ser/Thr kinase. Proc Natl Acad Sci USA 107:21418–23

Kim J, Cho YA, Choi IJ, Lee YS, Kim SY, Shin A, Cho SJ, Kook MC, Nam JH, Ryu KW, Lee JH, Kim YW (2012) Effects of interleukin-10 polymorphisms, *Helicobacter pylori* infection, and smoking on the risk of noncardia gastric cancer. PLoS ONE 7(1):e29643

Kim DJ, Park JH, Franchi L, Backert S, Núñez G (2013) The Cag pathogenicity island and interaction between TLR2/NOD2 and NLRP3 regulate 1L-1β production in *Helicobacter pylori* infected dendritic cells. Eur J Immunol 43:2650–2658

Kitajima Y, Ohtaka K, Mitsuno M, Tanaka M, Sato S, Nakafusa Y, Miyazaki K (2008) *Helicobacter pylori* infection is an independent risk factor for Runx3 methylation in gastric cancer. Oncol Rep 19(1):197–202

Koch KN, Hartung ML, Urban S, Kyburz A, Bahlmann AS, Lind J, Backert S, Taube C, Müller A (2015) *Helicobacter* urease-induced activation of the TLR2/NLRP3/1L-18 axis protects against asthma. J Clin Invest 125:3297–3302

Koch M, Mollenkopf HJ, Meyer TF (2016) Macrophages recognize the *Helicobacter pylori* type IV secretion system in the absence of toll-like receptor signalling. Cell Microbiol 18(1): 137–147

Kusters JG, van Vliet AH, Kuipers EJ (2006) Pathogenesis of *Helicobacter pylori* in-fection. Clin Microbiol Rev 19:449–490

Lee SG, Kim B, Choi W, Lee I, Choi J, Song K (2003) Lack of association between pro-inflammatory genotypes of the interleukin-1 (1L-1B -31 C/+ and 1L-1RN *2/*2) and gastric cancer/duodenal ulcer in Korean population. Cytokine 21:167–171

Lim B, Ju H, Kim M, Kang C (2011) Increased genetic susceptibility to intestinal-type gastric cancer is associated with increased activity of the RUNX3 distal promoter. Cancer 15:117 (22):5161–5171

Luther J, Dave M, Higgins PD, Kao JY (2010) Association between *Helicobacter pylori* infection and inflammatory bowel disease: a meta-analysis and systematic review of the literature. Inflamm Bowel Dis 16(6):1077–1084

Luther J, Owyang SY, Takeuchi T, Cole TS, Zhang M, Liu M, Erb-Downward J, Rubenstein JH, Chen CC, Pierzchala AV, Paul JA, Kao JY (2011) *Helicobacter pylori* DNA decreases pro-inflammatory cytokine production by dendritic cells and attenuates dextran sodium sulphate-induced colitis. Gut 60(11):1479–1486

Martinon F, Burns K, Tschopp J (2002) The inflammasome: a molecular platform triggering activation of inflammatory caspases and processing of proIL-beta. Mol Cell 10(2):417–426

Matusiak M, Van Opdenbosch N, Vande Walle L, Sirard JC, Kanneganti TD, Lamkanfi M (2015) Flagellin-induced NLRC4 phosphorylation primes the inflammasome for activation by NAIP5. Proc Natl Acad Sci USA 112:1541–1546

Mayerle J, den Hoed CM, Schurmann C, Stolk L, Homuth G, Peters MJ, Capelle LG, Zimmermann K, Rivadeneira F, Gruska S, Völzke H, de Vries AC, Völker U, Teumer A, van Meurs JB, Steinmetz I, Nauck M, Ernst F, Weiss FU, Hofman A, Zenker M, Kroemer HK, Prokisch H, Uitterlinden AG, Lerch MM, Kuipers EJ (2013) Identification of genetic loci associated with *Helicobacter pylori* serologic status. JAMA 309:1912–1920

Moodley Y, Linz B (2009) *Helicobacter pylori* sequences reflect past human migrations. Genome Dyn 6:62–74

Ng GZ, Menheniott TR, Every AL, Stent A, Judd LM, Chionh YT, Dhar P, Komen JC, Giraud AS, Wang TC, McGuckin MA, Sutton P (2015) The MUC1 mucin protects against *Helicobacter pylori* pathogenesis in mice by regulation of the NLRP3 inflammasome. Gut. doi:10.1136/gutjnl-2014-307175

Noach LA, Bosma NB, Jansen J, Hoek FJ, van Deventer SJ, Tytgat GN (1994) Mucosal tumor necrosis factor-alpha, interleukin-1 beta, and interleukin-8 production in patients with *Helicobacter pylori* infection. Scand J Gastroenterol 29(5):425–429

Oertli M, Sundquist M, Hitzler I, Engler DB, Arnold IC, Reuter S, Maxeiner J, Hansson M, Taube C, Quiding-Järbrink M, Müller A (2012) DC-derived 1L-18 drives Treg differentiation, murine *Helicobacter pylori*-specific immune tolerance, and asthma protection. J Clin Invest 122:1082–1096

Pachathundikandi SK, Tegtmeyer N, Backert S (2013) Signal transduction of *Helicobacter pylori* during interaction with host cell protein receptors of epithelial and immune cells. Gut Microbes 4(6):454–474

Pérez-Figueroa E, Torres J, Sánchez-Zauco N, Contreras-Ramos A, Alvarez-Arellano L, Maldonado-Bernal C (2016) Activation of NLRP3 inflammasome in human neutrophils by *Helicobacter pylori* infection. Innate Immun 22:103–112

Peterson AJ, Menheniott TR, O'Connor L et al (2010) *Helicobacter pylori* infection promotes methylation and silencing of trefoil factor 2, leading to gastric tumor development in mice and humans. Gastroenterology 139:2005–2017

Polk DB, Peek RM Jr (2010) *Helicobacter pylori*: gastric cancer and beyond. Nat Rev Cancer 10:403–414

Ravishankar Ram M, Goh KL, Leow AH, Poh BH, Loke MF, Harrison R, Shankar EM, Vadivelu J (2015) Polymorphisms at Locus 4p14 of Toll-Like Receptors TLR-1 and TLR-10 Confer Susceptibility to Gastric Carcinoma in *Helicobacter pylori* Infection. PLoS ONE 10 (11):e0141865

Resende C, Regalo G, Durães C, Pinto MT, Wen X, Figueiredo C, Carneiro F, Machado JC (2016) Interleukin-1B signalling leads to increased survival of gastric carcinoma cells through a CREB-C/EBPβ-associated mechanism. Gastric Cancer 19(1):74–84

Saeki N, Saito A, Choi IJ et al (2011) A functional single nucleotide polymorphism in Mucin 1, at chromosome 1q22, determines susceptibility to diffuse-type gastric cancer. Gastroenterology 140:892–902

Salama NR, Hartung ML, Müller A (2013) Life in the human stomach: persistence strategies of the bacterial pathogen *Helicobacter pylori*. Nat Rev Microbiol 11:385–399

Savage SA, Abnet CC, Mark SD, Qiao YL, Dong ZW, Dawsey SM, Taylor PR, Chanock SJ (2004) Variants of the IL8 and IL8RB genes and risk for gastric cardia adenocarcinoma and esophageal squamous cell carcinoma. Cancer Epidemiol Biomark Prev 13:2251–2257

Semper RP, Mejías-Luque R, Groß C, Anderl F, Müller A, Vieth M, Busch DH, Prazeres da Costa C, Ruland J, Groß O, Gerhard M (2014) *Helicobacter pylori*-induced 1L-1β secretion in innate immune cells is regulated by the NLRP3 inflammasome and requires the cag pathogenicity island. J Immunol 193:3566–3576

Shanks AM, El-Omar EM (2009) *Helicobacter pylori* infection, host genetics and gastric cancer. J Dig Dis 10:157–164

Shi J, Zhao Y, Wang K, Shi X, Wang Y, Huang H, Zhuang Y, Cai T, Wang F, Shao F (2015) Cleavage of GSDMD by inflammatory caspases determines pyroptotic cell death. Nature 526 (7575):660–665

Sonnenberg A, Genta RM (2012) Low prevalence of *Helicobacter pylori* infection among patients with inflammatory bowel disease. Aliment Pharmacol Ther 35(4):469–476

Tenguria S, Ansari SA, Khan N, Ranjan A, Devi S, Tegtmeyer N, Lind J, Backert S, Ahmed N (2014) *Helicobacter pylori* cell translocating kinase (CtkA/JHP0940) is pro-apoptotic in mouse macrophages and acts as auto-phosphorylating tyrosine kinase. Int J Med Microbiol 304:1066–1076

Tomita T, Jackson AM, Hida N, Hayat M, Dixon MF, Shimoyama T, Axon AT, Robinson PA, Crabtree JE (2001) Expression of Interleukin-18, a Th1 cytokine, in human gastric mucosa in increased in *Helicobacter pylori* infection. J Infect Dis 183:620–627

Tu S, Bhagat G, Cui G, Takaishi S, Kurt-Jones EA, Rickman B, Betz KS, Penz-Oesterreicher M, Bjorkdahl O, Fox JG, Wang TC (2008) Overexpression of interleukin-1beta induces gastric inflammation and cancer and mobilizes myeloid-derived suppressor cells in mice. Cancer Cell 14(5):408–419

Vinall LE, King M, Novelli M et al (2002) Altered expression and allelic association of the hypervariable membrane mucin MUC1 in *Helicobacter pylori* gastritis. Gastroenterology 123:41

von Moltke J, Ayres JS, Kofoed EM, Chavarría-Smith J, Vance RE (2013) Recognition of bacteria by inflammasomes. Annu Rev Immunol 31:73–106

Yamauchi K, Choi IJ, Lu H, Ogiwara H, Graham DY, Yamaoka Y (2008) Regulation of IL-18 in *Helicobacter pylori* infection. J Immunol 180:1207–1216

Ying HY, Yu BW, Yang Z, Yang SS, Bo LH, Shan XY, Wang HJ, Zhu YJ, Wu XS (2016) Interleukin-1B 31 C > T polymorphism combined with *Helicobacter pylori*-modified gastric cancer susceptibility: evidence from 37 studies. J Cell Mol Med 20(3):526–536

Yokota S, Ohnishi T, Muroi M, Tanamoto K, Fujii N, Amano K (2007) Highly-purified *Helicobacter pylori* LPS preparations induce weak inflammatory reactions and utilize Toll-like receptor 2 complex but not Toll-like receptor 4 complex. FEMS Immunol Med Microbiol 51:140–148

Listeria monocytogenes and the Inflammasome: From Cytosolic Bacteriolysis to Tumor Immunotherapy

Erin Theisen and John-Demian Sauer

Abstract Inflammasomes are cytosolic innate immune surveillance systems that recognize a variety of danger signals, including those from pathogens. *Listeria monocytogenes* is a Gram-positive intracellular bacterium evolved to live within the harsh environment of the host cytosol. Further, *L. monocytogenes* can activate a robust cell-mediated immune response that is being harnessed as an immunotherapeutic platform. Access to the cytosol is critical for both causing disease and inducing a protective immune response, and it is hypothesized that the cytosolic innate immune system, including the inflammasome, is critical for both host protection and induction of long-term immunity. *L. monocytogenes* can activate a variety of inflammasomes via its pore-forming toxin listeriolysin-O, flagellin, or DNA released through bacteriolysis; however, inflammasome activation attenuates *L. monocytogenes*, and as such, *L. monocytogenes* has evolved a variety of ways to limit inflammasome activation. Surprisingly, inflammasome activation also impairs the host cell-mediated immune response. Thus, understanding how *L. monocytogenes* activates or avoids detection by the inflammasome is critical to understand the pathogenesis of *L. monocytogenes* and improve the cell-mediated immune response generated to *L. monocytogenes* for more effective immunotherapies.

Contents

E. Theisen · J.-D. Sauer (✉)
Department of Medical Microbiology and Immunology, School of Medicine and Public Health, University of Wisconsin-Madison, 1550 Linden Dr. Rm 4203, Madison, WI 53706, USA
e-mail: sauer3@wisc.edu

S. Backert (ed.), *Inflammasome Signaling and Bacterial Infections*,
Current Topics in Microbiology and Immunology 397,
DOI 10.1007/978-3-319-41171-2_7

1 Introduction

1.1 Overview

Listeria monocytogenes is a Gram-positive facultative intracellular pathogen that has evolved to survive in a variety of severe environments and infect a wide range of hosts. These features combine to make *L. monocytogenes* an important food-borne pathogen most frequently associated with unpasteurized dairy products, deli meats, and more recently, fresh produce (Ferreira et al. 2014). Infection *with L. monocytogenes* can result in a noninvasive gastroenteritis that is likely severely underreported (Swaminathan and Gerner-Smidt 2007). More importantly, systemic listeriosis poses a severe risk to the immunocompromised, including the very young and old, as well as pregnant women. Complications of listeriosis include septicemia, meningitis, encephalitis, abortion, and still-borne births, resulting in a strikingly high mortality rate of 20–30 % even with proper antibiotic therapy (Swaminathan and Gerner-Smidt 2007).

While *L. monocytogenes* remains an important foodborne pathogen, there is an increasing interest in harnessing the robust cell-mediated immune response induced upon *L. monocytogenes* infection for use as a potential immunotherapy. Due to its almost exclusive intracellular lifecycle, described more in detail below, *L. monocytogenes* stimulates a robust $CD8^+$ T-cell response. Additionally, *L. monocytogenes* is genetically tractable and is able to break self-tolerance, key factors in making *L. monocytogenes* an extremely promising cancer immunotherapeutic platform (Le et al. 2012). As access to the cytosol is an essential prerequisite for both the pathogenesis and the induction of immunity following *L. monocytogenes* infection, understanding how bacteria are sensed by the innate immune system in this environment has been a focal point of *L. monocytogenes* research for the past decade. Particularly, activation and avoidance of the inflammasome, a critical innate immune signaling pathway that all cytosolic pathogens must deal with, has become a recent focus.

1.2 Life Cycle

Listeria monocytogenes can infect a variety of cell types either through phagocytosis by myeloid cells or through active invasion of epithelial cells or hepatocytes with its virulence factors, internalin A or internalin B, respectively (Mengaud et al. 1996; Shen et al. 2000). Upon entry into the host cell, *L. monocytogenes* is initially contained within a phagocytic vacuole. Using its cholesterol-dependent pore-forming toxin listeriolysin-O (LLO, encoded by the gene *hly*) and a pair of phospholipases (PlcA and PlcB), *L. monocytogenes* escapes from the phagosome and enters into the cytosol (Portnoy et al. 1988; Mengaud et al. 1991; Camilli et al. 1991; Vazquez-Boland et al. 1992; Hamon et al. 2012).

Unlike most bacterial pathogens, *L. monocytogenes* is able to not only survive but also flourish and replicate within the host cytosol. This is a property unique to *L. monocytogenes* and other cytosol-adapted pathogens as intracellular pathogens that mislocalize to the cytosol or non-intracellular pathogens that are placed within the cytosol are unable to survive or replicate (Beuzón et al. 2000; Goetz et al. 2001; Slaghuis et al. 2004; Creasey and Isberg 2012). These studies suggest that *L. monocytogenes* has evolved specific adaptations to deal with cytosolic stresses, cell autonomous defense mechanisms, and innate immune detection in the cytosol, including the inflammasome. For example, *L. monocytogenes* modifies its peptidoglycan through N-deacetylation and O-acetylation to avoid killing by lysozyme and detection by the innate immune system (Boneca et al. 2007; Rae et al. 2011). *L. monocytogenes* can also modulate the production of the innate signaling cytokine IL-6 through its virulence factor InlH (Personnic et al. 2010); misregulation of these factors results in the attenuation of *L. monocytogenes*, demonstrating the importance of *L. monocytogenes* carefully controlling its detection by the host (Boneca et al. 2007; Personnic et al. 2010; Rae et al. 2011).

Once in the cytosol, *L. monocytogenes* avoid exposure to the extracellular milieu and the host defenses found there by utilizing the virulence factor ActA to hijack host actin and propel itself through membrane protrusions into neighboring cells (Kocks et al. 1992). Both *Δhly* and *ΔactA L. monocytogenes* mutants are attenuated; however, of these, only *ΔactA L. monocytogenes* are able to mount a protective cell-mediated immune response (Portnoy et al. 1988; Goossens and Milon 1992; Bahjat et al. 2009), further highlighting the importance of the cytosol and cytosolic innate immune recognition in stimulating robust $CD8^+$ T cell-mediated immunity.

1.3 Innate Immune Response

Once *L. monocytogenes* accesses the cytosol, it is rarely exposed to extracellular host defenses; as such, detection by cytosolic innate immune sensors is likely critical for combating infection. Likewise, as *L. monocytogenes* must access the cytosol to cause disease, avoidance or direct inhibition of the cytosolic innate immune system is likely critical for *L. monocytogenes'* virulence. Indeed, multiple cytosolic innate immune pathways have been demonstrated to recognize *L. monocytogenes* infection (reviewed in Witte et al. 2012). Briefly, peptidoglycan fragments of *L. monocytogenes* can activate nucleotide binding oligomerization domain (NOD) proteins leading to the phosphorylation of p38 mitogen-activated protein kinase (MAPK) and transcription factor NF-κB activation, further leading to inflammatory cytokine production (O'Riordan et al. 2002; Opitz et al. 2006). Further, NOD signaling can synergize with interferon (IFN) signaling to enhance the host innate response (Leber et al. 2008), although it should be noted that type I IFN induction has been shown to inversely correlate with host protection (O'Connell et al. 2004; Auerbuch et al. 2004; Carrero et al. 2006). More recently, cyclic dinucleotides secreted by *L. monocytogenes* have been shown to be recognized by the STING/IRF3 pathway resulting in the production of type I IFNs (Woodward et al. 2010; Sauer et al. 2011b). Additionally, as will be highlighted in this chapter, *L. monocytogenes* has been reported to activate the inflammasome through a variety of receptors, ultimately resulting in the activation of caspase-1 (Fig. 1). Caspase-1 activation leads to an inflammatory cell death, pyroptosis, release of the pro-inflammatory cytokines IL-1β and IL-18, and the production and release of lipid mediators known as eicosanoids (Fig. 2). The engagement of these innate pathways resulting in the production of pro-inflammatory cytokines and the recruitment of innate immune cells is hypothesized to be critical in the development of adaptive immunity.

In this chapter, we will focus on how *L. monocytogenes* activates different inflammasomes. We will also discuss the downstream consequences of inflammasome activation on the pathogenesis of *L. monocytogenes*. Finally, we will discuss how engagement of the inflammasome influences the innate and adaptive immune responses to *L. monocytogenes*, with a focus on its impact on the development of *L. monocytogenes* as an immunotherapeutic platform.

2 Activation of Different Inflammasomes

2.1 NLRP3 Activation

NLRP3 belongs to the Nod-like receptor family of pattern recognition receptors and recognizes a variety of danger signals including ATP, uric acid crystals, and toxins (Mariathasan et al. 2006). *L. monocytogenes* was first observed to activate the NLRP3

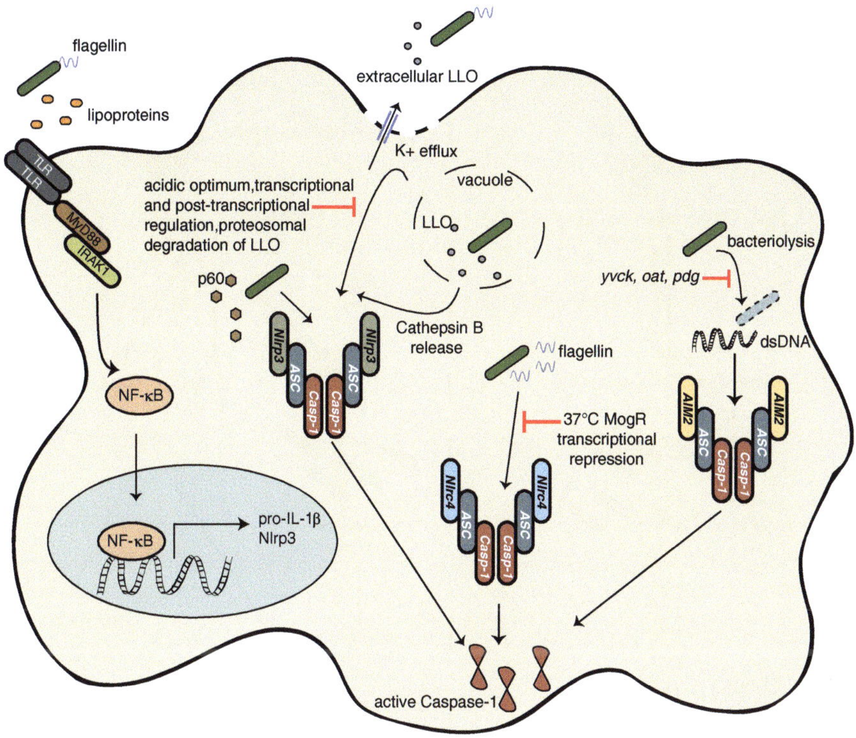

Fig. 1 In vitro mechanisms of activation and avoidance of the inflammasome by *L. monocytogenes*. *L. monocytogenes* lipoproteins and flagellin can signal through toll-like receptors (TLRs) via MyD88 and IRAK1 to activate the transcription factor NF-κB and prime cells for inflammasome activation by up-regulating IL-1β and NLRP3 transcripts. *L. monocytogenes* can then activate the NLRP3 inflammasome via a K^+ efflux secondary to extracellular or intracellular LLO-induced pores. Vacuole rupture by LLO can result in cathepsin B release that along with p60 can also activate the NLRP3 inflammasome. Flagellin can activate the NLRC4 inflammasome, while bacteriolysis and the subsequent DNA release can activate the AIM2 inflammasome. *L. monocytogenes* limits the activation of any of these inflammasomes via the tight control of LLO or flagellin, or by unknown mechanisms of combating cytosolic stress and limiting bacteriolysis

inflammasome in the seminal paper by Mariathasan et al. (2006) who demonstrated that cytosolic *L. monocytogenes* can activate the NLRP3 inflammasome, while vacuole-contained *L. monocytogenes* fail to activate the inflammasome, suggesting that the cytosolic recognition is critical. Concurrently, Kannengati et al. (2006) observed that *L. monocytogenes* total RNA could stimulate the NLRP3 inflammasome, further suggesting that *L. monocytogenes* could activate the NLRP3 inflammasome through a variety of mechanisms. More recently, activation of the Nrlp3 inflammasome specifically in unprimed bone marrow-derived macrophages infected with *L. monocytogenes* has been reported (Wu et al. 2010; Fernandes-Alnemri et al. 2013; Lin et al. 2014). However, the importance of NLRP3 in *L. monocytogenes* infection has also been questioned, as Franchi et al. (2007) failed to find any

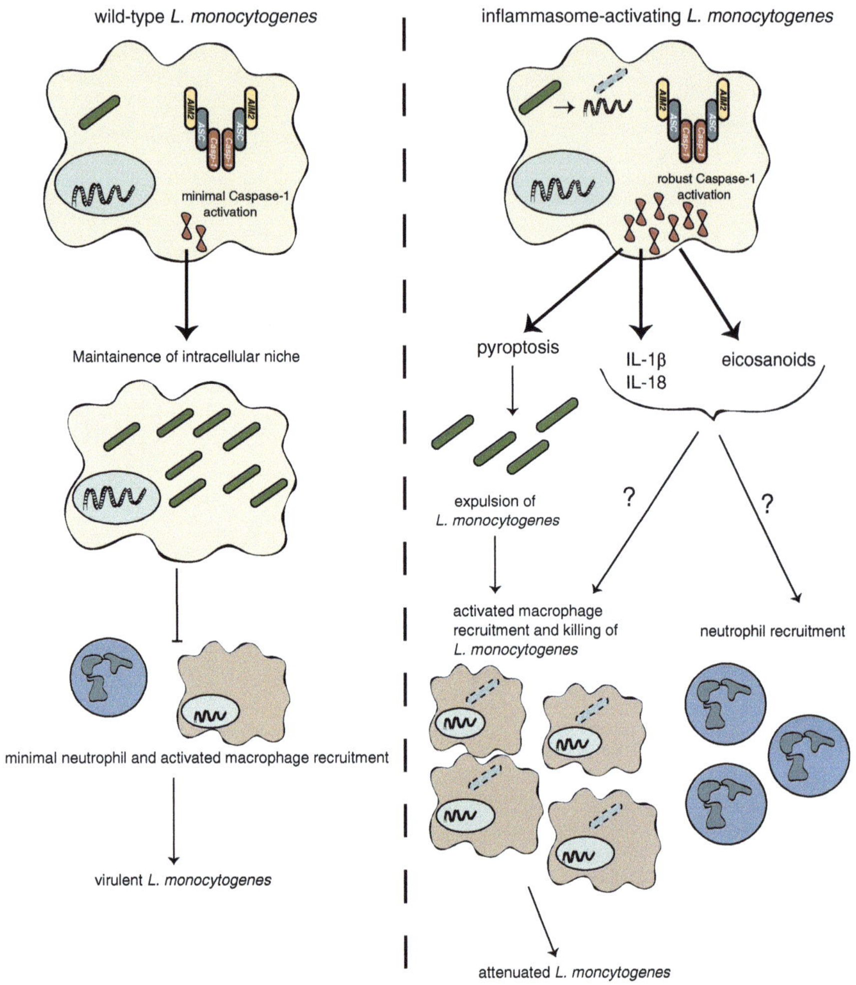

Fig. 2 Activation of the inflammasome attenuates *L. monocytogenes* through macrophage recruitment. In wild-type *L. monocytogenes* infection, *L. monocytogenes* minimally activates caspase-1 and maintains its intracellular niche, ultimately promoting a pathogenic infection. In contrast, inflammasome-activating *L. monocytogenes* results in robust caspase-1 activation and pyroptosis, release of IL-1β and IL-18, and eicosanoids. These effects result in expulsion of *L. monocytogenes,* increased activated macrophage recruitment, and potentially an influx of neutrophils. Increased macrophage recruitment restricts *L. monocytogenes* infection

difference in caspase-1 processing and IL-1β release in Nlrp3$^{-/-}$ cells following *L. monocytogenes* infection. Similarly, Sauer et al. (2010) failed to find any significant role for the NLRP3 inflammasome during *L. monocytogenes* infection.

These differing observations beg the questions: "Which components of *L. monocytogenes* stimulate the NLRP3 inflammasome and what role, if any, does it play under physiologic conditions?" Pore-forming toxins have been demonstrated to activate the NLRP3 inflammasome, and Mariathasan and colleagues hypothesized that LLO could act like many other NLRP3 agonists and result in intracellular K^+ disruptions to activate the NLRP3 inflammasome (Mariathasan et al. 2006). While their original interpretation was that the importance of LLO was to facilitate the access of *L. monocytogenes* to the cytosol, it is also possible that pore formation, and not cytosolic access, was the critical function of LLO in NLRP3 activation during their studies, particularly as these studies used a high MOI of 50. Additionally, Meixenberger et al. (2010) suggest that cathepsin B release from LLO-damaged phagosomes can stimulate the NLRP3 inflammasome. While some data have directly suggested that cytosolic access of intact bacteria is required for *L. monocytogenes* NLRP3 activation, more data are consistent with a role of extracellular LLO in NLRP3 activation (Meixenberger et al. 2010; Hamon and Cossart 2011; Sakhon et al. 2013). Extracellular LLO can induce caspase-1 processing and IL-1β release depending on NLRP3 (Meixenberger et al. 2010) by stimulating a K^+ efflux (Hamon and Cossart 2011). Blocking bacterial uptake with cytochalasin D still results in the activation of caspase-1, suggesting that low amounts of LLO from extracellular *L. monocytogenes* can activate the inflammasome (Hamon and Cossart 2011). Further, purified LLO that maintains pore-forming ability, but not hemolytically inactive LLO (LLO^{W492A}), can activate caspase-1 (Sakhon et al. 2013), and these results are independent of the cholesterol-binding activity (Hara et al. 2008; Hamon and Cossart 2011). Taken together, these results suggest that pore formation from extracellular LLO can trigger K^+ efflux resulting in NLRP3 inflammasome activation. However, LLO expression, activity, and stability are tightly regulated during *L. monocytogenes* infection (Glomski et al. 2002; Schnupf et al. 2006a, b), thus limiting the likelihood of high extracellular concentrations of LLO during infection in vivo and potentially suggesting that NLRP3 activation by LLO is an in vitro artifact.

While LLO is believed to be the primary molecule triggering the NLRP3 inflammasome, both bacterial RNA (Kanneganti et al. 2006) and the secreted virulence factor, p60 encoded by the gene invasion-associated secreted endopeptidase (*iap*), from *L. monocytogenes* have been shown to activate the NLRP3 inflammasome (Schmidt and Lenz 2012). The N-terminal LysM and SH3 domain region (L1S) of p60 stimulates IL-1β and IL-18 release, independent of the act of pyroptotic cell death, in bone marrow-derived dendritic cells (Schmidt and Lenz 2012). Given the recent reports about the importance of gasdermin D-dependent host cell death in cytokine secretion (Shi et al. 2015; Kayagaki et al. 2015), it is unclear how IL-1β and IL-18 are being secreted in this scenario; however, Schmidt and Lenz found that ROS inhibition with the inhibitor DPI impaired IL-1β secretion, while IL-18 secretion remained intact. Further, in BMDCs from mice on a 129S6 background that lack caspase-11, IL-1β secretion was impaired, while IL-18 was unaffected (Schmidt and Lenz 2012). These observations suggest different licensing mechanisms for IL-1β and IL-18, potentially allowing either the host cell or the invading pathogen to fine-tune the downstream consequences of inflammasome activation.

Most work examining *L. monocytogenes* and the NLRP3 inflammasome has been done with murine cells. However, human peripheral blood mononuclear cells (PBMCs) infected with *L. monocytogenes* can undergo LLO-induced NLRP3-dependent inflammasome activation that depends on phagosomal acidification and cathepsin B release (Meixenberger et al. 2010). These results suggest that the observed differences for the role of the NLRP3 inflammasome may also stem from the cell types used. More importantly, the state of the cells used in experiments matters. Multiple groups were able to find a role for the NLRP3 inflammasome in the first hour of *L. monocytogenes* infection when using unprimed cells (Wu et al. 2010; Fernandes-Alnemri et al. 2013; Lin et al. 2014). Interestingly, for *L. monocytogenes* to induce a rapid NLRP3-dependent inflammasome activation, intact toll-like receptor (TLR) signaling must be present as $MyD88^{-/-}$ cells fail to activate the inflammasome (Fernandes-Alnemri et al. 2013; Lin et al. 2014). Further, this response is dependent on the MyD88 downstream signaling molecule IL-1 receptor-associated kinase (IRAK1) as $IRAK1^{-/-}$ cells also fail to activate the inflammasome and secrete IL-1β and IL-18 (Thomas et al. 1999; Fernandes-Alnemri et al. 2013; Lin et al. 2014). In the presence of priming, however, IRAK1 is dispensable for inflammasome activation (Lin et al. 2014); additionally, a prolonged infection results in NLRP3-independent inflammasome activation (Lin et al. 2014). These results suggest that *L. monocytogenes* can rapidly activate the NLRP3 inflammasome early in infection, as long as an intact TLR pathway is present. Many groups do not use unprimed cells, as priming is generally required for inflammasome activation in vitro (Latz et al. 2013), potentially explaining the observed discrepancies in the importance of NLRP3.

2.2 NLRC4 Activation

The NLRC4 inflammasome recognizes flagellin and type III secretion system components (Miao et al. 2010b). As *L. monocytogenes* is a Gram-positive pathogen and lacks type III secretion components, its flagellin is likely the key activator of the NLRC4 inflammasome. In line with this, *L. monocytogenes* engineered to hyper-express flagellin can activate the NLRC4 inflammasome (Sauer et al. 2011a; Warren et al. 2011). Additionally, *ΔflgK L. monocytogenes* lack the adaptor molecule between the flagellar hook and flagellin monomers, and infection with this strain results in excess flagellin secretion and hyperactivation of the NLRC4 inflammasome (Warren et al. 2008). However, *L. monocytogenes* likely avoids the activation of the NLRC4 inflammasome by shutting off flagellin expression in mammalian hosts (Peel et al. 1988; Shen and Higgins 2006). As such, *L. monocytogenes* inflammasome activation was shown to be independent of NLRC4 (Franchi et al. 2007; Sauer et al. 2010). Similarly, infection with *L. monocytogenes* deficient in flagellin (*ΔflaA*) resulted in similar caspase-1 processing and IL-1β release (Warren et al. 2008) as infection with wild-type *L. monocytogenes*. Taken together, these results suggest that although *L. monocytogenes* flagellin can activate the NLRC4 inflammasome, likely in a manner similar to other bacterial flagellin being dependent on the adaptor molecule Naip5 (Kofoed and Vance 2011; Zhao et al. 2011), *L. monocytogenes* largely avoids the activation of this inflammasome through the control of its flagellin in vivo.

2.3 AIM2 Activation

As opposed to the NLR family members described above, AIM2 (absent in melanoma) is a member of the interferon-inducible class of genes (DeYoung et al. 1997). The AIM2 inflammasome through its HIN200 domain recognizes double-stranded DNA, rich in adenine and thymine that is mislocalized to the cytosol (Fernandes-Alnemri et al. 2009; Hornung et al. 2009). Initial evidence for the involvement of AIM2 in *L. monocytogenes* infection came from observations that *ΔflaA* mutant *L. monocytogenes* could still induce caspase-1 activation in the absence of NLRP3 (Warren et al. 2008; Wu et al. 2010). However, caspase-1 activation was abrogated in Aim2$^{-/-}$ cells, suggesting that the AIM2 inflammasome is responsible for the remaining caspase-1 activation (Wu et al. 2010; Kim et al. 2010). Additionally, Warren et al. (2010) observed that a component of lysed *L. monocytogenes* could activate the AIM2 inflammasome. Similarly, transfected *L. monocytogenes*' genomic DNA could also activate the AIM2 inflammasome (Warren et al. 2010). Using Hoechst-labeled *L. monocytogenes*, Wu et al. (2010) were able to observe colocalization of *L. monocytogenes* with GFP-labeled AIM2, further suggesting that *L. monocytogenes* DNA is able to activate the AIM2 inflammasome.

Finally, Sauer et al. (2010) demonstrated that bacteriolysis within the cytosol resulted in the activation of the AIM2 inflammasome, identifying a likely mechanism for how bacterial DNA accesses the cytosol to be recognized by AIM2. These experiments demonstrated that wild-type *L. monocytogenes* lyses at low levels throughout the course of infection resulting in the recognition of cytosolic DNA by AIM2. Treatment of cytosolically replicating *L. monocytogenes* with the bactericidal β-lactam antibiotic, ampicillin, but not the bacteriostatic antibiotic chloramphenicol, triggered the increased bacteriolysis and subsequently increased inflammasome activation (Sauer et al. 2010). Additionally, a *L. monocytogenes* mutant was identified that was defective for cytosolic survival leading to increased cytosolic bacteriolysis and ultimately increased AIM2-dependent inflammasome activation. Finally, *L. monocytogenes* engineered to lyse specifically within the host cytosol through the expression of the bacteriophage proteins holin and lysin resulted in a significant bacteriolysis and subsequent cell death (Sauer et al. 2010). Importantly, these data implicated cytosolic survival and avoidance of cell autonomous defenses as important mechanisms of avoiding detection by the AIM2 inflammasome. Further, a similar mechanism of cytosolic bacteriolysis was subsequently found to be the mechanism of AIM2 activation following *Francisella tularensis* infection using a similar reporter system (Peng et al. 2011).

2.4 Other NLR Engagement

The inflammasomes mentioned above are the traditionally described inflammasomes as well as the primary inflammasomes that *L. monocytogenes* has been demonstrated to engage. Beyond the receptors described above, other NLRs including NLRC5 and NLRP6 have been studied in regard to *L. monocytogenes* infection. The NLRC5 inflammasome largely regulates MHC class I-associated gene expression (Meissner et al. 2010). Nlrc5$^{-/-}$ mice infected with *L. monocytogenes* have impaired CD8^{+} T-cell responses, both in the number and in the ability to produce the effector cytokine, IFNγ. CD4^{+} T-cell responses are not impaired in these mice, consistent with the role of NLRC5 in regulating MHC I-associated gene expression (Yao et al. 2012; Biswas et al. 2012). NLRC5 has also been shown to associate with NLRP3 and contribute to NLRP3 inflammasome activation (Davis et al. 2011; Yao et al. 2012). As discussed above, *L. monocytogenes* has been shown, in some cases, to activate the NLRP3 inflammasome, although the specific contribution of NLRC5 to this activation will require additional studies. That said, infection of Nlrc5$^{-/-}$ mice with *L. monocytogenes* resulted in partial impairments in caspase-1 processing and IL-1β secretion, suggesting a role for this protein in a more traditional inflammasome-dependent response (Yao et al. 2012). Finally, Nlrc5$^{-/-}$ mice infected with *L. monocytogenes* have higher bacterial burdens (Yao et al. 2012; Biswas et al. 2012), although it is unclear whether this is due to deficient CD8^{+} T-cell responses or to a more canonical inflammasome-dependent infection control mechanism.

NLRP6 has been shown to have roles in altering the gut microbiota to promote gut inflammation and ultimately tumorigenesis (Elinav et al. 2011; Normand et al. 2011; Chen et al. 2011). Interestingly, Nlrp6$^{-/-}$ mice infected orally with *L. monocytogenes* are largely protected from lethal *L. monocytogenes* infection, with 75 % surviving lethal challenge. Further, Nlrp6$^{-/-}$ mice harbor lower bacterial burdens in their spleens and livers and fail to lose weight following infection (Anand et al. 2012). These effects appear to be independent of the gut microbiota composition, but instead depend on increases in circulating monocytes and neutrophils in Nlrp6$^{-/-}$ mice after *L. monocytogenes* infection both systemically in the blood and locally in the peritoneal cavity (Anand et al. 2012). Though NLRP6 seems to have important roles in *L. monocytogenes* infection, this appears to occur independent of canonical inflammasome activation as caspase-1 activation and IL-1β processing are not impaired following *L. monocytogenes* infection in Nlrp6$^{-/-}$ macrophages (Anand et al. 2012). Similarly, in human PBMCs, NLRP6 is dispensable for canonical inflammasome activation by *L. monocytogenes* (Meixenberger et al. 2010). Taken together, these results suggest that though NLRP6 has a role in *L. monocytogenes* pathogenesis, it is likely independent of the effects from inflammasome signaling.

Finally, among the remaining NLRs, a few have been tested for *L. monocytogenes*-dependent phenotypes. HEK293 cells reconstituted with NLRP2 or NLRP12 result in increases in IL-1β production when infected with *L. monocytogenes*, suggesting that *L. monocytogenes* can also engage these inflammasomes, although which components of *L. monocytogenes* are recognized by these receptors remains unknown (Tsuchiya et al. 2010). Conversely, in human PBMCs, NLRP1 and NLRP12 are dispensable for inflammasome activation by *L. monocytogenes* (Meixenberger et al. 2010). These results suggest that *L. monocytogenes* can potentially engage other NLRs; however, the effects of this engagement are unclear and the mechanisms by which they are activated require more in-depth future studies.

In conclusion, although under in vitro conditions *L. monocytogenes* has been found to activate the NLRP3, NLRC4, and AIM2 inflammasomes (Fig. 1), the in vivo relevance of each of inflammasomes is unclear. As discussed in more detail below and as shown in Fig. 1, molecules that engage the NLRP3 and NLRC4 inflammasomes are under the tight control in physiologic settings, thus limiting their ability to be sensed by their respective inflammasomes. Although AIM2 is the most likely relevant inflammasome in vivo, *L. monocytogenes* also has yet undefined ways of avoiding cytosolic bacteriolysis, suggesting that overall, *L. monocytogenes* limits inflammasome activation.

3 Avoidance of Inflammasome Activation

Avoidance of cell death is critical for the virulence of *L. monocytogenes* as mutants that misregulate LLO activity in the cytosol cause necrosis and are highly attenuated in vivo (Glomski et al. 2003). Similarly, *L. monocytogenes* that hyperactivate the inflammasome or induce apoptosis are also attenuated (Sauer et al. 2011a; Warren et al. 2011; Theisen and Sauer, unpublished results). Thus, although *L. monocytogenes* can activate multiple inflammasomes during infection in vitro, it is likely that it largely avoids doing so during infection in vivo to protect its intracellular niche. As a pathogen that utilizes a replication niche rich in inflammasome receptors, it is critical that *L. monocytogenes* has evolved mechanisms to avoid detection or actively inhibit the inflammasome.

3.1 Avoidance of NLRP3

As described above, LLO, through its pore-forming activity, is capable of activating the NLRP3 inflammasome, likely when found in high concentrations extracellularly. However, expression, stability, and activity of LLO are tightly regulated to prevent host cell toxicity, either directly through pore formation and necrosis or potentially through pore-induced NLRP3 activation. LLO expression is both transcriptionally and post-transcriptionally regulated through the activity of the PrfA transcription factor and the 5′ UTR, respectively (Shen and Higgins 2005; Schnupf et al. 2006a). Additionally, LLO in the host cytosol is ubiquitinated and targeted for proteosomal degradation (Schnupf et al. 2006b). Finally, and perhaps most importantly, LLO toxicity is limited at the level of pore-forming activity such that LLO is maximally active only at an acidic pH, a characteristic unique to LLO among the cholesterol-dependent cytolysins (Glomski et al. 2002). Each of these levels of regulation contributes to minimizing the toxicity of LLO to maximize virulence as misregulated LLO can lead to cytotoxicity and result in severe attenuation of virulence (Glomski et al. 2003). Taken together, these studies demonstrate that *L. monocytogenes* carefully regulates the expression of its pore-forming toxin to limit host cell damage and maintain its virulence, in the process likely avoiding a robust activation of the NLRP3 inflammasome under physiologic conditions.

3.2 Avoidance of NLRC4

Similar to the tight regulation of LLO, *L. monocytogenes* has developed exquisite control over expression of its flagellin, regulating the expression through temperature sensing as well as yet to be identified signals found in the host cell cytosol. At the physiologic temperature of 37 °C, *L. monocytogenes* limits the expression of its flagellin through the transcriptional repressor, MogR. However, at temperatures below 37 °C, such as those found in the external environment, *L. monocytogenes* must express flagellin (Peel et al. 1988), and MogR repression must be removed.

Repression by MogR is relieved by the anti-repressor GmaR (Shen et al. 2006). Thus, the balance between these two factors is important in the expression of flagellin. As an added measure to ensure that flagellin is not expressed at 37 °C, there is post-transcriptional regulation of GmaR so that MogR is the dominant component regulating flagellin expression (Kamp and Higgins 2009). Therefore, similar to what is observed for LLO, although flagellin can activate the NLRC4 inflammasome, *L. monocytogenes* tightly limits its expression in the presence of inflammasome components. Whether this level of regulation evolved to facilitate the avoidance of TLR5 detection or inflammasome activation is unclear; however given that the majority of *L. monocytogenes'* lifecycle in the host is in the cytosol, it is enticing to imagine a strong selective pressure exerted by inflammasome detection leading to the tight control of flagellin expression. Indeed, recent evidence suggests that, contrary to previous reports, primary human macrophages utilize a specific splice variant of NLR family, apoptosis inhibiting protein (NAIP) to facilitate the recognition of flagellin through the NLRC4 inflammasome (Kortmann et al. 2015).

3.3 Avoidance of AIM2

In addition to regulating LLO and flagellin to avoid detection by the inflammasome, as a pathogen that makes its life in the cytosol, fidelity during replication and evasion of host defense mechanisms is essential for *L. monocytogenes* to avoid detection by the AIM2 inflammasome. Work by Werner Goebel's laboratory suggests that cytosolic pathogens have developed unique defenses to survive the harsh environment, that is, the host cytosol. Non-cytosolic pathogens that are placed into the cytosol by microinjection fail to replicate and in some cases are killed, whereas cytosol-adapted pathogens survive and thrive (Goetz et al. 2001; Slaghuis et al. 2004). Further, additional studies demonstrated that even intracellular pathogens that mislocalize to the cytosol, such as *ΔsifA* mutant *Salmonella enterica* subsp. Typhimurium, are unable to survive (Beuzón et al. 2000). Finally, *Legionella pneumophila ΔsdhA* mutants mislocalize to the cytosol, subsequently lyse, and activate the inflammasome (Creasey and Isberg 2012; Ge et al. 2012). Taken together, these data suggest that bacteria such as *L. monocytogenes* that are able to survive and replicate within the harsh environment of the cytosol must have evolved mechanisms to deal with these stresses and/or active host defenses. Recent evidence suggests that the loss of these survival adaptations ultimately results in the lysis of *L. monocytogenes* in the cytosol, leading to AIM2 activation and virulence attenuation, further highlighting the importance of inflammasome avoidance in promoting *L. monocytogenes* virulence (Sauer et al. 2010, Daniel Pensinger, Grischa Chen and JD Sauer unpublished data).

The specific adaptations that allow *L. monocytogenes* to survive within the cytosol of cells and avoid AIM2 inflammasome activation are largely unknown. *L monocytogenes Δyvck* mutants lyse in the macrophage cytosol, resulting in hyperactivation of the AIM2 inflammasome and virulence attenuation (Sauer et al. 2010). The function of YvcK, a conserved protein of unknown function, in maintaining cell wall integrity specifically in the cytosol remains unknown, though the loss of *yvck* sensitizes *L. monocytogenes* to lysozyme and numerous cell wall-acting antibiotics in vitro, suggesting that *yvck* plays an integral role in dealing with cell wall stress (Burke et al. 2014, Daniel Pensinger and JD Sauer unpublished data). Similarly, *L. monocytogenes* deficient in either the N-deacetylation (*Δpgd*) or O-acetylation (*Δoat*) of peptidoglycan are sensitive to lysozyme and lyse within the host cytosol (Rae et al. 2011). Finally, Witte et al. (2012) reported a number of *L. monocytogenes* mutants that hyperinduce host cell death and undergo bacteriolysis in the host cell cytosol, though these mutants were not fully characterized as activating the AIM2 inflammasome. However, it is likely given their lysis phenotypes correlating with increased host cell death that these genes are required for the avoidance of host cytosolic defense mechanisms to promote the avoidance of the AIM2 inflammasome. Importantly, maintaining bacterial integrity in the cytosol as a mechanism of avoiding AIM2 detection is a conserved virulence mechanism as Peng et al. (2011) recently demonstrated that a series of *Francisella tularensis* mutants that hyperactivate the inflammasome do so due to cytosolic bacteriolysis. Taken together, these data suggest that maintenance of the cell wall is a critical determinant of *L. monocytogenes* virulence, in part due to the necessity of avoiding AIM2 activation.

3.4 Active Inhibition of the Inflammasome

Although pathogens such as *Yersinia* spp. and *Pseudomonas aeruginosa* express molecules that actively inhibit inflammasome activation (Sutterwala et al. 2007; Brodsky et al. 2010), there is little evidence of *L. monocytogenes* actively inhibiting the inflammasome. Instead, it is presumed that the lack of robust inflammasome activation observed during *L. monocytogenes* infection is due to avoiding detection. These data are supported by the studies demonstrating that *L. monocytogenes* engineered to activate the inflammasome through ectopic expression of flagellin are able to potently activate the inflammasome (Sauer et al. 2011a; Warren et al. 2011). Likewise, *L. monocytogenes*-infected macrophages are still capable of robustly responding to transfected DNA, suggesting that *L. monocytogenes* is not actively inhibiting AIM2 inflammasome activation (JD Sauer unpublished observations). While it is possible, and maybe even likely given its importance in host defense, that *L. monocytogenes* actively inhibits pyroptosis under certain conditions or for the activation of specific inflammasomes, studies to date have not identified such a mechanism. The lone potential example is during in vivo infection where *L. monocytogenes* can activate cholesterol 25-hydroxylase (Ch25h) downstream of liver X receptors (LXRs) through type I INF production (Zou et al. 2011). Ch25h results in the expression of Cd5l, a pro-survival gene that limits the activation of both caspase-3 and caspase-1, ultimately limiting host cell inflammasome activation and pyroptosis. Further, this response seems to be conserved among other downstream targets of LXRs (Zou et al. 2011), suggesting that *L. monocytogenes* may activate this pathway to limit caspase-1 activation and preserve its niche.

4 Role in Pathogenesis

4.1 Role of Caspase-1/11

As described above, *L. monocytogenes* can interact with many different inflammasomes in vitro, culminating in the activation of caspase-1. Initial studies of murine caspase-1 were actually studies of both caspase-1 and caspase-11 double deletions (Kayagaki et al. 2011). Caspase-11 has largely been implicated in Gram-negative infections recognizing the LPS moiety of many of these pathogens (Wang et al. 1998; Kayagaki et al. 2011). *L. monocytogenes* has been found to induce caspase-11 expression upon infection; however, genetic deletion of caspase-11 still allows caspase-1 to function normally (Akhter et al. 2012), suggesting that caspase-11 is dispensable in *L. monocytogenes* infection.

Initial reports suggested that mice deficient in caspase-1/11 had impaired clearance of *L. monocytogenes*, resulting in an increased death rate, particularly at late time points in infection (Tsuji et al. 2004). Mice were rescued with exogenous IL-18 or IFNγ, suggesting that the production of these critically important cytokines is impaired in caspase-1/11$^{-/-}$ mice (Tsuji et al. 2004). While these initial experiments suggest a critical role for caspase-1/11 in controlling *L. monocytogenes* infection, further experiments have failed to find a similar role. Both Sauer et al. (2011a) and Tsuchiya et al. (2014) observed no increased susceptibility to *L. monocytogenes* in caspase-1/11$^{-/-}$ mice at 2 days post infection. Similarly, there was no increased susceptibility to *L. monocytogenes* in zebrafish lacking caspase-1 (Vincent et al. 2015). These studies are consistent with the tight levels of regulation observed with potential inflammasome agonists as well as the critical nature of host cell survival that has previously been reported to promote *L. monocytogenes* virulence. The differences between the initial characterization of caspase-1 and more recent studies likely stem from the time points examined, the background of mice used in the initial studies (C57Bl/6 J × 129SV/J), as well as the differences in the strains of *L. monocytogenes* used for each of the studies. Taken together, the more recent studies suggest that caspase-1 is dispensable for the control of *L. monocytogenes* early during infection, likely due to *L. monocytogenes* avoiding the activation of the inflammasome and maintaining its intracellular niche. However, it is possible that later in infection when bacterial burdens are high, the inflammasome plays a role in host protection. Importantly, as briefly discussed above and in more detail below, under conditions in which *L. monocytogenes* does activate the inflammasome, due to ectopic flagellin expression or mutations that result in detection by AIM2, caspase-1 activation and the subsequent pyroptosis severely attenuate virulence, further highlighting the importance of inflammasome evasion for *L. monocytogenes* pathogenesis.

4.2 Role of ASC

ASC is an adaptor molecule required for cytokine secretion and pyroptosis downstream of both the NLRP3 and AIM2 inflammasomes (Broz et al. 2010). In the case of CARD-containing inflammasomes, such as the NLRC4 inflammasome, ASC is only required for cytokine secretion and is dispensable for pyroptosis (Broz et al. 2010). Thus, as discussed above, the majority of inflammasome activation by *L. monocytogenes* requires signaling through ASC for both pyroptosis and cytokine secretion (Fig. 1). As such, ASC$^{-/-}$ peritoneal macrophages fail to activate caspase-1 or secrete IL-1β and IL-18 (Ozören et al. 2006). Consistent with observations that the loss of caspase-1 has no effect on *L. monocytogenes* virulence in vivo (Sauer et al. 2011a; Tsuchiya et al. 2014), the loss of ASC results in similar susceptibility to *L. monocytogenes* (Sauer et al. 2011a). Interestingly, recent work by Tsuchiya suggests that ASC deficiency may in some cases result in enhanced host protection during *L. monocytogenes* infection (Tsuchiya et al. 2014). ASC$^{-/-}$ mice are able to withstand a bacterial load of 1×10^6 colony forming units (CFU) of *L. monocytogenes*, a log higher than the LD_{50} for wild-type C57Bl/6 mice, and harbor lower bacteria burdens in their livers. The investigators report that ASC promotes the production of the anti-inflammatory cytokine IL-10 and that protection in ASC-deficient mice is likely due to reduced IL-10 production (Tsuchiya et al. 2014). Taken together, these results suggest that similar to the loss of caspase-1, the loss of ASC does not promote the virulence of *L. monocytogenes*, and in fact may result in enhanced protection, although this phenotype may be independent of ASC's role in caspase-1 activation (Tsuchiya et al. 2014).

4.3 Role of IL-1β and IL-18

IL-1β and IL-18 have been examined in the context of *L. monocytogenes* infection well before these cytokines were known to be downstream of the inflammasome. Initial reports of IL-1 receptor-deficient (IL-1R1$^{-/-}$) mice that lack signaling of IL-1α and IL-1β suggested that these mice are more susceptible to *L. monocytogenes* (Labow et al. 1997); however, this was done in a mixed background (B6/129) and is likely confounded by the fact that 129 mice are more susceptible to *L. monocytogenes* at baseline (Gahan and Collins 1995). Further examination on a clean C57Bl/6 background suggests that there is no alteration in acute virulence in the absence of IL-1 signaling (Glaccum et al. 1997). Similarly, mice deficient just in IL-1β show no differences in susceptibility (Zheng et al. 1995), suggesting that IL-1 signaling is dispensable during initial infection.

Priming of cells via signal one, largely through NF-κB signaling, is critical in up-regulating IL-1β transcripts (Latz et al. 2013). In TLR2$^{-/-}$ macrophages, IL-1β secretion is impaired following *L. monocytogenes* infection likely through transcriptional impairment (Ozören et al. 2006). Further, the loss of tumor progression locus 2 (TPL2), a factor critical in orchestrating cytokine responses downstream of TLR signaling, impairs IL-1β secretion following *L. monocytogenes* infection through impairments of IL-1β transcript up-regulation (Mielke et al. 2009). Interestingly, the loss of TPL2 increases the susceptibility of mice to *L. monocytogenes*, though the regulation of other key innate cytokines such as TNF-α and IFN-γ is similar regardless of whether TPL2 is present or not (Mielke et al. 2009). Even though TPL2 influences the levels of IL-1β, the increased susceptibility of TPL2$^{-/-}$ mice does not recapitulate the phenotype of IL-1R1$^{-/-}$ mice; thus, the increased susceptibility of TPL2$^{-/-}$ mice is not likely due to the changes in IL-1β but instead to some other undescribed effect of loss of TPL2.

Characterization of the role of IL-18, particularly in acute infection, has also resulted in conflicting reports. Initial studies suggested that IL-18 is critical in acute infection, and the lack of IL-18 leads to an increased susceptibility (Neighbors et al. 2001). IL-18 is critical in stimulating IFN-γ, a crucial cytokine for controlling *L. monocytogenes* infection, and thus, decreases in IL-18 production result in less IFN-γ production, leading mice to succumb to infection. While these observations were made in initial reports, the authors also observed decreases in TNF-α production, another critical cytokine in innate susceptibility, and suggest that the corresponding lack of TNF-α from IL-18-deprived mice is largely what is responsible for susceptibility to infection. More recent studies in both IL-18$^{-/-}$ and

IL-18Rα$^{-/-}$ contradict these earlier results and suggest that IL-18$^{-/-}$ mice are actually more resistant to *L. monocytogenes*, particularly at high doses (Lochner et al. 2008; Tsuchiya et al. 2014). These results are in line with reports that suggest that ASC$^{-/-}$ mice, which also lack IL-18 secretion, are less susceptible to *L. monocytogenes* (Tsuchiya et al. 2014). The reason for this decreased susceptibility could stem from an increased influx of leukocytes in the initial 72 h postinfection that are resistant to apoptosis (Lochner et al. 2008).

While most IL-1β and IL-18 signaling is believed to occur downstream of the inflammasome, there have been caspase-1-independent observations of IL-1β and IL-18 secretion (Uchiyama et al. 2013; Tsuchiya et al. 2014). One such mechanism relies on Fas (CD95/Apo-1) signaling. Peritoneal exudate cells (PECs) deficient in Fas secreted less IL-1β and IL-18 after *L. monocytogenes* infection (Uchiyama et al. 2013), suggesting that Fas is required for cytokine secretion. Interestingly, this mechanism is caspase-1, caspase-11, NLRP3, and NLRC4 independent, while it is dependent on ASC (Uchiyama et al. 2013). In line with the reports that suggest IL-1β maturation can occur through caspase-8 activation (Gringhuis et al. 2012; Bossaller et al. 2012), *L. monocytogenes* can activate caspase-8 in a Fas-dependent manner, suggesting that in this instance, IL-1β and IL-18 could be processed through caspase-8 and not caspase-1. Additionally, IL-18 can be processed to its mature form through neutrophil serine proteases (Sugawara et al. 2001). In caspase-1/11$^{-/-}$ mice infected with lethal doses of *L. monocytogenes*, elevated neutrophil serine protease activity was observed that correlated with the increased IL-18 production (Tsuchiya et al. 2014). Taken together, while not all studies are consistent, and differences in mouse backgrounds, *L. monocytogenes* strains, doses, and infection kinetics could account for discrepancies, the preponderance of data suggests that the inflammatory cytokines downstream of inflammasome activation are dispensable for host defense during *L. monocytogenes* infection.

4.4 Innate Immune Cell Infiltrate

During murine *L. monocytogenes* infection, there is a robust, initial infiltrate of myelomonocytic cells (MMCs) consisting of neutrophils and Ly6C^{hi} monocytes into the splenic white pulp (Waite et al. 2011; Williams et al. 2013). Because *L. monocytogenes* largely avoids activating the inflammasome, much of the work examining the host response to pyroptosis has used an engineered strain of *L. monocytogenes* that ectopically expresses flagellin resulting in hyperactivation of the inflammasome and attenuation (Sauer et al. 2011a; Warren et al. 2011; Williams et al. 2013; Vincent et al. 2015). Infection with inflammasome-activating *L. monocytogenes* resulted in an earlier MMC infiltrate that penetrates into the deep T-cell zones of the spleen; however, this infiltrate is less robust and clears earlier than with wild-type infection, corresponding to a drop in burden (Williams et al. 2013). Similar early innate immune cell infiltrates consisting of predominantly macrophages were observed in zebrafish infected with inflammasome-activating

L. monocytogenes ultimately resulting in attenuation (Fig. 2) (Vincent et al. 2015). Previous work with *Salmonella enterica* subsp. Typhimurium similarly engineered to robustly activate the inflammasome suggests that pyroptosis expels bacteria for uptake by bactericidal neutrophils ultimately resulting in phagocyte oxidase-dependent attenuation (Miao et al. 2010a). However, neutrophil depletion does not specifically rescue virulence defects following the activation of pyroptosis by *L. monocytogenes* (Sauer et al. 2011a; Vincent et al. 2015). Instead, depletion of macrophages rescues the virulence defect of inflammasome-activating *L. monocytogenes* (Vincent et al. 2015). Why the attenuation of *L. monocytogenes* is independent of neutrophils while *S. typhimurium* is dependent on neutrophils is unclear, although it is important to note that the depletion of neutrophils was not directly assessed in the *Salmonella* studies; rather, the loss of phagocyte oxidase implicated the role for neutrophils (Miao et al. 2010a). In wild-type *L. monocytogenes* infection, neutrophils are dispensable for defense, while inflammatory monocytes are critical (Shi et al. 2011), suggesting that attenuation following inflammasome activation likely depends on the cell type critical in clearing the pathogen normally.

5 Adaptive Immune Response to *L. monocytogenes*

The specific mechanisms by which *L. monocytogenes* infection triggers robust $CD8^+$ T cell-dependent cell-mediated immune responses are largely unknown. Generally, the initial innate response is posited to inform the type and robustness of the adaptive immune response (Janeway 1989). It was hypothesized that activation of the inflammasome and its resulting robust innate inflammatory response would contribute to mounting a better adaptive immune response. As discussed below, surprising results suggest that activation of the inflammasome is detrimental for the activation of *L. monocytogenes*-triggered adaptive immunity (Sauer et al. 2011a). Importantly, understanding how $CD8^+$ T-cell responses form following *L. monocytogenes* infection and how we can improve them is critical in the development of *L. monocytogenes* as a cancer immunotherapeutic platform.

5.1 Protective Immunity

To understand how the inflammasome impacts the generation of adaptive immunity, multiple groups independently generated strains of *L. monocytogenes* that hyperactivate the inflammasome (Sauer et al. 2011a; Warren et al. 2011). Both groups utilized expression systems whereby flagellin expression was activated exclusively in the cytosol of infected macrophages, limiting flagellin detection to the NAIP5/NLRC4 inflammasome while limiting the potential impact of TLR5-mediated flagellin recognition. Warren et al. (2011) observed that mice immunized with flagellin-expressing *L. monocytogenes* could survive a subsequent lethal challenge, suggesting that the activation of the inflammasome allows a protective immune response to form. While all mice immunized with a high dose of inflammasome-activating bacteria survived the subsequent lethal dose challenge, no wild-type control mice were analyzed as a comparison, and neither T-cell responses nor bacterial burdens after lethal challenge were examined to determine whether the response was better or worse than that activated by non-inflammasome-activating *L. monocytogenes*.

In contrast, similar immunization studies with inflammasome-activating *L. monocytogenes* found that adaptive immune responses were impaired. While immunization of mice with high doses of attenuated inflammasome-activating or nonactivating *L. monocytogenes* both conferred some degree of protection from the subsequent lethal challenge, mice were significantly less protected when immunized with strains that activated the inflammasome. More strikingly, when immunized at low doses, inflammasome-activating strains were unable to confer any protection above mock-immunized mice, whereas non-inflammasome-activating immunizations led to a significant protection from the subsequent lethal challenge (Sauer et al. 2011a). Similarly, immunization with inflammasome-activating *L. monocytogenes* resulted in an impaired antigen-specific primary $CD8^+$ T-cell response, with about half as many antigen-specific $CD8^+$ T cells forming compared to immunization with non-inflammasome-activating strains (Sauer et al. 2011a). Further, these $CD8^+$ T-cell deficits persist into the memory stage (Sauer et al. 2011a), and levels of the recall $CD8^+$ T cells at 2 days after challenge are also impaired (Theisen and Sauer, unpublished results). Interestingly, the recall response examined at 5 days post challenge results in similar levels of expanded antigen-specific $CD8^+$ T cells (Sauer et al. 2011a), suggesting that the rate of expansion upon recall is slower and likely contributes to the differences seen in bacterial burdens upon challenge. Finally, caspase-1 mice immunized with *L. monocytogenes* mount more robust protective immune responses compared to the wild-type (Sauer et al. 2011a, JD Sauer unpublished data). Taken together, these data suggest that at high immunizing doses, some protection can be conferred from *L. monocytogenes* strains that activate the inflammasome, but that this response is impaired compared to strains that avoid inflammasome activation, and at lower immunizing doses, T-cell activation is insufficient to confer the protection from a subsequent lethal challenge (Fig. 3).

How inflammasome activation inhibits *L. monocytogenes*-stimulated immunity remains unclear. Inflammasome-activating strains of *L. monocytogenes* are attenuated in vivo (Sauer et al. 2011a; Warren et al. 2011; Vincent et al. 2015); however, some of the adaptive immunity studies have been done with *Δact L. monocytogenes* to normalize the differences in bacterial burden (Sauer et al. 2011a; Williams et al. 2013). Further, the depletion of MMCs, which partially rescues inflammasome-activating *L. monocytogenes* virulence defects, does not affect the defects in the development of $CD8^+$ T cells or protective immunity (Fig. 3) (Williams et al. 2013), suggesting that deficits in adaptive immunity are independent of virulence defects and are consistent with evidence that immunity is independent of bacterial burden (Mercado et al. 2000). Interestingly, inflammasome activation is required

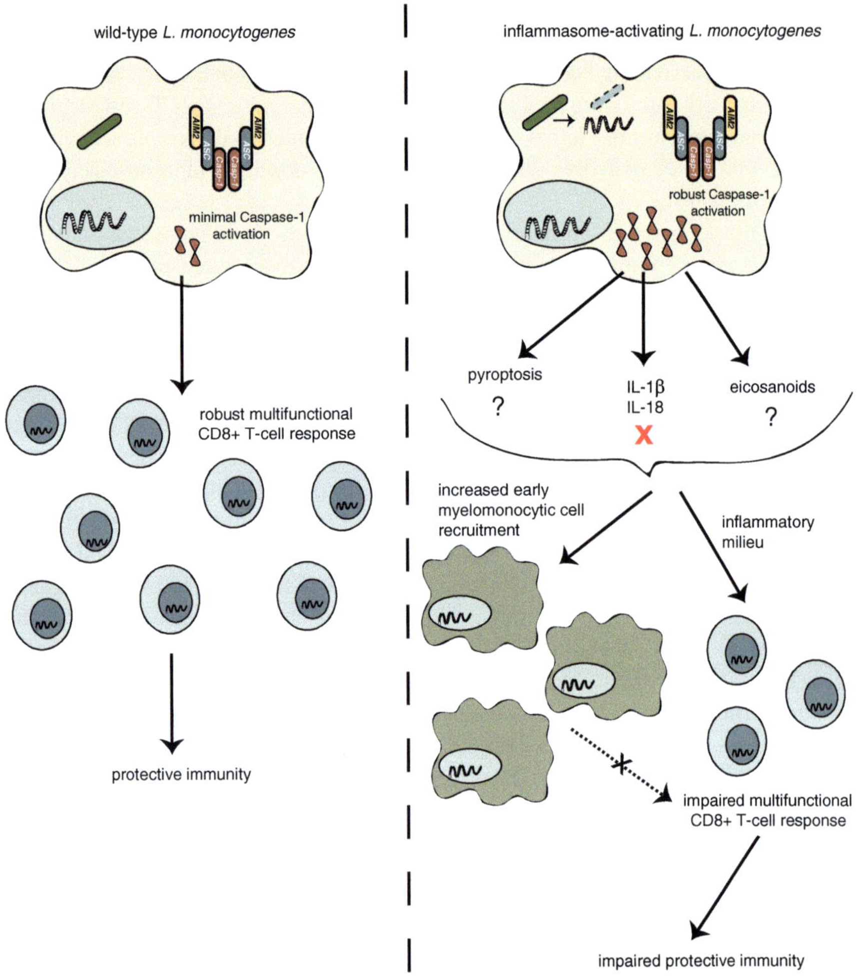

Fig. 3 Inflammasome-associated inflammation impairs host cell-mediated immune responses to *L. monocytogenes*. In wild-type *L. monocytogenes* infection, *L. monocytogenes* minimally activates caspase-1 and, through a not fully understood mechanism, induces a robust multifunctional CD8$^+$ T-cell response. In contrast, activation of the inflammasome by *L. monocytogenes* results in robust caspase-1 activation and impairment of multifunctional CD8$^+$ T-cell generation through the inflammasome-associated inflammatory milieu, independent of IL-1β and IL-18. Inflammasome activation also results in a robust, early recruitment of myelomonocytic cells that while attenuating *L. monocytogenes* do not impair CD8$^+$ T-cell generation

for T-cell responses to a variety of other pathogens, including *Candida albicans, Schistosoma mansoni*, and *Bordetella pertussis* (Dunne et al. 2010; Ritter et al. 2010; van de Veerdonk et al. 2011). However, these responses largely require CD4+ T cell-dependent Th1 or Th17 responses, not the CD8$^+$ T-cell response that

L. monocytogenes induces, suggesting that inflammasome activation may have different roles in regulating the adaptive immune response pending the type of response required for pathogen control.

5.2 *Influence of Cytokines*

Preliminary data suggest that inflammasome-dependent inhibition of immunity is largely mediated by the inflammatory milieu and not deficits in antigen presentation or costimulation. However, the specific inflammatory mediators that inhibit the generation of robust cell-mediated immunity remain unknown (Erin Theisen and JD Sauer, unpublished data). We and others have examined how deficiency of either IL-1β or IL-18 influences the adaptive immune response. In our hands, the loss of IL-1β, IL-18, or both still results in impairments in the protective immune response following the immunization with inflammasome-activating *L. monocytogenes*, suggesting that these cytokines do not negatively influence the development of adaptive immunity. Further, IL-$18^{-/-}$ or IL-$18R\alpha^{-/-}$ mice do not have impairments in generating antigen-specific $CD8^+$ T cells that have effector functions (Lochner et al. 2008; Haring and Harty 2009). Similarly, the loss of IL-18Rα does not impact the rate of contraction or formation or maintenance of memory T cells (Haring and Harty 2009). Interestingly, it seems that the loss of IL-18 does result in fewer $CD4^+$ T cells; however, this ultimately does not impact the generation of protective immunity as IL-$18^{-/-}$ or IL-$18R\alpha^{-/-}$ mice harbor similar burdens of *L. monocytogenes* following the lethal challenge (Lochner et al. 2008; Haring and Harty 2009).

Taken together, these results suggest that inflammasome activation is detrimental to the influence of adaptive immunity; however, the reason for this inhibition remains largely unknown. We hypothesize that an undescribed inflammatory component is responsible for these defects.

6 Concluding Remarks

In this chapter, we have discussed the mechanisms by which *L. monocytogenes* can activate the inflammasome. However, in vivo data suggest that under physiologic conditions, *L. monocytogenes* largely avoids activating the inflammasome as evidenced by multiple studies that demonstrate no change in virulence in caspase-1-deficient mice and attenuation when *L. monocytogenes* is engineered to robustly activate the inflammasome (Sauer et al. 2011a; Warren et al. 2011; Vincent et al. 2015).

Attenuated *L. monocytogenes* is currently being developed as a cancer immunotherapeutic platform, and early returns in multiple phase I and II clinical trials have been exceptionally positive (Le et al. 2015). How *L. monocytogenes* is able to mount a $CD8^+$ T-cell response that is able to break self-tolerance remains unknown. It was initially hypothesized that inflammasome activation and its associated inflammation would positively influence cell-mediated immunity; however, inflammasome activation ultimately impairs the adaptive immune response generated to *L. monocytogenes* (Sauer et al. 2011a; Williams et al. 2013). It is possible that although the inflammasome is detrimental in the context of strong non-self-antigens, it could be beneficial in the context of less robust or self-antigens. Studies to assess this possibility are currently underway.

While *L. monocytogenes* largely avoids inflammasome activation, immunization of caspase-1/$11^{-/-}$ mice results in better protective immune responses than wild-type mice, suggesting that even low-level inflammasome activation is detrimental to the development of adaptive immunity (Sauer et al. 2011a). Additionally, current platforms of *L. monocytogenes* immunotherapy are limited in the complexity of antigens they can produce, and for the production of more complex antigens, the host must be able to transcribe and translate the antigens—similar to the idea of immunization with plasmid DNA. Platforms of *L. monocytogenes* designed to release plasmid DNA encoding complex tumor antigens through engineered bacteriolysis or antibiotic-mediated lysis (van Pijkeren et al. 2010; Sauer et al. 2010) are severely inhibited due to inflammasome activation. Thus, understanding and modulating inflammasome activation by *L. monocytogenes* is critical for its success as an immunotherapeutic platform.

Acknowledgments The authors would like to thank Dr. Laurie Ristow and Grischa Chen for critical reading of this manuscript. The authors would also like to thank other members of the Sauer laboratory for commentary on the figures. This work is funded by the NIH (R01 CA188034) to J-D.S., and an AAI Careers in Immunology Fellowship to E.T.

References

Akhter A, Caution K, Abu Khweek A et al (2012) Caspase-11 promotes the fusion of phagosomes harboring pathogenic bacteria with lysosomes by modulating actin polymerization. Immunity 37:35–47. doi:10.1016/j.immuni.2012.05.001

Anand PK, Malireddi RKS, Lukens JR et al (2012) NLRP6 negatively regulates innate immunity and host defence against bacterial pathogens. Nature 488:389–393. doi:10.1038/nature11250

Auerbuch V, Brockstedt DG, Meyer-Morse N et al (2004) Mice lacking the type I interferon receptor are resistant to *Listeria monocytogenes*. J Exp Med 200:527–533. doi:10.1084/jem.20040976

Bahjat KS, Meyer-Morse N, Lemmens EE et al (2009) Suppression of cell-mediated immunity following recognition of phagosome-confined bacteria. PLoS Pathog. doi:10.1371/journal.ppat.1000568

Beuzón CR, Méresse S, Unsworth KE et al (2000) Salmonella maintains the integrity of its intracellular vacuole through the action of SifA. EMBO J 19:3235–3249. doi:10.1093/emboj/19.13.3235

Biswas A, Meissner TB, Kawai T, Kobayashi KS (2012) Cutting edge: impaired MHC class I expression in mice deficient for Nlrc5/class I transactivator. J Immunol 189:516–520. doi:10.4049/jimmunol.1200064

Boneca IG, Dussurget O, Cabanes D et al (2007) A critical role for peptidoglycan N-deacetylation in *Listeria* evasion from the host innate immune system. Proc Natl Acad Sci USA 104:997–1002. doi:10.1073/pnas.0609672104

Bossaller L, Chiang P-I, Schmidt-Lauber C et al (2012) Cutting edge: FAS (CD95) mediates noncanonical IL-1β and IL-18 maturation via caspase-8 in an RIP3-independent manner. J Immunol 189:5508–5512. doi:10.4049/jimmunol.1202121

Brodsky IE, Palm NW, Sadanand S et al (2010) A *Yersinia* effector protein promotes virulence by preventing inflammasome recognition of the type III secretion system. Cell Host Microbe 7:376–387. doi:10.1016/j.chom.2010.04.009

Broz P, von Moltke J, Jones JW et al (2010) Differential requirement for caspase-1 autoproteolysis in pathogen-induced cell death and cytokine processing. Cell Host Microbe 8:471–483. doi:10.1016/j.chom.2010.11.007

Burke TP, Loukitcheva A, Zemansky J, et al (2014) Listeria monocytogenes is resistant to lysozyme through the regulation, not the acquisition, of cell wall-modifying enzymes. J Bacteriol 196:3756–3767. doi: 10.1128/JB.02053-14

Camilli A, Goldfine H, Portnoy DA (1991) *Listeria monocytogenes* mutants lacking phosphatidylinositol-specific phospholipase C are avirulent. J Exp Med 173:751–754

Carrero JA, Calderon B, Unanue ER (2006) Lymphocytes are detrimental during the early innate immune response against *Listeria monocytogenes*. J Exp Med 203:933–940. doi:10.1084/jem.20060045

Chen GY, Liu M, Wang F et al (2011) A functional role for Nlrp6 in intestinal inflammation and tumorigenesis. J Immunol 186:7187–7194. doi:10.4049/jimmunol.1100412

Creasey EA, Isberg RR (2012) The protein SdhA maintains the integrity of the *Legionella*-containing vacuole. Proc Natl Acad Sci USA 109:3481–3486. doi:10.1073/pnas.1121286109

Davis BK, Roberts RA, Huang MT et al (2011) Cutting edge: NLRC5-dependent activation of the inflammasome. J Immunol 186:1333–1337. doi:10.4049/jimmunol.1003111

DeYoung KL, Ray ME, Su YA et al (1997) Cloning a novel member of the human interferon-inducible gene family associated with control of tumorigenicity in a model of human melanoma. Oncogene 15:453–457. doi:10.1038/sj.onc.1201206

Dunne A, Ross PJ, Pospisilova E et al (2010) Inflammasome activation by adenylate cyclase toxin directs Th17 responses and protection against *Bordetella pertussis*. J Immunol 185:1711–1719. doi:10.4049/jimmunol.1000105

Elinav E, Strowig T, Kau AL et al (2011) NLRP6 inflammasome regulates colonic microbial ecology and risk for colitis. Cell 145:745–757. doi:10.1016/j.cell.2011.04.022

Fernandes-Alnemri T, Yu J-W, Datta P et al (2009) AIM2 activates the inflammasome and cell death in response to cytoplasmic DNA. Nature 458:509–513. doi:10.1038/nature07710

Fernandes-Alnemri T, Kang S, Anderson C et al (2013) Cutting edge: TLR signaling licenses IRAK1 for rapid activation of the NLRP3 inflammasome. J Immunol 191:3995–3999. doi:10.4049/jimmunol.1301681

Ferreira V, Wiedmann M, Teixeira P, Stasiewicz MJ (2014) *Listeria monocytogenes* persistence in food-associated environments: epidemiology, strain characteristics, and implications for public health. J Food Prot 77:150–170. doi:10.4315/0362-028X.JFP-13-150

Franchi L, Kanneganti T-D, Dubyak GR, Núñez G (2007) Differential requirement of P2X7 receptor and intracellular K+ for caspase-1 activation induced by intracellular and extracellular bacteria. J Biol Chem 282:18810–18818. doi:10.1074/jbc.M610762200

Gahan CG, Collins JK (1995) Non-dystrophic 129 REJ mice are susceptible to i.p. infection with *Listeria monocytogenes* despite an ability to recruit inflammatory neutrophils to the peritoneal cavity. Microb Pathog 18:355–364. doi:10.1006/mpat.1995.0032

Ge J, Gong Y-N, Xu Y, Shao F (2012) Preventing bacterial DNA release and absent in melanoma 2 inflammasome activation by a Legionella effector functioning in membrane trafficking. Proc Natl Acad Sci USA 109:6193–6198. doi:10.1073/pnas.1117490109

Glaccum MB, Stocking KL, Charrier K et al (1997) Phenotypic and functional characterization of mice that lack the type I receptor for IL-1. J Immunol 159:3364–3371

Glomski IJ, Gedde MM, Tsang AW et al (2002) The *Listeria monocytogenes* hemolysin has an acidic pH optimum to compartmentalize activity and prevent damage to infected host cells. J Cell Biol 156:1029–1038. doi:10.1083/jcb.200201081

Glomski IJ, Decatur AL, Portnoy DA (2003) *Listeria monocytogenes* mutants that fail to compartmentalize Listerolysin O activity are cytotoxic, avirulent, and unable to evade host extracellular defenses. Infect Immun 71:6754–6765. doi:10.1128/IAI.71.12.6754-6765.2003

Goetz M, Bubert A, Wang G et al (2001) Microinjection and growth of bacteria in the cytosol of mammalian host cells. Proc Natl Acad Sci USA 98:12221–12226. doi:10.1073/pnas.211106398

Goossens PL, Milon G (1992) Induction of protective CD8+ T lymphocytes by an attenuated *Listeria monocytogenes* actA mutant. Int Immunol 4:1413–1418

Gringhuis SI, Kaptein TM, Wevers BA et al (2012) Dectin-1 is an extracellular pathogen sensor for the induction and processing of IL-1β via a noncanonical caspase-8 inflammasome. Nat Immunol 13:246–254. doi:10.1038/ni.2222

Hamon MA, Cossart P (2011) K+ efflux is required for histone H3 dephosphorylation by *Listeria monocytogenes* listeriolysin O and other pore-forming toxins. Infect Immun 79:2839–2846. doi:10.1128/IAI.01243-10

Hamon MA, Ribet D, Stavru F, Cossart P (2012) Listeriolysin O: the Swiss army knife of Listeria. Trends Microbiol 20:360–368. doi:10.1016/j.tim.2012.04.006

Hara H, Tsuchiya K, Nomura T et al (2008) Dependency of caspase-1 activation induced in macrophages by *Listeria monocytogenes* on cytolysin, listeriolysin O, after evasion from phagosome into the cytoplasm. J Immunol 180:7859–7868

Haring JS, Harty JT (2009) Interleukin-18-related genes are induced during the contraction phase but do not play major roles in regulating the dynamics or function of the T-cell response to *Listeria monocytogenes* infection. Infect Immun 77:1894–1903. doi:10.1128/IAI.01315-08

Hornung V, Ablasser A, Charrel-Dennis M et al (2009) AIM2 recognizes cytosolic dsDNA and forms a caspase-1-activating inflammasome with ASC. Nature 458:514–518. doi:10.1038/nature07725

Janeway CA (1989) Approaching the asymptote? Evolution and revolution in immunology. Cold Spring Harb Symp Quant Biol 54(Pt 1):1–13

Kamp HD, Higgins DE (2009) Transcriptional and post-transcriptional regulation of the GmaR antirepressor governs temperature-dependent control of flagellar motility in *Listeria monocytogenes*. Mol Microbiol 74:421–435. doi:10.1111/j.1365-2958.2009.06874.x

Kanneganti T-D, Ozören N, Body-Malapel M et al (2006) Bacterial RNA and small antiviral compounds activate caspase-1 through cryopyrin/Nalp3. Nature 440:233–236. doi:10.1038/nature04517

Kayagaki N, Warming S, Lamkanfi M et al (2011) Non-canonical inflammasome activation targets caspase-11. Nature 479:117–121. doi:10.1038/nature10558

Kayagaki N, Stowe IB, Lee BL et al (2015) Caspase-11 cleaves gasdermin D for non-canonical inflammasome signalling. Nature 526:666–671. doi:10.1038/nature15541

Kim S, Bauernfeind F, Ablasser A et al (2010) *Listeria monocytogenes* is sensed by the NLRP3 and AIM2 inflammasome. Eur J Immunol 40:1545–1551. doi:10.1002/eji.201040425

Kocks C, Gouin E, Tabouret M et al (1992) *L. monocytogenes*-induced actin assembly requires the actA gene product, a surface protein. Cell 68:521–531

Kofoed EM, Vance RE (2011) Innate immune recognition of bacterial ligands by NAIPs determines inflammasome specificity. Nature 477:592–595. doi:10.1038/nature10394

Kortmann J, Brubaker SW, Monack DM (2015) Cutting edge: inflammasome activation in primary human macrophages is dependent on flagellin. J Immunol 195:815–819. doi:10.4049/jimmunol.1403100

Labow M, Shuster D, Zetterstrom M et al (1997) Absence of IL-1 signaling and reduced inflammatory response in IL-1 type I receptor-deficient mice. J Immunol 159:2452–2461

Latz E, Xiao TS, Stutz A (2013) Activation and regulation of the inflammasomes. Nat Rev Immunol 13:397–411. doi:10.1038/nri3452

Le DT, Dubensky TW, Brockstedt DG (2012) Clinical development of *Listeria monocytogenes*-based immunotherapies. Semin Oncol 39:311–322. doi:10.1053/j.seminoncol.2012.02.008
Le DT, Wang-Gillam A, Picozzi V et al (2015) Safety and survival with GVAX pancreas prime and *Listeria monocytogenes*-expressing mesothelin (CRS-207) boost vaccines for metastatic pancreatic cancer. J Clin Oncol 33:1325–1333. doi:10.1200/JCO.2014.57.4244
Leber JH, Crimmins GT, Raghavan S et al (2008) Distinct TLR- and NLR-mediated transcriptional responses to an intracellular pathogen. PLoS Pathog 4:e6. doi:10.1371/journal.ppat.0040006
Lin K-M, Hu W, Troutman TD et al (2014) IRAK-1 bypasses priming and directly links TLRs to rapid NLRP3 inflammasome activation. Proc Natl Acad Sci USA 111:775–780. doi:10.1073/pnas.1320294111
Lochner M, Kastenmüller K, Neuenhahn M et al (2008) Decreased susceptibility of mice to infection with *Listeria monocytogenes* in the absence of interleukin-18. Infect Immun 76:3881–3890. doi:10.1128/IAI.01651-07
Mariathasan S, Weiss DS, Newton K et al (2006) Cryopyrin activates the inflammasome in response to toxins and ATP. Nature 440:228–232. doi:10.1038/nature04515
Meissner TB, Li A, Biswas A et al (2010) NLR family member NLRC5 is a transcriptional regulator of MHC class I genes. Proc Natl Acad Sci USA 107:13794–13799. doi:10.1073/pnas.1008684107
Meixenberger K, Pache F, Eitel J et al (2010) *Listeria monocytogenes*-infected human peripheral blood mononuclear cells produce IL-1beta, depending on listeriolysin O and NLRP3. J Immunol 184:922–930. doi:10.4049/jimmunol.0901346
Mengaud J, Braun-Breton C, Cossart P (1991) Identification of phosphatidylinositol-specific phospholipase C activity in *Listeria monocytogenes*: a novel type of virulence factor? Mol Microbiol 5:367–372
Mengaud J, Ohayon H, Gounon P et al (1996) E-cadherin is the receptor for internalin, a surface protein required for entry of *L. monocytogenes* into epithelial cells. Cell 84:923–932
Mercado R, Vijh S, Allen SE et al (2000) Early programming of T cell populations responding to bacterial infection. J Immunol 165:6833–6839
Miao EA, Leaf IA, Treuting PM et al (2010a) Caspase-1-induced pyroptosis is an innate immune effector mechanism against intracellular bacteria. Nat Immunol 11:1136–1142. doi:10.1038/ni.1960
Miao EA, Mao DP, Yudkovsky N et al (2010b) Innate immune detection of the type III secretion apparatus through the NLRC4 inflammasome. Proc Natl Acad Sci USA 107:3076–3080. doi:10.1073/pnas.0913087107
Mielke LA, Elkins KL, Wei L et al (2009) Tumor progression locus 2 (Map3k8) is critical for host defense against *Listeria monocytogenes* and IL-1 beta production. J Immunol 183:7984–7993. doi:10.4049/jimmunol.0901336
Neighbors M, Xu X, Barrat FJ et al (2001) A critical role for interleukin 18 in primary and memory effector responses to *Listeria monocytogenes* that extends beyond its effects on interferon production. J Exp Med 194:343–354
Normand S, Delanoye-Crespin A, Bressenot A et al (2011) Nod-like receptor pyrin domain-containing protein 6 (NLRP6) controls epithelial self-renewal and colorectal carcinogenesis upon injury. Proc Natl Acad Sci USA 108:9601–9606. doi:10.1073/pnas.1100981108
O'Connell RM, Saha SK, Vaidya SA et al (2004) Type I interferon production enhances susceptibility to *Listeria monocytogenes* infection. J Exp Med 200:437–445. doi:10.1084/jem.20040712
O'Riordan M, Yi CH, Gonzales R et al (2002) Innate recognition of bacteria by a macrophage cytosolic surveillance pathway. Proc Natl Acad Sci 99:13861–13866. doi:10.1073/pnas.202476699
Opitz B, Püschel A, Beermann W et al (2006) *Listeria monocytogenes* activated p38 MAPK and induced IL-8 secretion in a nucleotide-binding oligomerization domain 1-dependent manner in endothelial cells. J Immunol 176:484–490

Ozören N, Masumoto J, Franchi L et al (2006) Distinct roles of TLR2 and the adaptor ASC in IL-1beta/IL-18 secretion in response to *Listeria monocytogenes*. J Immunol 176:4337–4342

Peel M, Donachie W, Shaw A (1988) Temperature-dependent expression of flagella of *Listeria monocytogenes* studied by electron microscopy, SDS-PAGE and western blotting. J Gen Microbiol 134:2171–2178. doi:10.1099/00221287-134-8-2171

Peng K, Broz P, Jones J et al (2011) Elevated AIM2-mediated pyroptosis triggered by hypercytotoxic *Francisella* mutant strains is attributed to increased intracellular bacteriolysis. Cell Microbiol 13:1586–1600. doi:10.1111/j.1462-5822.2011.01643.x

Personnic N, Bruck S, Nahori M-A et al (2010) The stress-induced virulence protein InlH controls interleukin-6 production during murine listeriosis. Infect Immun 78:1979–1989. doi:10.1128/IAI.01096-09

Portnoy DA, Jacks PS, Hinrichs DJ (1988) Role of hemolysin for the intracellular growth of *Listeria monocytogenes*. J Exp Med 167:1459–1471

Rae CS, Geissler A, Adamson PC, Portnoy DA (2011) Mutations of the *Listeria monocytogenes* peptidoglycan N-deacetylase and O-acetylase result in enhanced lysozyme sensitivity, bacteriolysis, and hyperinduction of innate immune pathways. Infect Immun 79:3596–3606. doi:10.1128/IAI.00077-11

Ritter M, Gross O, Kays S et al (2010) *Schistosoma mansoni* triggers Dectin-2, which activates the Nlrp3 inflammasome and alters adaptive immune responses. Proc Natl Acad Sci USA 107:20459–20464. doi:10.1073/pnas.1010337107

Sakhon OS, Victor KA, Choy A et al (2013) NSD1 mitigates caspase-1 activation by listeriolysin O in macrophages. PLoS ONE 8:e75911. doi:10.1371/journal.pone.0075911

Sauer J-D, Witte CE, Zemansky J et al (2010) *Listeria monocytogenes* triggers AIM2-mediated pyroptosis upon infrequent bacteriolysis in the macrophage cytosol. Cell Host Microbe 7:412–419. doi:10.1016/j.chom.2010.04.004

Sauer J-D, Pereyre S, Archer KA et al (2011a) *Listeria monocytogenes* engineered to activate the Nlrc4 inflammasome are severely attenuated and are poor inducers of protective immunity. Proc Natl Acad Sci USA 108:12419–12424. doi:10.1073/pnas.1019041108

Sauer J-D, Sotelo-Troha K, von Moltke J et al (2011b) The N-ethyl-N-nitrosourea-induced Goldenticket mouse mutant reveals an essential function of Sting in the in vivo interferon response to *Listeria monocytogenes* and cyclic dinucleotides. Infect Immun 79:688–694. doi:10.1128/IAI.00999-10

Schmidt RL, Lenz LL (2012) Distinct licensing of IL-18 and IL-1β secretion in response to NLRP3 inflammasome activation. PLoS ONE 7:e45186. doi:10.1371/journal.pone.0045186

Schnupf P, Hofmann J, Norseen J et al (2006a) Regulated translation of listeriolysin O controls virulence of *Listeria monocytogenes*. Mol Microbiol 61:999–1012. doi:10.1111/j.1365-2958.2006.05286.x

Schnupf P, Portnoy DA, Decatur AL (2006b) Phosphorylation, ubiquitination and degradation of listeriolysin O in mammalian cells: role of the PEST-like sequence. Cell Microbiol 8:353–364. doi:10.1111/j.1462-5822.2005.00631.x

Shen A, Higgins DE (2005) The 5' untranslated region-mediated enhancement of intracellular listeriolysin O production is required for *Listeria monocytogenes* pathogenicity. Mol Microbiol 57:1460–1473. doi:10.1111/j.1365-2958.2005.04780.x

Shen A, Higgins DE (2006) The MogR transcriptional repressor regulates nonhierarchal expression of flagellar motility genes and virulence in *Listeria monocytogenes*. PLoS Pathog 2:e30. doi:10.1371/journal.ppat.0020030

Shen Y, Naujokas M, Park M, Ireton K (2000) InlB-dependent internalization of *Listeria* is mediated by the met receptor tyrosine kinase. Cell 103:501–510. doi:10.1016/S0092-8674(00)00141-0

Shen A, Kamp HD, Grundling A, Higgins DE (2006) A bifunctional O-GlcNAc transferase governs flagellar motility through anti-repression. Genes Dev 20:3283–3295. doi:10.1101/gad.1492606

Shi C, Hohl TM, Leiner I et al (2011) Ly6G+ neutrophils are dispensable for defense against systemic *Listeria monocytogenes* infection. J Immunol 187:5293–5298. doi:10.4049/jimmunol.1101721

Shi J, Zhao Y, Wang K et al (2015) Cleavage of GSDMD by inflammatory caspases determines pyroptotic cell death. Nature 526:660–665. doi:10.1038/nature15514

Slaghuis J, Goetz M, Engelbrecht F, Goebel W (2004) Inefficient replication of *Listeria innocua* in the cytosol of mammalian cells. J Infect Dis 189:393–401. doi:10.1086/381206

Sugawara S, Uehara A, Nochi T et al (2001) Neutrophil proteinase 3-mediated induction of bioactive IL-18 secretion by human oral epithelial cells. J Immunol 167:6568–6575

Sutterwala FS, Mijares LA, Li L et al (2007) Immune recognition of *Pseudomonas aeruginosa* mediated by the IPAF/NLRC4 inflammasome. J Exp Med 204:3235–3245. doi:10.1084/jem.20071239

Swaminathan B, Gerner-Smidt P (2007) The epidemiology of human listeriosis. Microbes Infect 9:1236–1243. doi:10.1016/j.micinf.2007.05.011

Thomas JA, Allen JL, Tsen M et al (1999) Impaired cytokine signaling in mice lacking the IL-1 receptor-associated kinase. J Immunol 163:978–984

Tsuchiya K, Hara H, Kawamura I et al (2010) Involvement of absent in melanoma 2 in inflammasome activation in macrophages infected with *Listeria monocytogenes*. J Immunol 185:1186–1195. doi:10.4049/jimmunol.1001058

Tsuchiya K, Hara H, Fang R et al (2014) The adaptor ASC exacerbates lethal *Listeria monocytogenes* infection by mediating IL-18 production in an inflammasome-dependent and - independent manner. Eur J Immunol 44:3696–3707. doi:10.1002/eji.201444673

Tsuji NM, Tsutsui H, Seki E et al (2004) Roles of caspase-1 in *Listeria* infection in mice. Int Immunol 16:335–343

Uchiyama R, Yonehara S, Tsutsui H (2013) Fas-mediated inflammatory response in *Listeria monocytogenes* infection. J Immunol (Baltimore, Md 1950) 190:4245–4254. doi:10.4049/jimmunol.1203059

van de Veerdonk FL, Joosten LAB, Shaw PJ et al (2011) The inflammasome drives protective Th1 and Th17 cellular responses in disseminated candidiasis. Eur J Immunol 41:2260–2268. doi:10.1002/eji.201041226

van Pijkeren JP, Morrissey D, Monk IR et al (2010) A novel *Listeria monocytogenes*-based DNA delivery system for cancer gene therapy. Hum Gene Ther 21:405–416. doi:10.1089/hum.2009.022

Vazquez-Boland JA, Kocks C, Dramsi S et al (1992) Nucleotide sequence of the lecithinase operon of *Listeria monocytogenes* and possible role of lecithinase in cell-to-cell spread. Infect Immun 60:219–230

Vincent WJB, Freisinger CM, Lam P-Y et al (2015) Macrophages are required for inflammasome-dependent host defense in vivo. Cell Microbiol. doi:10.1111/cmi.12536

Waite JC, Leiner I, Lauer P et al (2011) Dynamic imaging of the effector immune response to listeria infection in vivo. PLoS Pathog 7:e1001326. doi:10.1371/journal.ppat.1001326

Wang S, Miura M, Jung YK et al (1998) Murine caspase-11, an ICE-interacting protease, is essential for the activation of ICE. Cell 92:501–509

Warren SE, Mao DP, Rodriguez AE et al (2008) Multiple Nod-like receptors activate caspase 1 during *Listeria monocytogenes* infection. J Immunol 180:7558–7564

Warren SE, Armstrong A, Hamilton MK et al (2010) Cutting edge: cytosolic bacterial DNA activates the inflammasome via Aim2. J Immunol 185:818–821. doi:10.4049/jimmunol.1000724

Warren SE, Duong H, Mao DP et al (2011) Generation of a *Listeria* vaccine strain by enhanced caspase-1 activation. Eur J Immunol 41:1934–1940. doi:10.1002/eji.201041214

Williams CR, Dustin ML, Sauer J-D (2013) Inflammasome-mediated inhibition of *Listeria monocytogenes*-stimulated immunity is independent of myelomonocytic function. PLoS ONE 8:e83191. doi:10.1371/journal.pone.0083191

Witte CE, Archer KA, Rae CS et al (2012) Innate immune pathways triggered by *Listeria monocytogenes* and their role in the induction of cell-mediated immunity. Adv Immunol 113:135–156. doi:10.1016/B978-0-12-394590-7.00002-6

Woodward JJ, Iavarone AT, Portnoy DA (2010) c-di-AMP secreted by intracellular *Listeria monocytogenes* activates a host type I interferon response. Science 80(328):1703–1705

Wu J, Fernandes-Alnemri T, Alnemri ES (2010) Involvement of the AIM2, NLRC4, and NLRP3 inflammasomes in Caspase-1 activation by *Listeria monocytogenes*. J Clin Immunol 30:693–702. doi:10.1007/s10875-010-9425-2

Yao Y, Wang Y, Chen F et al (2012) NLRC5 regulates MHC class I antigen presentation in host defense against intracellular pathogens. Cell Res 22:836–847. doi:10.1038/cr.2012.56

Zhao Y, Yang J, Shi J et al (2011) The NLRC4 inflammasome receptors for bacterial flagellin and type III secretion apparatus. Nature 477:596–600. doi:10.1038/nature10510

Zheng H, Fletcher D, Kozak W et al (1995) Resistance to fever induction and impaired acute-phase response in interleukin-1β-deficient mice. Immunity 3:9–19. doi:10.1016/1074-7613(95)90154-X

Zou T, Garifulin O, Berland R, Boyartchuk VL (2011) *Listeria monocytogenes* infection induces prosurvival metabolic signaling in macrophages. Infect Immun 79:1526–1535. doi:10.1128/IAI.01195-10

Inflammasome Recognition and Regulation of the *Legionella* Flagellum

Ursula Schell, Sylvia Simon and Hubert Hilbi

Abstract The Gram-negative bacterium *Legionella pneumophila* colonizes extracellular environmental niches and infects free-living protozoa. Upon inhalation into the human lung, the opportunistic pathogen grows in macrophages and causes a fulminant pneumonia termed Legionnaires' disease. *L. pneumophila* employs a biphasic life cycle, comprising a replicative, non-virulent, and a stationary, virulent form. In the latter phase, the pathogen produces a plethora of so-called effector proteins, which are injected into host cells, where they subvert pivotal processes and promote the formation of a distinct membrane-bound compartment, the *Legionella*-containing vacuole. In the stationary phase, the bacteria also produce a single monopolar flagellum and become motile. *L. pneumophila* flagellin is recognized by and triggers the host's NAIP5 (Birc1e)/NLRC4 (Ipaf) inflammasome, which leads to caspase-1 activation, pore formation, and pyroptosis. The production of *L. pneumophila* flagellin and pathogen–host interactions are controlled by a complex stationary phase regulatory network, detecting nutrient availability as well as the *Legionella* quorum sensing (Lqs) signaling compound LAI-1 (3-hydroxypentadecane-4-one). Thus, the small molecule LAI-1 coordinates *L. pneumophila* flagellin production and motility, inflammasome activation, and virulence.

U. Schell
Max von Pettenkofer Institute, Ludwig-Maximilians University,
Pettenkoferstraße 9a, 80336 Munich, Germany

S. Simon · H. Hilbi (✉)
Institute of Medical Microbiology, University of Zürich,
Gloriastrasse 30/32, 8006 Zürich, Switzerland
e-mail: hilbi@imm.uzh.ch

S. Backert (ed.), *Inflammasome Signaling and Bacterial Infections*,
Current Topics in Microbiology and Immunology 397,
DOI 10.1007/978-3-319-41171-2_8

Contents

1 Introduction: The Facultative Intracellular Pathogen *Legionella pneumophila*

1.1 Environmental Niches, Human Infection and Mouse Models

Legionella pneumophila is a ubiquitous environmental bacterium, which is found in a wide range of natural and artificial water systems (Newton et al. 2010; Hilbi et al. 2011). In the environment, the Gram-negative bacterium colonizes complex aquatic biofilms composed of prokaryotic and eukaryotic organisms (Declerck 2010). Yet, the major niche of *Legionella* spp. are free-living protozoa, wherein the bacteria naturally replicate (Fields 1996) (Fig. 1). These include amoeba (*Acanthamoeba, Hartmanella, Naegleria,* or *Vahlkampfia* spp.), ciliates (*Tetrahymena* spp.), and—at least under laboratory conditions—the social soil amoeba *Dictyostelium discoideum* (Solomon and Isberg 2000; Steinert 2011). Moreover, nematodes such as *Caenorhabditis elegans* have been discussed as a potential environmental niche of *L. pneumophila* (Brassinga et al. 2010; Komura et al. 2010). Upon respiration of bacteria-containing aerosols, *L. pneumophila* can replicate in alveolar macrophages and cause a severe pneumonia known as Legionnaires' disease (McDade et al. 1977; Horwitz and Silverstein 1980).

Legionnaires' disease preferentially strikes elderly or immuno-compromised persons and can spread in outbreaks comprising as many as 450 cases (Garcia-Fulgueiras et al. 2003). For the last 40 years, it was believed that *L. pneumophila* cannot be transmitted via interpersonal contact and that humans represent a dead end in the life cycle of this bacterium (Hilbi et al. 2010; Newton et al. 2010). However, recently the first case of a probable person-to-person transmission of Legionnaires' disease has been reported (Correia et al. 2016). In this

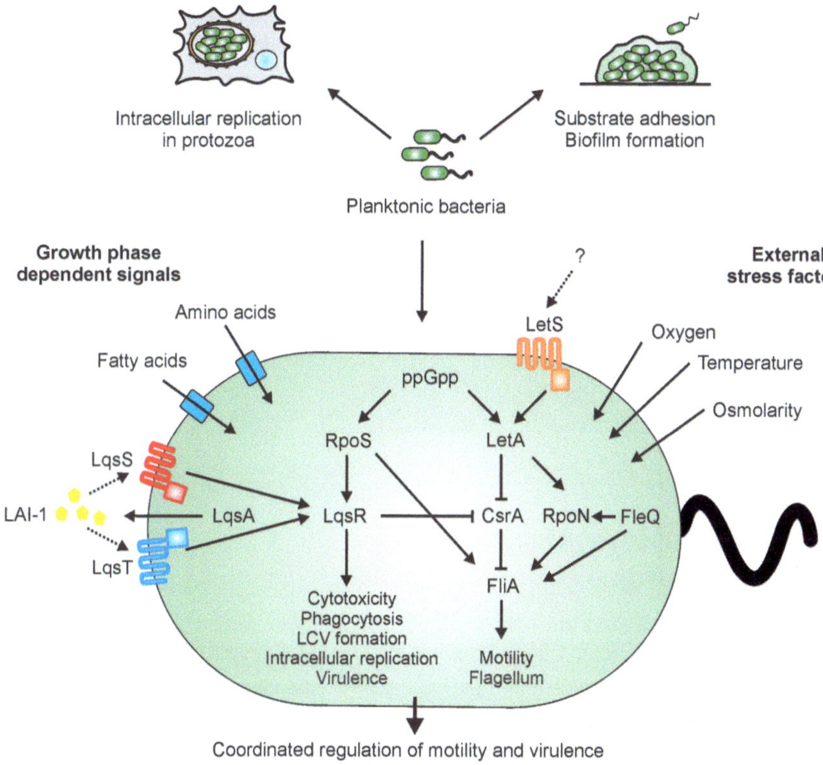

Fig. 1 Environmental niches and the biphasic life cycle of *L. pneumophila*. In the environment, planktonic *L. pneumophila* can replicate intracellularly in protozoa or colonize biofilms. *L. pneumophila* adopts a growth phase-dependent, biphasic life cycle, comprising a replicative, non-virulent, and a transmissive, virulent phase. In the latter (stationary) phase, *L. pneumophila* becomes flagellated and is highly motile and virulent. Cues triggering the switch between the two phases include (i) availability of nutrients (amino acids, fatty acids) as gauged by the second messenger guanosine 3′,5′-bispyrophosphate (ppGpp); (ii) the cell-density-dependent *Legionella* quorum sensing system (Lqs) and its autoinducer signal LAI-1 (3-hydroxypentadecane-4-one); and (iii) external stress factors such as oxygen, temperature, and osmolarity. The LetAS two-component system, in concert with the alternative sigma factors RpoS (σ38), RpoN (σ54), and FliA (σ28), negatively regulates the sRNA-binding global regulator CsrA and promotes the production of FlaA, the major flagellin subunit

tragic incident, a highly pathogenic *L. pneumophila* strain was apparently transmitted within a household located 300 km away from the original outbreak area, and both patients involved died from pneumonia caused by clonal isolates.

Inbred mouse strains such as C57BL/6 or BALB/c restrict *L. pneumophila* due to the activation of a programmed cell death pathway in macrophages as an ultimate line of defense against the intracellular pathogen (Newton et al. 2010; Massis and Zamboni 2011). In contrast, the A/J strain fails to restrict *L. pneumophila*, and consequently, this strain was established as a mouse model of Legionnaires' disease

(Brieland et al. 1994). Mimicking human disease, *L. pneumophila* triggers in A/J mice an acute inflammatory reaction, including production of the cytokines tumor necrosis factor (TNF)-α, interferon (IFN)-γ, interleukin (IL)-12, and IL-18 (Newton et al. 2010; Massis and Zamboni 2011). IFN-γ is particularly important to inhibit growth of *L. pneumophila* in monocytes and alveolar macrophages, thus limiting the infection (Bhardwaj et al. 1986; Nash et al. 1988; Byrd and Horwitz 1989). On a cellular level, type I (IFN-α/β) and type II (IFN-γ) interferons profoundly modify the intracellular pathogen compartment, termed the *Legionella*-containing vacuole (LCV) (Naujoks et al. 2016). Comparative quantitative mass spectrometry of intact LCVs purified from *L. pneumophila*-infected resting or IFN-treated bone marrow-derived macrophages revealed that type I as well as type II IFNs upregulate immune responsive gene (IRG)1. The enzyme IRG1 catalyzes the production of the bactericidal metabolite itaconate, which restricts *L. pneumophila* and other intravacuolar pathogens (Naujoks et al. 2016).

1.2 Intracellular Replication Within the Legionella*-Containing Vacuole*

After uptake by metazoan phagocytes or protozoa, *L. pneumophila* evades the canonical bactericidal endocytic pathway and instead forms a replication-permissive membrane-bound compartment, the LCV (Isberg et al. 2009; Hilbi and Haas 2012). In this highly dynamic process, the nascent LCV communicates with the endosomal, secretory, and retrograde vesicle trafficking pathways (Personnic et al. 2016) and finally fuses with the endoplasmic reticulum (Swanson and Isberg 1995; Lu and Clarke 2005; Robinson and Roy 2006). The mechanism of LCV formation and intracellular replication in metazoan and protozoan phagocytes appears to be evolutionarily conserved, as many bacterial effectors target conserved eukaryotic components.

On the pathogen side, the crucial virulence factor is the Icm/Dot type IV secretion system (T4SS) (Nagai and Kubori 2011), a multicomponent molecular "syringe" that injects the stunning number of 300 different so-called effector proteins into eukaryotic host cells (Hubber and Roy 2010; Sherwood and Roy 2013). In eukaryotic cells, the effector proteins subvert crucial processes, including signal transduction, gene expression, vesicle trafficking, cell motility, and cell death pathways. To this end, the effectors target pivotal host factors, such as small GTPases and phosphoinositide (PI) lipids.

Only rather few Icm/Dot-translocated effectors have been thoroughly characterized to date (Finsel and Hilbi 2015). These include the histone methyltransferase RomA (Rolando et al. 2013), the GTPase activators RalF (Nagai et al. 2002), SidM (DrrA) (Machner and Isberg 2006; Murata et al. 2006; Brombacher et al. 2009; Müller et al. 2010) and LegG1 (Rothmeier et al. 2013; Simon et al. 2014), the PI-binding E3 ubiquitin ligase SidC (Weber et al. 2006; Hsu et al. 2014), the PI phosphatase SidF (Hsu et al. 2012), the retromer interactor RidL (Finsel et al. 2013), the Atg8 protease RavZ (Choy et al. 2012), or the phytase LppA (Weber et al. 2014).

1.3 The Biphasic Life Cycle of L. pneumophila

L. pneumophila is a facultative intracellular microorganism that survives and replicates both in extracellular and intracellular niches. The pathogen employs a biphasic life cycle consisting of a replicative (non-virulent) and a transmissive (virulent) phase (Molofsky and Swanson 2004). In the replicative phase, *L. pneumophila* is non-motile; in the stationary growth phase, e.g., at the end of intracellular growth, the pathogen becomes flagellated and exits the host cell (Rowbotham 1986; Byrne and Swanson 1998). Moreover, toward the end of the replication period, *L. pneumophila* upregulates its resistance to extracellular stress factors (Hales and Shuman 1999) and seems to exhibit pore-forming activity to escape the host (Alli et al. 2000). Collectively, the transmissive and virulence traits enable the pathogen to evade its protozoan hosts, survive in the environment as a motile planktonic cell and re-establish a replicative niche within a new host, wherein the cycle repeats.

Virulence and motility traits are expressed in response to nutrient shortage, i.e., amino acid starvation (Byrne and Swanson 1998; Sauer et al. 2005) and inhibition of fatty acid biosynthesis (Dalebroux et al. 2009). Concomitantly, the intracellular second messenger guanosine 3′,5′-bispyrophosphate (ppGpp) is produced (Hammer and Swanson 1999; Molofsky and Swanson 2004) by the synthetase RelA and the bifunctional synthetase/hydrolase SpoT (Zusman et al. 2002; Dalebroux et al. 2009) (Fig. 1). The level of ppGpp regulates a complex network that controls transmissive traits through transcription factors such as the sigma factors RpoS (σ^{38}) (Hales and Shuman 1999; Bachman and Swanson 2001), RpoN (σ^{54}) (Jacobi et al. 2004) and FliA (σ^{28}) (Heuner et al. 1997; Hammer et al. 2002; Molofsky et al. 2005), as well as LetAS and other two-component systems (Hammer et al. 2002). The signal recognized by LetA in stationary growth phase is unknown (Rasis and Segal 2009; Sahr et al. 2009).

The response regulator LetA directly upregulates the expression of the small non-coding sRNAs *rsmY* and *rsmZ* (Rasis and Segal 2009; Sahr et al. 2009). The sRNAs bind and sequester the conserved RNA-binding, global regulatory protein CsrA (Fettes et al. 2001; Molofsky and Swanson 2003; Forsbach-Birk et al. 2004), and thus, the repression of transmissive traits by CsrA is relieved. CsrA is an essential activator of intracellular replication and a global repressor of transmission traits. Accordingly, the overproduction of CsrA in *L. pneumophila* leads to a reduction of flagellation (Suzuki et al. 2006). The *L. pneumophila* stationary phase regulatory network controlled by RpoS and LetAS also includes the *Legionella* quorum sensing (Lqs) system (Tiaden et al. 2007, 2010a) (Fig. 1). Taken together, in the transmissive growth phase, *L. pneumophila* represses replication and instead exhibits motility and virulence traits, such as the flagellar apparatus, the type IV pilus machinery, and Icm/Dot-dependent and Icm/Dot-independent virulence factors (Molofsky and Swanson 2004; Molofsky et al. 2005).

2 Modulation of Inflammasome Activity by *L. pneumophila*

2.1 Pattern Recognition Receptors Implicated in L. pneumophila *Infection*

Various leukocytes such as neutrophils, dendritic cells, and macrophages produce so-called pattern recognition receptors (PRRs), which are activated by pathogen-associated molecular patterns (PAMPs), such as bacterial lipopolysaccharide (LPS), lipoproteins, peptidoglycan, nucleic acids, or flagellin. By recognizing PAMPs or cellular danger-associated molecular patterns (DAMPs), the infected host cell is alerted and induces cytokine production and cell death pathways, eventually leading to clearance of invading pathogens (Creagh and O'Neill 2006; Chow et al. 2015; Sellge and Kufer 2015).

PRRs include the membrane-bound Toll-like receptors (TLRs) such as TLR4 and TLR5, which recognize bacterial LPS and flagellin, respectively, and activate the transcription factor NF-κB through the adaptor MyD88 (Creagh and O'Neill 2006) (Fig. 2). Compared to enterobacterial LPS, the *L. pneumophila* lipid A (endotoxin) membrane anchor is much more hydrophobic, which might account for its low endotoxic activity (Neumeister et al. 1998; Albers et al. 2007). Cytosolic PRRs include Nod-like receptors (NLRs), such as the neuronal apoptosis inhibitory protein (NAIP) and NRLC4 (Ipaf) subfamilies, or RIG-I like receptors (RLRs), which trigger inflammation and programmed cell death (Zhao and Shao 2015). Noteworthy, the NLR family member NLRX1 (Lei et al. 2012) was identified by a proteomics approach on LCVs from infected RAW 264.7 macrophages (Hoffmann et al. 2014a).

Distinct NLRs (likely recognizing different PAMPs) are components of cytoplasmic multiprotein complexes termed inflammasomes, which are pivotal modules of the innate immune system (Sutterwala et al. 2007; Schroder and Tschopp 2010; Vance 2015). Through inflammasomes, many microbial PAMPs activate the cysteine protease caspase-1, a key factor of the innate immune response, which promotes processing and release of the pro-inflammatory cytokines IL-1β and IL-18, and leads to a form of cell death termed pyroptosis (Fig. 2). Noteworthy, neither inflammasomes and caspases nor NF-κB are evolutionary conserved in amoeba (Hoffmann et al. 2014b). Since *L. pneumophila* underwent co-evolution predominantly if not exclusively in protozoa rather than in mammalian (human) hosts, the bacteria likely did not experience a rigorous selection against recognition by

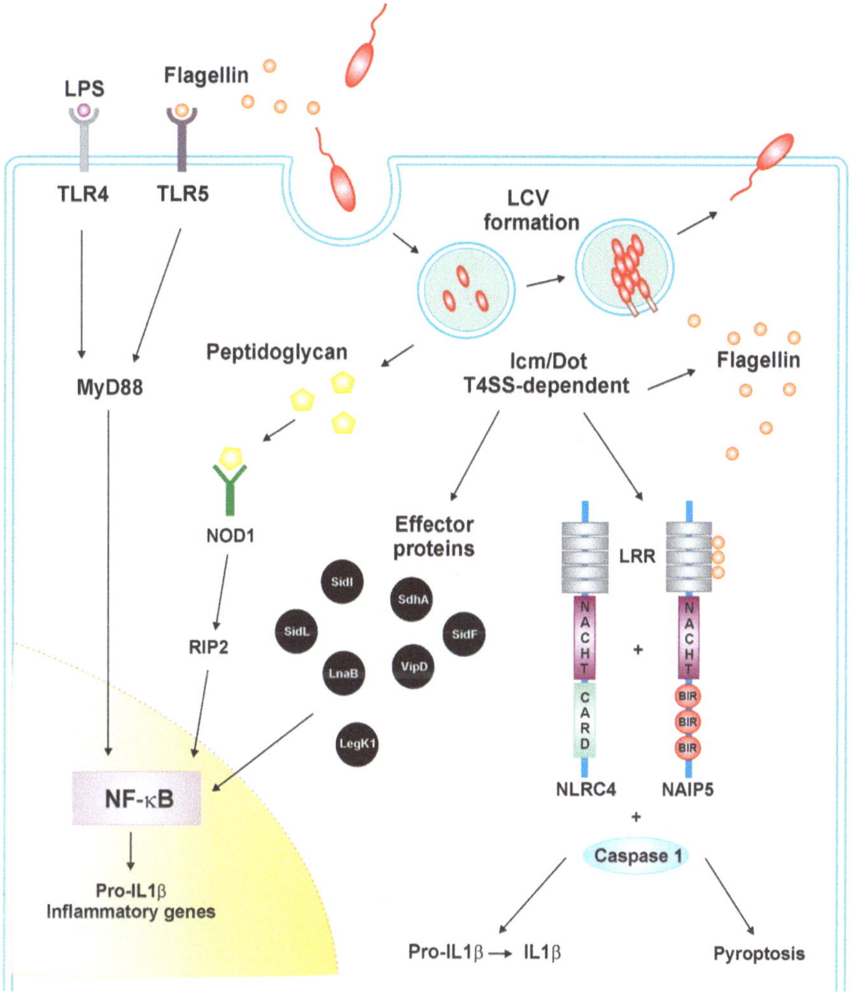

Fig. 2 Inflammasome activation by *L. pneumophila* flagellin and Icm/Dot substrates. In mammalian phagocytes, *L. pneumophila* PAMPs (LPS, flagellin) are (weakly) recognized by membrane-bound PRRs, such as TLR4 or TLR5, and activate immune gene expression through the adaptor molecule MyD88 and the transcription factor NF-κB. *L. pneumophila* peptidoglycan might be recognized by the PRR Nod1, which signals through RIP2 kinase to NF-κB. *L. pneumophila* producing the Icm/Dot T4SS exposes flagellin to the cytosol and thus triggers cytoplasmic NLRs, in particular the NAIP5/NLRC4 inflammasome. This induces caspase-1 activation, the production of mature IL-1β and IL-18, and cell death by pyroptosis. Several Icm/Dot substrates (SdhA, VipD, SidF, SidI, SidL, LnaB, and LegK1) are implicated in host cell death by determining the integrity of the pathogen vacuole or by modulating NF-κB activation. *Abbreviations*: *BIR* baculo virus inhibitor of apoptosis repeat; *CARD* caspase recruitment domain; *LCV Legionella*-containing vacuole; *LPS* lipopolysaccharide; *LRR* leucine rich repeat; *MyD88* myeloid differentiation primary response gene 88; NACHT, domain present in neuronal apoptosis inhibitory protein (NAIP), CIITA, HET-E and TP1; *NF-κB* nuclear factor kappa-light-chain-enhancer of activated B cells; *NOD* nucleotide-binding oligomerization domain-containing protein; *NLR* nod-like receptor; *RIP2* receptor interacting protein 2; *T4SS* type IV secretion system; *TLR* Toll-like receptor; *IL* interleukin

mammalian PRRs. Accordingly, the intracellular pathogen triggers the activation of all known PRR families (Newton et al. 2010; Massis and Zamboni 2011; Simon and Hilbi 2015).

2.2 *Inflammasome Activation by* L. pneumophila *Flagellin*

Macrophages from most inbred mouse strains are non-permissive for intracellular replication of *L. pneumophila*, with the notable exception of the A/J strain (Yamamoto et al. 1988; Brieland et al. 1994). The genetic locus conferring sensitivity against *L. pneumophila* was identified in the *Lgn1* region present on the mouse chromosome 13 (Dietrich et al. 1995). Considerably later, the corresponding allele was mapped to the *Naip5* gene (also termed *Birc1e*) (Diez et al. 2003; Wright et al. 2003). Through the NLR Naip5, non-permissive macrophages sense cytosolic flagellin and trigger caspase-1-dependent pyroptosis, thus restricting intracellular bacterial growth (Molofsky et al. 2006; Ren et al. 2006; Zamboni et al. 2006) (Fig. 2). In contrast, A/J macrophages harbor a defective *Naip5* allele and fail to restrict *L. pneumophila*. A conserved carboxy-terminal domain of flagellin triggers Naip5 inflammasome activation, as was revealed by *Naip5*$^{-/-}$ mice (Lightfield et al. 2008). In addition to its role in inflammasome activation, Naip5 also negatively affects LCV formation, such that in presence or absence of functional Naip5 more LCVs acquire lysosomal markers or ER markers, respectively (Fortier et al. 2007). In agreement with this notion, Naip5 also restricts intracellular growth of *L. pneumophila* independently of caspase-1 (Lamkanfi et al. 2007). Finally, the NAIP5 and NLRC4 (see below) inflammasomes control the replication of flagellated *L. pneumophila* not only in mouse but also in human cells (Vinzing et al. 2008).

NLRC4 (also called Ipaf) is another member of the NLR family (Sutterwala et al. 2007). Interestingly, different NAIP paralogues determine the specificity of NLRC4 for distinct bacterial PAMPs, e.g., NAIP5 activates the NLRC4 inflammasome specifically in response to flagellin by ligand-dependent oligomerization (Kofoed and Vance 2011). *Nlrc4*$^{-/-}$ macrophages are permissive for *L. pneumophila*, the corresponding LCV acquires ER markers, and flagellin-dependent activation of caspase-1 as well as pyroptosis is abrogated (Amer et al. 2006; Pereira et al. 2011a, b). To clear *L. pneumophila*, NLRC4-dependent pyroptosis and rapid pore formation (Coers et al. 2007; Case et al. 2009; Silveira and Zamboni 2010) depends on caspase-1 but not on IL-1β and IL-18 (Miao et al. 2010) (Fig. 2). Moreover, the adaptor protein Asc appears to dampen caspase-1-dependent pyroptosis (Case and Roy 2011).

In addition to caspase-1, caspase-7 is activated by flagellin downstream of the NAIP5/NLRC4 inflammasome (Akhter et al. 2009), and independently of caspase-1, caspase-11 is also activated (Akhter et al. 2012), but not required for NLRC4-dependent pyroptosis (Cerqueira et al. 2015). Interestingly, caspase-11 promotes a rapid, flagellin-independent pyroptotic pathway requiring the NLRP3/Asc inflammasome (Case and Roy 2011; Case et al. 2013). Both, caspase-7 and caspase-11 contribute to restrict intracellular replication of *L. pneumophila* by promoting the fusion of the pathogen vacuole with lysosomes (Akhter et al. 2009; Amer 2010; Akhter et al. 2012). Collectively, flagellin-dependent NAIP5/NLRC4 inflammasome activation is a crucial mechanism of *L. pneumophila* restriction in macrophages as well as in dendritic cells (Nogueira et al. 2009).

Interestingly, flagellin-dependent NAIP5/NLRC4 inflammasome activation by *L. pneumophila* requires a functional Icm/Dot T4SS, and accordingly, wild-type but not Icm/Dot-deficient bacteria trigger pyroptotic cell death (Molofsky et al. 2006; Ren et al. 2006; Zamboni et al. 2006). While it was postulated that the T4SS forms pores through which flagellin might get access to the host cytoplasm, the mechanism of pathogen vacuole exit of this PAMP remains poorly understood.

2.3 Regulation of Cell Death by Icm/Dot-Translocated Effectors

Numerous cytoplasmic PRRs detect and respond to pathogenic bacteria residing in this cellular compartment. Thus, maintenance of LCV integrity and modulation of programmed host cell death are critical for avoiding detection of *L. pneumophila* and preserving its intracellular replication niche. Accordingly, Icm/Dot-translocated effectors that control pathogen vacuole integrity play a pivotal role for *L. pneumophila*–phagocyte interactions (Fig. 2). The SidH family of Icm/Dot substrates (Luo and Isberg 2004)—in particular SdhA—is implicated in maintaining the LCV membrane integrity and prevention of pyroptotic cell death (Laguna et al. 2006; Creasey and Isberg 2012). *L. pneumophila* lacking *sdhA* localizes to the macrophage cytoplasm and triggers caspase-1 activation and IL-1β secretion as well as pyroptosis. These processes do not involve the flagellin-NAIP5-NLRC4 pathway (discussed above), but rather the DNA-sensing absent in melanoma 2 (AIM2) inflammasome (Ge et al. 2012). The induction of cell death by an *sdhA* mutant is suppressed by the secreted bacterial phospholipase A, PlaA, through an unknown mechanism (Creasey and Isberg 2012). Furthermore, *L. pneumophila* lacking *sdhA* triggers the production of type I IFNs (IFN-α/β) through the RNA-sensing RIG-I receptor (Monroe et al. 2009). Interestingly, the *sdhA* mutant also activates caspase-11 in a NLRC4-, NLRP3-, and Asc-independent manner, thereby enhancing clearance of cytoplasmic pathogens (Aachoui et al. 2013). Taken together, the Icm/Dot-translocated effector SdhA is a key suppressor of inflammatory and pyroptotic host responses to *L. pneumophila* mediated through cytoplasmic nucleic acid-sensing PRRs.

A number of other Icm/Dot substrates are also implicated in the inhibition of programmed host cell death (Fig. 2). The *L. pneumophila* PI 3-phosphatase SidF (Hsu et al. 2012) inactivates the pro-apoptotic factors BNIP3 and Bcl-rambo by an unknown mechanism and thus counteracts cell death induction (Banga et al. 2007). Moreover, several *L. pneumophila* effectors activate the transcription factor NF-κB, a master regulator of the mammalian innate immune response, which controls the production of anti-apoptotic and pro-survival factors, as well as inflammatory mediators (Gilmore 2006). The Icm/Dot T4SS and its substrates, e.g., LnaB and LegK1, strongly induce NF-κB (Losick and Isberg 2006; Abu-Zant et al. 2007; Bartfeld et al. 2009; Losick et al. 2010). LnaB is a novel effector of unknown function. LegK1 is a Ser/Thr kinase that phosphorylates the NF-κB inhibitor IκBα. This leads to IκBα degradation and hence, a robust NF-κB activation due to its release from the inhibitor and nuclear translocation (Ge et al. 2009). Furthermore, the Icm/Dot substrates Lgt1-3, SidI, and SidL are cytotoxic translation inhibitors, which selectively decrease the production of IκB (Shen et al. 2009; Fontana et al. 2011). To shut down translation, the UDP-glucosyl-transferases Lgt1-3 modify the elongation factor eEF1A (Belyi et al. 2006, 2008), and SidI inactivates eEF1A/eEF1Bγ by an unknown mechanism (Shen et al. 2009).

The phospholipase C, PlcC, is an Icm/Dot substrate that hydrolyzes several lipids, which might destabilize target membranes and cause cell toxicity (Aurass et al. 2013). VipD is an Icm/Dot-translocated phospholipase A that destabilizes mitochondrial membranes, thus potentially releasing cytochrome *c* into the cytosol and promoting caspase-3-dependent programmed cell death (Zhu et al. 2013). Interestingly, VipD is a Rab5(GTP)-dependent phospholipase, the activity of which requires activated small GTPase in its GTP-bound form (Gaspar and Machner 2014). The stimulation of apoptosis by *L. pneumophila* seems counter-intuitive, but might reflect a tight spatial and temporal control of anti- and pro-apoptotic processes during LCV maturation, release of the bacteria from the pathogen compartment and exit from the host cell at the end of an infection cycle.

3 Regulation of *Legionella* Flagellin and Motility

3.1 Legionella *Flagellin as a Transmissive and Virulence Factor*

The *L. pneumophila* flagellin is a potent activator of the host NAIP5/NLRC4 inflammasome and caspase-1. Therefore, the regulation of this major flagellum component and motility at large need to be tightly regulated and coordinated with the pathogen's biphasic life cycle. The switch of *L. pneumophila* from the replicative to the virulent/transmissive phase is triggered by various external and internal stimuli, including (i) nutrient availability, (ii) a density-dependent autoinducer signal, and (iii) external stress factors such as oxygen, temperature and osmolarity (Fig. 1).

Under conditions of nutrient shortage, the bacterial replication rate drops, and the flagellar promoter P_{flaA} as well as virulence traits are expressed in response to the increase of the "alarmone" ppGpp (Hammer and Swanson 1999; Molofsky and Swanson 2004). Transcriptomic studies revealed that the upregulation of the *flaA* and *fliA* genes is dependent on the alternative sigma factor RpoS (Hovel-Miner et al. 2009). Moreover, the expression of P_{flaA} and the production of flagellin as well as bacterial motility are regulated by the *L. pneumophila* small signaling molecule LAI-1 (*Legionella* autoinducer-1; see below) (Tiaden et al. 2007, 2010a).

L. pneumophila is a strictly aerobic bacterium (Pine et al. 1979). External physical stress factors such as oxygen limitation inhibit the bacterial metabolism and lead to loss of the flagellum (Mauchline et al. 1992). Furthermore, an increase of temperature also negatively influences the production and assembly of the flagellum (Ott et al. 1991; Mauchline et al. 1992; Heuner et al. 1995). Consequently, *L. pneumophila* grown at high temperatures (up to 41 °C) is less motile and virulent, compared to lower temperatures (25 °C) (Edelstein et al. 1987). Finally, using a *flaA*::*luxAB* gene fusion, flagellin expression was found to be upregulated in stationary growth phase in media with high osmolarity, whereas expression was downregulated during exponential growth in medium of high viscosity (Ott et al. 1991; Heuner et al. 1999).

3.2 *Regulation of* Legionella *Motility by the Signaling Molecule LAI-1*

L. pneumophila employs a sole autoinducer compound, LAI-1 (3-hydroxypentadecane-4-one), for cell–cell communication and density-dependent gene regulation known as quorum sensing (Tiaden et al. 2010a). LAI-1 is produced and detected by the Lqs (*Legionella* quorum sensing) system (Tiaden et al. 2008), comprising the autoinducer synthase LqsA (Spirig et al. 2008), the homologous sensor kinases LqsS (Tiaden et al. 2010b) and LqsT (Kessler et al. 2013), and the response regulator LqsR (Tiaden et al. 2007) (Fig. 1). Phosphorylation signaling through LqsS and LqsT converges on LqsR, which dimerizes upon phosphorylation (Schell et al. 2014).

Lqs-regulated processes include pathogen–host interactions, expression of a 133 kb chromosomal “fitness island”, production of extracellular filaments, and natural competence for DNA uptake (Tiaden and Hilbi 2012). An *L. pneumophila lqsR* mutant strain shows a shortened lag phase and starts to grow earlier than the parental strain, indicating that LqsR controls the induction of virulence traits and negatively regulates the transition from the transmissive (stationary) to the replicative phase (Tiaden et al. 2007). Strikingly, *L. pneumophila* mutant strains lacking individual *lqs* genes are impaired for motility, as determined by tracking single bacterial cells, and the expression of the flagellar promoter P_{flaA} in *L. pneumophila lqs* mutant strains was reduced, indicating that components of the Lqs system positively regulate P_{flaA} (Schell et al. 2015).

Synthctic LAI-1 promotes the motility of *L. pneumophila* by signaling through LqsS/LqsT and LqsR (Schell et al. 2015). Upon addition of LAI-1, autophosphorylation of LqsS/LqsT by [γ-^{32}P]-ATP was inhibited in a dose-dependent manner;

yet, LAI-1 did neither affect the stability of phospho-LqsS or phospho-LqsT, nor the dephosphorylation by LqsR. Transcriptome analysis of *L. pneumophila* treated with LAI-1 revealed that the compound positively regulates a number of genes, including the sRNAs *rsmY* and *rsmZ*, and negatively regulates the RNA-binding global regulator *crsA* (Schell et al. 2015). Accordingly, LAI-1 controls the switch from the replicative to the transmissive growth phase of *L. pneumophila*, and the compound specifically regulates bacterial motility through LqsS- and LqsT-dependent phosphorylation signaling. Interestingly, synthetic LAI-1 also inhibited the migration of eukaryotic cells through a pathway comprising the scaffold protein IQGAP1, the small GTPase Cdc42 and a Cdc42 activator, the guanine nucleotide exchange factor (GEF) ARHGEF9 (Simon et al. 2015). Thus, LAI-1 modulates the motility of *L. pneumophila* as well as the migration of eukaryotic cells. While the bacterial receptors of LAI-1 are the sensor kinases LqsS and LqsT, the eukaryotic receptor of LAI-1 is currently not known.

3.3 The Flagellar Regulon of Legionella

L. pneumophila surface structures such as the flagellum and pili play important roles in bacterial pathogenicity and survival in the environment. Flagellum formation and invasion capacity of the pathogen are closely linked, as *flaA* mutant strains are defective for infection of amoeba and macrophages (Pruckler et al. 1995; Bosshardt et al. 1997; Dietrich et al. 2001). In contrast, the intracellular replication rate of *L. pneumophila* is not affected by FlaA or bacterial motility (Dietrich et al. 2001). Flagellin is mainly produced at the end of an infection, as growth of *L. pneumophila* slows down, before highly motile and virulent bacteria exit the host cells (Rowbotham 1986; Albert-Weissenberger et al. 2010).

The single monopolar flagellum of *L. pneumophila* adopts a canonical architecture (Fig. 3a). The flagellar secretion apparatus exports the majority of flagellar components, including the rod, hook, flagellin, and capping proteins, to the bacterial periplasm or the external milieu. The basal body stabilizes the flagellum, and the rotational forces generated by the Mot ion channels are transmitted from the basal body through the hook to the external filament, causing its rotation. FlaA plays a special role in flagellum biosynthesis and assembly, as the flagellin subunit forms the flagellar filament, thus completing the complex structure (Heuner et al. 1995; Albert-Weissenberger et al. 2010).

The genes corresponding to individual flagellum components are organized in the complex flagellar regulon (Fig. 3b). The regulon comprises more than 40 different genes, divided into four classes, whose temporal expression pattern is highly coordinated (Ott et al. 1991; Molofsky et al. 2005; Albert-Weissenberger et al. 2010). The network is controlled directly by hierarchically organized regulators such as FleQ, RpoN (σ^{54}) and FliA (σ^{28}), but also indirectly by RpoS (σ^{38}) (Bachman and Swanson 2001; Albert-Weissenberger et al. 2010). The sigma factor σ^{54}, RpoN, initiates gene transcription together with the enhancer-binding proteins

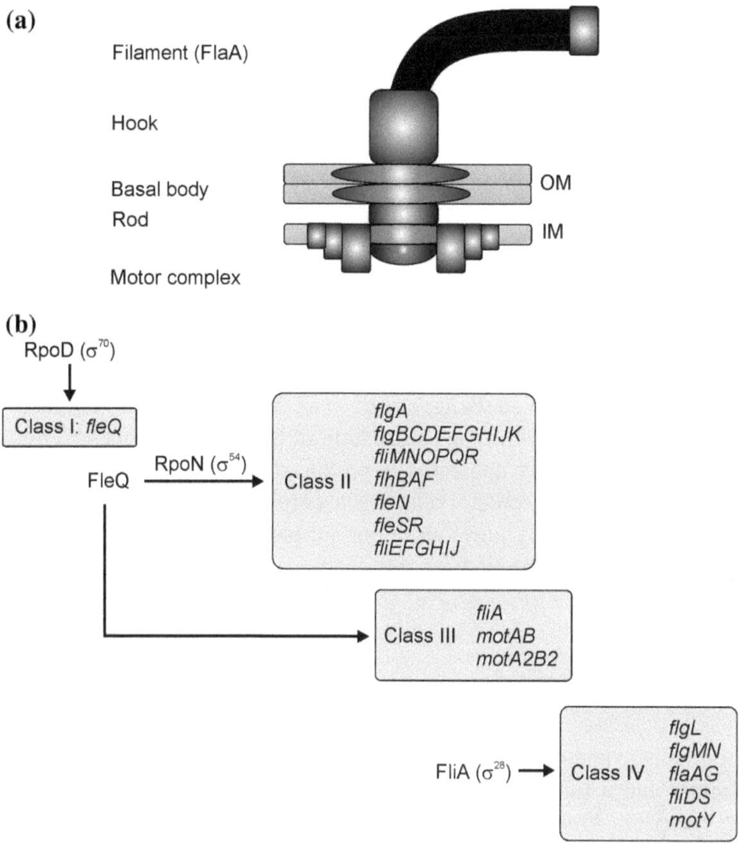

Fig. 3 *L. pneumophila* flagellum architecture and flagellar regulon. **a** Schematic architecture of the flagellum of Gram-negative bacteria. The flagellum is composed of the motor complex, rod, basal body, hook, and flagellin subunits forming the filament. **b** The complex flagellar regulon comprises more than 40 different genes, divided into four classes (I–IV). The network is controlled in a temporal manner directly by hierarchically organized regulators such as FleQ, RpoD (σ^{70}), RpoN (σ^{54}), or FliA (σ^{28}), but also indirectly by the alternative sigma factor RpoS (σ^{38}). FleQ (regulated by RpoD), together with RpoN, induces the expression of the bulk of flagellar genes (class II). The *flaA* gene is upregulated up to 100 times in transmissive bacteria compared to replicative bacteria. The flagellin subunit FlaA plays a special role in flagellum biosynthesis and assembly, as the protein forms the flagellar filament, thus completing the flagellum

FleQ or FleR (Jacobi et al. 2004; Wigneshweraraj et al. 2008). The expression of FleQ is transcriptionally controlled by the sigma factor RpoD (σ^{70}) (Albert-Weissenberger et al. 2010). FleQ and RpoN coordinately regulate nearly all flagellar class II genes emphasizing their major role in the timely control of the flagellar regulon (Albert-Weissenberger et al. 2010). Late flagellar genes (class III and IV) are expressed dependent on FleQ but independent of RpoN or FleR (Albert-Weissenberger et al. 2010). Thus, FleQ represents the master regulator for motility genes, which controls flagellar gene expression both dependent and independent of RpoN (Albert-Weissenberger et al. 2010). As the last step in flagellar biosynthesis, the sigma factor σ^{28}, FliA, induces the expression of flagellar class IV genes such as *flaA*, encoding the flagellin subunit (Albert-Weissenberger et al. 2010).

Belonging to the class IV in the hierarchy of the flagellar genes, the *flaA* gene represents a good indicator for flagellum formation in the *L. pneumophila* intracellular replication cycle (Albert-Weissenberger et al. 2010). Transcriptomic analysis of *L. pneumophila* shows that *flaA* is upregulated up to 100 times in transmissive bacteria compared to replicative bacteria. Similarly, in the late growth phase, several genes encoding proteins implicated in flagellum biosynthesis (e.g., *fliS, fliD, fliN, flgBCDEFGHIJKL, fhF, fleN*) and *fliA* encoding the sigma factor FliA are strongly upregulated (Brüggemann et al. 2006; Jules and Buchrieser 2007). In summary, the tight and complex regulation of flagellar components ensures that this bioenergetically costly apparatus is produced and assembled at precisely the right time during the infection cycle. As a corollary, the intricate regulation of flagellum production likely plays an important role for NAIP5/NLRC4 inflammasome activation and pyroptotic host cell death.

4 Concluding Remarks

The opportunistic pathogen *L. pneumophila* adopts a biphasic live cycle comprising a replicative and a transmissive phase. This strategy allows the facultative intracellular bacterium to alter between intracellular replication in protozoa and extracellular transmission between hosts. Transmissive phase *L. pneumophila* is not only motile but also highly virulent and rapidly forms a specific replication-permissive pathogen vacuole upon contact with host cells. To this end, the pathogen employs the Icm/Dot T4SS, which translocates as many as 300 different effector proteins into host cells, where they subvert crucial processes. While the flagellum plays a crucial role for dissemination and niche colonization of *L. pneumophila*, its major component, the highly conserved flagellin, is efficiently recognized as a PAMP by metazoan phagocytes. Indeed, flagellin activates caspase-1 through the cytoplasmic NAIP5/NLRC4 inflammasome in *L. pneumophila*-infected cells. Caspase-1 activation generates mature IL-1β and IL-18 and triggers pyroptotic cell death, thus eliminating the pathogen's replicative niche. Accordingly, maintenance of pathogen vacuole integrity is critical for preserving the intracellular niche. In agreement with this notion, the *L. pneumophila* Icm/Dot substrate SdhA stabilizes the pathogen vacuole and prevents pyroptosis. Due to its role in *L. pneumophila* pathogenicity, environmental dissemination and immune recognition, the production of flagella needs to be tightly regulated and coordinated with the pathogen's biphasic life cycle. Flagellum production is regulated by various stimuli, including nutrient availability, a density-dependent autoinducer signal, and external stress factors such as oxygen, temperature, and osmolarity. Future studies should address the intriguing question of how and when during the infection cycle *L. pneumophila* flagellin and other PAMPs are exposed in the host cytoplasm and trigger inflammasomes. Research along this line will shed light on the mechanism of vacuolar replication of a major human pathogen, as well as on the detection and elimination of the bacterial invader by host cells.

Acknowledgments This work was supported by the Swiss National Science Foundation (SNF; 31003A_153200), the German Research Foundation (DFG; SPP 1617), and the Bundesministerium für Bildung und Forschung (BMBF; 031A410A; Infect-ERA project EUGENPATH).

References

Aachoui Y, Leaf IA, Hagar JA et al (2013) Caspase-11 protects against bacteria that escape the vacuole. Science 339:975–978

Abu-Zant A, Jones S, Asare R, Suttles J, Price C, Graham J, Kwaik YA (2007) Anti-apoptotic signalling by the Dot/Icm secretion system of *L. pneumophila*. Cell Microbiol 9:246–264

Akhter A, Gavrilin MA, Frantz L et al (2009) Caspase-7 activation by the Nlrc4/Ipaf inflammasome restricts *Legionella pneumophila* infection. PLoS Pathog 5:e1000361

Akhter A, Caution K, Abu Khweek A et al (2012) Caspase-11 promotes the fusion of phagosomes harboring pathogenic bacteria with lysosomes by modulating actin polymerization. Immunity 37:35–47

Albers U, Tiaden A, Spirig T, Al Alam D, Goyert SM, Gangloff SC, Hilbi H (2007) Expression of *Legionella pneumophila* paralogous lipid A biosynthesis genes under different growth conditions. Microbiology 153:3817–3829

Albert-Weissenberger C, Sahr T, Sismeiro O, Hacker J, Heuner K, Buchrieser C (2010) Control of flagellar gene regulation in *Legionella pneumophila* and its relation to growth phase. J Bacteriol 192:446–455

Alli OA, Gao LY, Pedersen LL, Zink S, Radulic M, Doric M, Abu Kwaik Y (2000) Temporal pore formation-mediated egress from macrophages and alveolar epithelial cells by *Legionella pneumophila*. Infect Immun 68:6431–6440

Amer AO (2010) Modulation of caspases and their non-apoptotic functions by *Legionella pneumophila*. Cell Microbiol 12:140–147

Amer A, Franchi L, Kanneganti TD et al (2006) Regulation of *Legionella* phagosome maturation and infection through flagellin and host Ipaf. J Biol Chem 281:35217–35223

Aurass P, Schlegel M, Metwally O, Harding CR, Schroeder GN, Frankel G, Flieger A (2013) The *Legionella pneumophila* Dot/Icm-secreted effector PlcC/CegC1 together with PlcA and PlcB promotes virulence and belongs to a novel zinc metallophospholipase C family present in bacteria and fungi. J Biol Chem 288:11080–11092

Bachman MA, Swanson MS (2001) RpoS co-operates with other factors to induce *Legionella pneumophila* virulence in the stationary phase. Mol Microbiol 40:1201–1214

Banga S, Gao P, Shen X, Fiscus V, Zong WX, Chen L, Luo ZQ (2007) *Legionella pneumophila* inhibits macrophage apoptosis by targeting pro-death members of the Bcl2 protein family. Proc Natl Acad Sci USA 104:5121–5126

Bartfeld S, Engels C, Bauer B, Aurass P, Flieger A, Brüggemann H, Meyer TF (2009) Temporal resolution of two-tracked NF-kappaB activation by *Legionella pneumophila*. Cell Microbiol 11:1638–1651

Belyi Y, Niggeweg R, Opitz B, Vogelsgesang M, Hippenstiel S, Wilm M, Aktories K (2006) *Legionella pneumophila* glucosyltransferase inhibits host elongation factor 1A. Proc Natl Acad Sci USA 103:16953–16958

Belyi Y, Tabakova I, Stahl M, Aktories K (2008) Lgt: a family of cytotoxic glucosyltransferases produced by *Legionella pneumophila*. J Bacteriol 190:3026–3035
Bhardwaj N, Nash TW, Horwitz MA (1986) Interferon-gamma-activated human monocytes inhibit the intracellular multiplication of *Legionella pneumophila*. J Immunol 137:2662–2669
Bosshardt SC, Benson RF, Fields BS (1997) Flagella are a positive predictor for virulence in *Legionella*. Microb Pathog 23:107–112
Brassinga AK, Kinchen JM, Cupp ME, Day SR, Hoffman PS, Sifri CD (2010) *Caenorhabditis* is a metazoan host for *Legionella*. Cell Microbiol 12:343–361
Brieland J, Freeman P, Kunkel R, Chrisp C, Hurley M, Fantone J, Engleberg C (1994) Replicative *Legionella pneumophila* lung infection in intratracheally inoculated A/J mice. A murine model of human Legionnaires' disease. Am J Pathol 145:1537–1546
Brombacher E, Urwyler S, Ragaz C, Weber SS, Kami K, Overduin M, Hilbi H (2009) Rab1 guanine nucleotide exchange factor SidM is a major phosphatidylinositol 4-phosphate-binding effector protein of *Legionella pneumophila*. J Biol Chem 284:4846–4856
Brüggemann H, Hagman A, Jules M et al (2006) Virulence strategies for infecting phagocytes deduced from the in vivo transcriptional program of *Legionella pneumophila*. Cell Microbiol 8:1228–1240
Byrd TF, Horwitz MA (1989) Interferon gamma-activated human monocytes downregulate transferrin receptors and inhibit the intracellular multiplication of *Legionella pneumophila* by limiting the availability of iron. J Clin Invest 83:1457–1465
Byrne B, Swanson MS (1998) Expression of *Legionella pneumophila* virulence traits in response to growth conditions. Infect Immun 66:3029–3034
Case CL, Roy CR (2011) Asc modulates the function of NLRC4 in response to infection of macrophages by *Legionella pneumophila*. MBio 2:e00117-11
Case CL, Shin S, Roy CR (2009) Asc and Ipaf Inflammasomes direct distinct pathways for caspase-1 activation in response to *Legionella pneumophila*. Infect Immun 77:1981–1991
Case CL, Kohler LJ, Lima JB et al (2013) Caspase-11 stimulates rapid flagellin-independent pyroptosis in response to *Legionella pneumophila*. Proc Natl Acad Sci USA 110:1851–1856
Cerqueira DM, Pereira MS, Silva AL, Cunha LD, Zamboni DS (2015) Caspase-1 but not Caspase-11 is required for NLRC4-mediated pyroptosis and restriction of infection by flagellated *Legionella* species in mouse macrophages and in vivo. J Immunol 195:2303–2311
Chow J, Franz KM, Kagan JC (2015) PRRs are watching you: localization of innate sensing and signaling regulators. Virology 479–480C:104–109
Choy A, Dancourt J, Mugo B, O'Connor TJ, Isberg RR, Melia TJ, Roy CR (2012) The *Legionella* effector RavZ inhibits host autophagy through irreversible Atg8 deconjugation. Science 338:1072–1076
Coers J, Vance RE, Fontana MF, Dietrich WF (2007) Restriction of *Legionella pneumophila* growth in macrophages requires the concerted action of cytokine and Naip5/Ipaf signalling pathways. Cell Microbiol 9:2344–2357
Correia AM, Ferreira JS, Borges V et al (2016) Probable person-to-person transmission of Legionnaires' disease. N Engl J Med 374:497–498
Creagh EM, O'Neill LA (2006) TLRs, NLRs and RLRs: a trinity of pathogen sensors that co-operate in innate immunity. Trends Immunol 27:352–357
Creasey EA, Isberg RR (2012) The protein SdhA maintains the integrity of the *Legionella*-containing vacuole. Proc Natl Acad Sci USA 109:3481–3486
Dalebroux ZD, Edwards RL, Swanson MS (2009) SpoT governs *Legionella pneumophila* differentiation in host macrophages. Mol Microbiol 71:640–658
Declerck P (2010) Biofilms: the environmental playground of *Legionella pneumophila*. Environ Microbiol 12:557–566
Dietrich WF, Damron DM, Isberg RR, Lander ES, Swanson MS (1995) Lgn1, a gene that determines susceptibility to *Legionella pneumophila*, maps to mouse chromosome 13. Genomics 26:443–450

Dietrich C, Heuner K, Brand BC, Hacker J, Steinert M (2001) Flagellum of *Legionella pneumophila* positively affects the early phase of infection of eukaryotic host cells. Infect Immun 69:2116–2122

Diez E, Lee SH, Gauthier S, Yaraghi Z, Tremblay M, Vidal S, Gros P (2003) Birc1e is the gene within the Lgn1 locus associated with resistance to *Legionella pneumophila*. Nat Genet 33:55–60

Edelstein PH, Beer KB, DeBoynton ED (1987) Influence of growth temperature on virulence of *Legionella pneumophila*. Infect Immun 55:2701–2705

Fettes PS, Forsbach-Birk V, Lynch D, Marre R (2001) Overexpresssion of a *Legionella pneumophila* homologue of the *E. coli* regulator *csrA* affects cell size, flagellation, and pigmentation. Int J Med Microbiol 291:353–360

Fields BS (1996) The molecular ecology of *Legionellae*. Trends Microbiol 4:286–290

Finsel I, Hilbi H (2015) Formation of a pathogen vacuole according to *Legionella pneumophila*: how to kill one bird with many stones. Cell Microbiol 17:935–950

Finsel I, Ragaz C, Hoffmann C et al (2013) The *Legionella* effector RidL inhibits retrograde trafficking to promote intracellular replication. Cell Host Microbe 14:38–50

Fontana MF, Banga S, Barry KC, Shen X, Tan Y, Luo ZQ, Vance RE (2011) Secreted bacterial effectors that inhibit host protein synthesis are critical for induction of the innate immune response to virulent *Legionella pneumophila*. PLoS Pathog 7:e1001289

Forsbach-Birk V, McNealy T, Shi C, Lynch D, Marre R (2004) Reduced expression of the global regulator protein CsrA in *Legionella pneumophila* affects virulence-associated regulators and growth in *Acanthamoeba castellanii*. Int J Med Microbiol 294:15–25

Fortier A, de Chastellier C, Balor S, Gros P (2007) Birc1e/Naip5 rapidly antagonizes modulation of phagosome maturation by *Legionella pneumophila*. Cell Microbiol 9:910–923

Garcia-Fulgueiras A, Navarro C, Fenoll D et al (2003) Legionnaires' disease outbreak in Murcia, Spain. Emerg Infect Dis 9:915–921

Gaspar AH, Machner MP (2014) VipD is a Rab5-activated phospholipase A1 that protects *Legionella pneumophila* from endosomal fusion. Proc Natl Acad Sci USA 111:4560–4565

Ge J, Xu H, Li T et al (2009) A *Legionella* type IV effector activates the NF-κB pathway by phosphorylating the IκB family of inhibitors. Proc Natl Acad Sci USA 106:13725–13730

Ge J, Gong YN, Xu Y, Shao F (2012) Preventing bacterial DNA release and absent in melanoma 2 inflammasome activation by a *Legionella* effector functioning in membrane trafficking. Proc Natl Acad Sci USA 109:6193–6198

Gilmore TD (2006) Introduction to NF-κB: players, pathways, perspectives. Oncogene 25: 6680–6684

Hales LM, Shuman HA (1999) The *Legionella pneumophila rpoS* gene is required for growth within *Acanthamoeba castellanii*. J Bacteriol 181:4879–4889

Hammer BK, Swanson MS (1999) Co-ordination of *Legionella pneumophila* virulence with entry into stationary phase by ppGpp. Mol Microbiol 33:721–731

Hammer BK, Tateda ES, Swanson MS (2002) A two-component regulator induces the transmission phenotype of stationary-phase *Legionella pneumophila*. Mol Microbiol 44: 107–118

Heuner K, Bender-Beck L, Brand BC et al (1995) Cloning and genetic characterization of the flagellum subunit gene (*flaA*) of *Legionella pneumophila* serogroup 1. Infect Immun 63: 2499–2507

Heuner K, Hacker J, Brand BC (1997) The alternative sigma factor sigma28 of *Legionella pneumophila* restores flagellation and motility to an *Escherichia coli fliA* mutant. J Bacteriol 179:17–23

Heuner K, Brand BC, Hacker J (1999) The expression of the flagellum of *Legionella pneumophila* is modulated by different environmental factors. FEMS Microbiol Lett 175:69–77

Hilbi H, Haas A (2012) Secretive bacterial pathogens and the secretory pathway. Traffic 13: 1187–1197

Hilbi H, Jarraud S, Hartland E, Buchrieser C (2010) Update on Legionnaires' disease: pathogenesis, epidemiology, detection and control. Mol Microbiol 76:1–11

Hilbi H, Hoffmann C, Harrison CF (2011) *Legionella* spp. outdoors: colonization, communication and persistence. Environ Microbiol Rep 3:286–296

Hoffmann C, Finsel I, Otto A et al (2014a) Functional analysis of novel Rab GTPases identified in the proteome of purified *Legionella*-containing vacuoles from macrophages. Cell Microbiol 16:1034–1052

Hoffmann C, Harrison CF, Hilbi H (2014b) The natural alternative: protozoa as cellular models for *Legionella* infection. Cell Microbiol 16:15–26

Horwitz MA, Silverstein SC (1980) Legionnaires' disease bacterium (*Legionella pneumophila*) multiplies intracellularly in human monocytes. J Clin Invest 66:441–450

Hovel-Miner G, Pampou S, Faucher SP et al (2009) SigmaS controls multiple pathways associated with intracellular multiplication of *Legionella pneumophila*. J Bacteriol 191:2461–2473

Hsu F, Zhu W, Brennan L, Tao L, Luo ZQ, Mao Y (2012) Structural basis for substrate recognition by a unique *Legionella* phosphoinositide phosphatase. Proc Natl Acad Sci USA 109:13567–13572

Hsu F, Luo X, Qiu J et al (2014) The *Legionella* effector SidC defines a unique family of ubiquitin ligases important for bacterial phagosomal remodeling. Proc Natl Acad Sci USA 111: 10538–10543

Hubber A, Roy CR (2010) Modulation of host cell function by *Legionella pneumophila* type IV effectors. Annu Rev Cell Dev Biol 26:261–283

Isberg RR, O'Connor TJ, Heidtman M (2009) The *Legionella pneumophila* replication vacuole: making a cosy niche inside host cells. Nat Rev Microbiol 7:13–24

Jacobi S, Schade R, Heuner K (2004) Characterization of the alternative sigma factor sigma54 and the transcriptional regulator FleQ of *Legionella pneumophila*, which are both involved in the regulation cascade of flagellar gene expression. J Bacteriol 186:2540–2547

Jules M, Buchrieser C (2007) *Legionella pneumophila* adaptation to intracellular life and the host response: clues from genomics and transcriptomics. FEBS Lett 581:2829–2838

Kessler A, Schell U, Sahr T, Tiaden A, Harrison C, Buchrieser C, Hilbi H (2013) The *Legionella pneumophila* orphan sensor kinase LqsT regulates competence and pathogen-host interactions as a component of the LAI-1 circuit. Environ Microbiol 15:646–662

Kofoed EM, Vance RE (2011) Innate immune recognition of bacterial ligands by NAIPs determines inflammasome specificity. Nature 477:592–595

Komura T, Yasui C, Miyamoto H, Nishikawa Y (2010) *Caenorhabditis elegans* as an alternative model host for *Legionella pneumophila*, and protective effects of *Bifidobacterium infantis*. Appl Environ Microbiol 76:4105–4108

Laguna RK, Creasey EA, Li Z, Valtz N, Isberg RR (2006) A *Legionella pneumophila*-translocated substrate that is required for growth within macrophages and protection from host cell death. Proc Natl Acad Sci USA 103:18745–18750

Lamkanfi M, Amer A, Kanneganti TD et al (2007) The Nod-like receptor family member Naip5/Birc1e restricts *Legionella pneumophila* growth independently of caspase-1 activation. J Immunol 178:8022–8027

Lei Y, Wen H, Yu Y et al (2012) The mitochondrial proteins NLRX1 and TUFM form a complex that regulates type I interferon and autophagy. Immunity 36:933–946

Lightfield KL, Persson J, Brubaker SW et al (2008) Critical function for Naip5 in inflammasome activation by a conserved carboxy-terminal domain of flagellin. Nat Immunol 9:1171–1178

Losick VP, Isberg RR (2006) NF-κB translocation prevents host cell death after low-dose challenge by *Legionella pneumophila*. J Exp Med 203:2177–2189

Losick VP, Haenssler E, Moy MY, Isberg RR (2010) LnaB: a *Legionella pneumophila* activator of NF-kappaB. Cell Microbiol 12:1083–1097

Lu H, Clarke M (2005) Dynamic properties of *Legionella*-containing phagosomes in *Dictyostelium* amoebae. Cell Microbiol 7:995–1007

Luo ZQ, Isberg RR (2004) Multiple substrates of the *Legionella pneumophila* Dot/Icm system identified by interbacterial protein transfer. Proc Natl Acad Sci USA 101:841–846

Machner MP, Isberg RR (2006) Targeting of host Rab GTPase function by the intravacuolar pathogen *Legionella pneumophila*. Dev Cell 11:47–56

Massis LM, Zamboni DS (2011) Innate immunity to *Legionella pneumophila*. Front Microbiol 2:109

Mauchline WS, Araujo R, Wait R, Dowsett AB, Dennis PJ, Keevil CW (1992) Physiology and morphology of *Legionella pneumophila* in continuous culture at low oxygen concentration. J Gen Microbiol 138:2371–2380

McDade JE, Shepard CC, Fraser DW, Tsai TR, Redus MA, Dowdle WR (1977) Legionnaires' disease: isolation of a bacterium and demonstration of its role in other respiratory disease. N Engl J Med 297:1197–1203

Miao EA, Leaf IA, Treuting PM et al (2010) Caspase-1-induced pyroptosis is an innate immune effector mechanism against intracellular bacteria. Nat Immunol 11:1136–1142

Molofsky AB, Swanson MS (2003) *Legionella pneumophila* CsrA is a pivotal repressor of transmission traits and activator of replication. Mol Microbiol 50:445–461

Molofsky AB, Swanson MS (2004) Differentiate to thrive: lessons from the *Legionella pneumophila* life cycle. Mol Microbiol 53:29–40

Molofsky AB, Shetron-Rama LM, Swanson MS (2005) Components of the *Legionella pneumophila* flagellar regulon contribute to multiple virulence traits, including lysosome avoidance and macrophage death. Infect Immun 73:5720–5734

Molofsky AB, Byrne BG, Whitfield NN, Madigan CA, Fuse ET, Tateda K, Swanson MS (2006) Cytosolic recognition of flagellin by mouse macrophages restricts *Legionella pneumophila* infection. J Exp Med 203:1093–1104

Monroe KM, McWhirter SM, Vance RE (2009) Identification of host cytosolic sensors and bacterial factors regulating the type I interferon response to *Legionella pneumophila*. PLoS Pathog 5:e1000665

Müller MP, Peters H, Blumer J, Blankenfeldt W, Goody RS, Itzen A (2010) The *Legionella* effector protein DrrA AMPylates the membrane traffic regulator Rab1b. Science 329:946–949

Murata T, Delprato A, Ingmundson A, Toomre DK, Lambright DG, Roy CR (2006) The *Legionella pneumophila* effector protein DrrA is a Rab1 guanine nucleotide-exchange factor. Nat Cell Biol 8:971–977

Nagai H, Kubori T (2011) Type IVB secretion systems of *Legionella* and other Gram-negative bacteria. Front Microbiol 2:136

Nagai H, Kagan JC, Zhu X, Kahn RA, Roy CR (2002) A bacterial guanine nucleotide exchange factor activates ARF on *Legionella* phagosomes. Science 295:679–682

Nash TW, Libby DM, Horwitz MA (1988) IFN-gamma-activated human alveolar macrophages inhibit the intracellular multiplication of *Legionella pneumophila*. J Immunol 140:3978–3981

Naujoks J, Tabeling C, Dill BD et al (2016) IFNs modify the proteome of *Legionella*-containing vacuoles and restrict infection via IRG1-derived itaconic acid. PLoS Pathog 12:e1005408

Neumeister B, Faigle M, Sommer M et al (1998) Low endotoxic potential of *Legionella pneumophila* lipopolysaccharide due to failure of interaction with the monocyte lipopolysaccharide receptor CD14. Infect Immun 66:4151–4157

Newton HJ, Ang DK, van Driel IR, Hartland EL (2010) Molecular pathogenesis of infections caused by *Legionella pneumophila*. Clin Microbiol Rev 23:274–298

Nogueira CV, Lindsten T, Jamieson AM, Case CL, Shin S, Thompson CB, Roy CR (2009) Rapid pathogen-induced apoptosis: a mechanism used by dendritic cells to limit intracellular replication of *Legionella pneumophila*. PLoS Pathog 5:e1000478

Ott M, Messner P, Heesemann J, Marre R, Hacker J (1991) Temperature-dependent expression of flagella in *Legionella*. J Gen Microbiol 137:1955–1961

Pereira MS, Marques GG, Dellama JE, Zamboni DS (2011a) The Nlrc4 inflammasome contributes to restriction of pulmonary infection by flagellated *Legionella* spp. that trigger pyroptosis. Front Microbiol 2:33

Pereira MS, Morgantetti GF, Massis LM, Horta CV, Hori JI, Zamboni DS (2011b) Activation of NLRC4 by flagellated bacteria triggers caspase-1-dependent and -independent responses to restrict *Legionella pneumophila* replication in macrophages and *in vivo*. J Immunol 187:6447–6455

Personnic N, Bärlocher K, Finsel I, Hilbi H (2016) Subversion of retrograde trafficking by translocated pathogen effectors. Trends Microbiol 24:450–462

Pine L, George JR, Reeves MW, Harrell WK (1979) Development of a chemically defined liquid medium for growth of *Legionella pneumophila*. J Clin Microbiol 9:615–626

Pruckler JM, Benson RF, Moyenuddin M, Martin WT, Fields BS (1995) Association of flagellum expression and intracellular growth of *Legionella pneumophila*. Infect Immun 63:4928–4932

Rasis M, Segal G (2009) The LetA-RsmYZ-CsrA regulatory cascade, together with RpoS and PmrA, post-transcriptionally regulates stationary phase activation of *Legionella pneumophila* Icm/Dot effectors. Mol Microbiol 72:995–1010

Ren T, Zamboni DS, Roy CR, Dietrich WF, Vance RE (2006) Flagellin-deficient *Legionella* mutants evade caspase-1- and Naip5-mediated macrophage immunity. PLoS Pathog 2:e18

Robinson CG, Roy CR (2006) Attachment and fusion of endoplasmic reticulum with vacuoles containing *Legionella pneumophila*. Cell Microbiol 8:793–805

Rolando M, Sanulli S, Rusniok C et al (2013) *Legionella pneumophila* effector RomA uniquely modifies host chromatin to repress gene expression and promote intracellular bacterial replication. Cell Host Microbe 13:395–405

Rothmeier E, Pfaffinger G, Hoffmann C et al (2013) Activation of Ran GTPase by a *Legionella* effector promotes microtubule polymerization, pathogen vacuole motility and infection. PLoS Pathog 9:e1003598

Rowbotham TJ (1986) Current views on the relationships between amoebae, legionellae and man. Isr J Med Sci 22:678–689

Sahr T, Brüggemann H, Jules M, Lomma M, Albert-Weissenberger C, Cazalet C, Buchrieser C (2009) Two small ncRNAs jointly govern virulence and transmission in *Legionella pneumophila*. Mol Microbiol 72:741–762

Sauer JD, Bachman MA, Swanson MS (2005) The phagosomal transporter A couples threonine acquisition to differentiation and replication of *Legionella pneumophila* in macrophages. Proc Natl Acad Sci USA 102:9924–9929

Schell U, Kessler A, Hilbi H (2014) Phosphorylation signalling through the *Legionella* quorum sensing histidine kinases LqsS and LqsT converges on the response regulator LqsR. Mol Microbiol 92:1039–1055

Schell U, Simon S, Sahr T et al (2015) The α-hydroxyketone LAI-1 regulates motility, Lqs-dependent phosphorylation signaling and gene expression of *Legionella pneumophila*. Mol Microbiol 99:778–793

Schroder K, Tschopp J (2010) The inflammasomes. Cell 140:821–832

Sellge G, Kufer TA (2015) PRR-signaling pathways—learning from microbial tactics. Semin Immunol 27:75–84

Shen X, Banga S, Liu Y et al (2009) Targeting eEF1A by a *Legionella pneumophila* effector leads to inhibition of protein synthesis and induction of host stress response. Cell Microbiol 11:911–926

Sherwood RK, Roy CR (2013) A Rab-centric perspective of bacterial pathogen-occupied vacuoles. Cell Host Microbe 14:256–268

Silveira TN, Zamboni DS (2010) Pore formation triggered by *Legionella* spp. is an Nlrc4 inflammasome-dependent host cell response that precedes pyroptosis. Infect Immun 78:1403–1413

Simon S, Hilbi H (2015) Subversion of cell-autonomous immunity and cell migration by *Legionella pneumophila* effectors. Front Immunol 6:447

Simon S, Wagner MA, Rothmeier E, Müller-Taubenberger A, Hilbi H (2014) Icm/Dot-dependent inhibition of phagocyte migration by *Legionella* is antagonized by a translocated Ran GTPase activator. Cell Microbiol 16:977–992

Simon S, Schell U, Heuer N et al (2015) Inter-kingdom signaling by the *Legionella* quorum sensing molecule LAI-1 modulates cell migration through an IQGAP1-Cdc42-ARHGEF9-dependent pathway. PLoS Pathog 11:e1005307

Solomon JM, Isberg RR (2000) Growth of *Legionella pneumophila* in *Dictyostelium discoideum*: a novel system for genetic analysis of host-pathogen interactions. Trends Microbiol 8:478–480

Spirig T, Tiaden A, Kiefer P, Buchrieser C, Vorholt JA, Hilbi H (2008) The *Legionella* autoinducer synthase LqsA produces an α-hydroxyketone signaling molecule. J Biol Chem 283:18113–18123
Steinert M (2011) Pathogen-host interactions in *Dictyostelium*, *Legionella*, *Mycobacterium* and other pathogens. Semin Cell Dev Biol 22:70–76
Sutterwala FS, Ogura Y, Flavell RA (2007) The inflammasome in pathogen recognition and inflammation. J Leukoc Biol 82:259–264
Suzuki K, Babitzke P, Kushner SR, Romeo T (2006) Identification of a novel regulatory protein (CsrD) that targets the global regulatory RNAs CsrB and CsrC for degradation by RNase E. Genes Dev 20:2605–2617
Swanson MS, Isberg RR (1995) Association of *Legionella pneumophila* with the macrophage endoplasmic reticulum. Infect Immun 63:3609–3620
Tiaden A, Hilbi H (2012) α-Hydroxyketone synthesis and sensing by *Legionella* and *Vibrio*. Sensors 12:2899–2919
Tiaden A, Spirig T, Weber SS, Brüggemann H, Bosshard R, Buchrieser C, Hilbi H (2007) The *Legionella pneumophila* response regulator LqsR promotes host cell interactions as an element of the virulence regulatory network controlled by RpoS and LetA. Cell Microbiol 9:2903–2920
Tiaden A, Spirig T, Carranza P et al (2008) Synergistic contribution of the *Legionella pneumophila lqs* genes to pathogen-host interactions. J Bacteriol 190:7532–7547
Tiaden A, Spirig T, Hilbi H (2010a) Bacterial gene regulation by α-hydroxyketone signaling. Trends Microbiol 18:288–297
Tiaden A, Spirig T, Sahr T, Wälti MA, Boucke K, Buchrieser C, Hilbi H (2010b) The autoinducer synthase LqsA and putative sensor kinase LqsS regulate phagocyte interactions, extracellular filaments and a genomic island of *Legionella pneumophila*. Environ Microbiol 12:1243–1259
Vance RE (2015) The NAIP/NLRC4 inflammasomes. Curr Opin Immunol 32:84–89
Vinzing M, Eitel J, Lippmann J et al (2008) NAIP and Ipaf control *Legionella pneumophila* replication in human cells. J Immunol 180:6808–6815
Weber SS, Ragaz C, Reus K, Nyfeler Y, Hilbi H (2006) *Legionella pneumophila* exploits PI(4) *P* to anchor secreted effector proteins to the replicative vacuole. PLoS Pathog 2:e46
Weber S, Stirnimann CU, Wieser M et al (2014) A type IV translocated *Legionella* cysteine phytase counteracts intracellular growth restriction by phytate. J Biol Chem 289:34175–34188
Wigneshweraraj S, Bose D, Burrows PC et al (2008) Modus operandi of the bacterial RNA polymerase containing the sigma54 promoter-specificity factor. Mol Microbiol 68:538–546
Wright EK, Goodart SA, Growney JD et al (2003) Naip5 affects host susceptibility to the intracellular pathogen *Legionella pneumophila*. Curr Biol 13:27–36
Yamamoto Y, Klein TW, Newton CA, Widen R, Friedman H (1988) Growth of *Legionella pneumophila* in thioglycolate-elicited peritoneal macrophages from A/J mice. Infect Immun 56:370–375
Zamboni DS, Kobayashi KS, Kohlsdorf T et al (2006) The Birc1e cytosolic pattern-recognition receptor contributes to the detection and control of *Legionella pneumophila* infection. Nat Immunol 7:318–325
Zhao Y, Shao F (2015) Diverse mechanisms for inflammasome sensing of cytosolic bacteria and bacterial virulence. Curr Opin Microbiol 29:37–42
Zhu W, Hammad LA, Hsu F, Mao Y, Luo ZQ (2013) Induction of caspase 3 activation by multiple *Legionella pneumophila* Dot/Icm substrates. Cell Microbiol 15:1783–1795
Zusman T, Gal-Mor O, Segal G (2002) Characterization of a *Legionella pneumophila relA* insertion mutant and roles of RelA and RpoS in virulence gene expression. J Bacteriol 184:67–75

Inflammasome Activation and Function During Infection with *Mycobacterium tuberculosis*

Andrea Ablasser and Anca Dorhoi

Abstract Tuberculosis (TB) is caused by infection with *Mycobacterium tuberculosis (Mtb)* and represents one of the most relevant bacterial diseases worldwide. Recent advances have yielded new insights into the molecular basis of the immune response required for restriction of the pathogen and also highlighted determinants of immunopathology in TB. Several innate immune mediators including soluble proteins as well as lipid molecules participate in both processes, and their mechanisms of action during TB have been extensively studies over the past years. Among those mediators, inflammasomes are essential signaling platforms that execute crucial functions in several areas of immunology and beyond. This chapter aims to summarize what is known about the roles of the inflammasome during infection with *Mtb* from both in vitro studies as well as from in vivo work. A better understanding of the complex interactions between *Mtb* and the host immune system could reveal novel therapeutic approaches and improve current vaccination protocols in TB.

Contents

A. Ablasser (✉)
Global Health Institute, Ecole Polytechnique Fédérale de Lausanne (EPFL), Station 19, 1015 Lausanne, Switzerland
e-mail: andrea.ablasser@epfl.ch

A. Dorhoi
Max Planck Institute for Infection Biology, 10117 Berlin, Germany

S. Backert (ed.), *Inflammasome Signaling and Bacterial Infections*, Current Topics in Microbiology and Immunology 397,
DOI 10.1007/978-3-319-41171-2_9

1 Introduction

Tuberculosis (TB) is a chronic bacterial disease most frequently affecting the lung. Despite available chemotherapy and vaccine programs, TB continues to cause high death tolls worldwide. In 2014, WHO estimated that 1.5 million people died from TB and about 9.6 million new cases have been diagnosed. The majority (90–95 %) of individuals exposed to *Mycobacterium tuberculosis (Mtb)*, the causative agent of TB, do not develop active disease. Instead, most individuals contain *Mtb*, likely within tightly organized accumulation of immune cells termed granulomas [so-called latent infection (LTBI)] (O'Garra et al. 2013). Active TB occurs either upon disease reactivation in LTBI cases or as progressive pulmonary disease in ca. 5–10 % of infected individuals. In these cases, unrestricted mycobacterial propagation leads to a protracted disease course, which is accompanied by destruction of the lung parenchyma.

The innate immune response is critical for host resistance against TB. This has been demonstrated by numerous observations in humans and experimental studies. Children with deficiencies in the IL-12/IFN-γ pathway or adults under perturbed TNF-α abundance or signaling are susceptible to TB (Abel et al. 2014; Dorhoi and Kaufmann 2014). Similarly, investigations in murine models of TB have unveiled an essential role for these cytokines in defense against *Mtb* (O'Garra et al. 2013). More recently, the IL-1 family of cytokines has emerged as indispensable for control of the disease. Indeed IL-1R knockout mice succumb rapidly following low-dose aerosol infection with *Mtb* (Di Paolo et al. 2015; Fremond et al. 2007; Mayer-Barber et al. 2011). Both IL-1R ligands, IL-1α and IL-1β, independently exert essential functions in anti-mycobacterial defense. While native IL-1α can exert a signal, the production of bioactive IL-1, especially IL-1β, is subject to complex layers of regulation. The recognition of the bacilli promotes transcription of the pro-form of the cytokine, whereas protease-mediated cleavage generates mature IL-1β.

During pulmonary TB, the most prevalent form of the disease, *Mtb* is transmitted via aerosols and recognized within the bronchoalveolar space by cells of the innate immune system, including alveolar macrophages, neutrophils, inflammatory monocytes, and dendritic cells (DCs) (Mayer-Barber and Barber 2015). Multiple pattern recognition receptors (PRRs) participate in the detection of *Mtb* during this early stage (Philips and Ernst 2012; Stamm et al. 2015). For example, cell surface expressed Toll-like receptors (TLR) and C-type lectin receptors (CLR) as well as cytosolic receptors including nucleotide oligomerization domain (NOD)-like receptors (NLR), cyclic GMP-AMP synthase (cGAS), and the aryl hydrocarbon receptor (AhR) have been implicated in the recognition of mycobacteria (see details below) (Philips and Ernst 2012; Stamm et al. 2015). Stimulation of these sensors collectively induces the expression of pro-inflammatory cytokines, including IL-1 members, chemokines, and cell adhesion molecules, which orchestrate the local and systemic immune response to *Mtb* (O'Garra et al. 2013; Philips and Ernst 2012). In macrophages, the preferred cellular host of mycobacteria, production of IL-1β depends on the activation of the inflammasome. However, studies with gene-deficient mice indicate that the role of inflammasomes in the context of pulmonary TB is more complex than anticipated. Determining how exactly inflammasomes affect anti-bacterial immunity or immunopathology at different stages of the disease remains a challenge. In-depth understanding of these biological processes will enrich our knowledge of fundamental pathophysiological mechanisms and could contribute to development of novel therapies or prevention strategies against TB. In this chapter, the role of inflammasomes as sensors of *Mtb* and their impact on the pathogenesis of TB will be discussed.

2 Early Events in *M. tuberculosis*-Infected Phagocytes

Macrophages represent the most studied phagocyte type in TB. Alveolar macrophages are among the first responders within the lung tissue and readily sense and phagocytose *Mtb* through a variety of surface-expressed receptors including scavenger receptors, Fc receptors, complement receptors, and CLR [reviewed in Philips and Ernst (2012) and Stamm et al. (2015)]. In addition to mediating bacilli uptake, such receptors are also critical for the early activation of macrophages, for example, by triggering signaling pathways that culminate in the production of immunomodulatory cytokines and chemokines. Numerous membrane-bound PRRs, including members of the TLR family (TLR2, 4, 9) and CLRs (Mincle, Dectin-1, mannose receptor, and dendritic cell-specific intercellular adhesion molecule-3-grabbing non-integrin) [for details see review Stamm et al. (2015)] and cytosolic PRRs (e.g., NOD2) contribute to the regulation of cytokines during TB. These PRRs recognize distinct agonists, some of which are common to many bacteria (e.g., peptidoglycan), whereas others are unique to *mycobacteria* (e.g., lipoarabinomannan). Because many PRRs are coexpressed by the same cell, the resulting cytokine signature emerges from the cooperative action of multiple signal transduction cascades. On the other hand, given the diverse mechanism of *Mtb* recognition, redundancy in mycobacterial detection has been observed. As such, deletion of single or several receptors does not impact on host susceptibility to TB (Court et al. 2010; Heitmann et al. 2013; Holscher et al. 2008). However, absence of adaptors positioned at hubs of PRR signaling network induces lethality of experimentally infected animals. Mice deficient in MyD88 or Card9, the common downstream adaptors for TLRs/IL-1R or CLRs, succumb rapidly after aerosol infection with *Mtb* (Dorhoi et al. 2010; Fremond et al. 2004; Mayer-Barber et al. 2010). These adaptors and their upstream

PRRs regulate transcript abundance of various immune mediators, including IL-1R ligands.

The exact sequence of molecular events that follow entry of *Mtb* into host macrophages is complex and still not entirely clear. However, it appears that through the coordinate action of key virulence factors, *Mtb* restricts phagosomal maturation to permit replication within macrophages. In addition, the type VII secretion system ESX-1 (see below) allows *Mtb* to escape into the cytosol. Further dissemination depends upon macrophage cytolysis, which is triggered upon reaching a critical threshold of intracellular bacteria load (Repasy et al. 2013). Although most research has focused on macrophages as prime targets for *Mtb*, various myeloid cells such as DCs, neutrophils, and monocytes also engulf *Mtb* and contribute to early defense against mycobacteria (Dorhoi and Kaufmann 2015; Lowe et al. 2012; Srivastava et al. 2014).

3 Regulation of Inflammasome Activation via the Mycobacterial Type VII Secretion System (T7SS)

An important feature of virulent *Mtb* strains is their ability to export proteins via specialized secretory systems, among which the T7SS called ESX-1 is most well studied. It is responsible for mediating the release of major virulence factors into infected cells, and it is also implicated in the rupture of the phagosomal membrane, an important prerequisite for the cytosolic escape of *Mtb* (de Jonge et al. 2007; Houben et al. 2012; Hsu et al. 2003; Simeone et al. 2012; van der Wel et al. 2007). Indeed, the ESX-1 system is regarded as a key mediator of the pathogenicity of *Mtb* and mutants lacking ESX-1 are defective for replication within macrophages and are severely attenuated in experimental murine TB (Hsu et al. 2003). The region of difference 1 (RD1) locus, which encodes many components of the ESX-1 apparatus, is deleted in the attenuated vaccine strain *M. bovis bacillus Calmette-Guerin* (BCG) (Pym et al. 2002). How the individual components of the ESX-1 system work together to generate an intact secretion machinery is incompletely understood, as is the precise function/nature of its effector proteins [see for review Majlessi et al. (2015), Simeone et al. (2009)]. Most of the investigations addressing interaction of the ESX-1 system with immune elements have focused on the substrates EsxA (ESAT-6) and EsxB (CFP-10), which are dominant T cell antigens in TB. These proteins are secreted as a heterodimer and require presence of auxiliary, co-secreted proteins in order to pass through the secretion apparatus (Majlessi et al. 2015; Renshaw et al. 2002). Interestingly, EsxA has membrane-permeabilizing ability and bacterial mutants lacking EsxA are attenuated (de Jonge et al. 2007). Apart from EsxA and EsxB, several other ESX-1 secretion-associated proteins (Esp) have been proposed as candidate substrates, which together with EsxA induce cell host lysis in the favor of *Mtb* persistence. Due to the importance of ESX-1 for translocation of bacterial effectors into the host cytosol, its expression has a strong impact on activation of intracellular PRR, including the inflammasome. As such, early reports document that mutant strains of *Mtb* and *M. marinum*, which lack the RD-1 locus, or *M. bovis* BCG are not able to induce IL-1β secretion in mouse and human macrophages in vitro (Dorhoi et al. 2012; Koo et al. 2008; Mishra et al. 2010; Novikov et al. 2011). Similarly, an EsxA mutant strain of *Mtb* does not induce IL-1β release in bone marrow-derived macrophages (BMDMs) (Mishra et al. 2010). Purified EsxA protein is also reported to trigger macrophage-dependent IL-1β production in the absence of mycobacteria (Mishra et al. 2010). Although EsxA represents an important virulence factor of *Mtb*, more recent findings suggest that the interaction between individual ESX-1 substrates and the host immune response is more complex than initially anticipated. For example, various attenuated strains, which are compromised in EsxA secretion (e.g., H37Ra, ΔEspC, ΔEspA), still provoke IL-1β production from mouse and human macrophages in vitro (McElvania Tekippe et al. 2010; Wassermann et al. 2015). These findings indicate that other ESX-1 substrates can perforate the phagosomal membrane or otherwise access the cytosol. Experiments employing mycobacterial knockin or knockout mutants will enable a more detailed understanding of the role of individual substrates in the host response toward *Mtb*. Aside from ESX-1, the ESX-5 system has also been implicated in macrophage activation and IL-1β production upon infection with *M. marinum* (Abdallah et al. 2011) and *Mtb* (Shah et al. 2015). Similar to ESX-1, the ESX-5 T7SS is responsible for the secretion of mycobacterial proteins, including EsxN/M and several PE and PPE proteins (Stoop et al. 2012). In addition, it also seems to play a role in the virulence of mycobacteria, since a *Mtb* ESX-5 mutant showed reduced ability to grow in vivo (Ates et al. 2016). However, the nature of the ESX-5 substrate(s) responsible for manipulation of the immune response of macrophages and the mechanism of action has not yet been resolved.

4 *M. tuberculosis* Activates the Inflammasome in Macrophages

Several reports have documented inflammasome activation after in vitro infection of macrophages with *Mtb*. Human monocytic cells (THP-1), BMDMs, and thioglycolate elicited macrophages readily secreted IL-1β upon *Mtb* infection (Koo et al. 2008; Mayer-Barber et al. 2010; McElvania Tekippe et al. 2010; Mishra et al. 2010). These initial studies showed that IL-1β secretion requires both ASC and caspase-1. Several studies suggest NLRP3 as a putative inflammasome sensor operating upstream of ASC and the inflammasome response upon *Mtb* infection (Dorhoi et al. 2012; Mishra et al. 2010; Walter et al. 2010) (Fig. 1). In support of this, macrophages from individuals that carry a combined polymorphism (Q705K) in the *NLRP3* gene together with a variant in *CARD8*, which is associated with up-regulated IL-1β production, better restrict mycobacterial growth ex vivo (Eklund et al. 2014). While the mechanism responsible for the antimycobacterial effects of

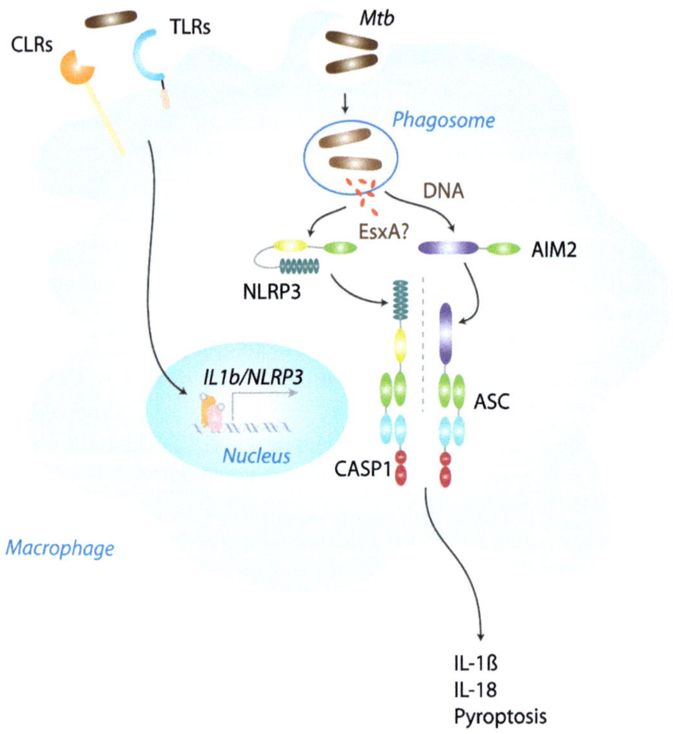

Fig. 1 Pattern recognition pathways relevant for *Mtb*-triggered inflammasome activation in macrophages. Cell surface-expressed TLRs and CLRs as well as cytosolic NLRs (NOD2) induce up-regulation of IL1B and NLRP3 mRNA following recognition of *Mtb*-derived molecules. In turn, ESX-1-mediated activation of the cytosolic inflammasome sensors NLRP3 and AIM2 leads to the maturation of IL-1 cytokines and to the induction of pyroptosis

NLRP3 was not examined in this study, it might be that NLRP3 does so indirectly through an IL-1β—TNFR feed-forward loop (Jayaraman et al. 2013). Similar to other NLRP3 agonists, the exact mechanism underlying mycobacterial-induced NLRP3 activation is not resolved. As highlighted above, the mycobacterial ESX-1 secretion system is an important trigger of the IL-1β response in macrophages. It has been suggested that through its membrano-lytic capability EsxA can directly promote activation of NLRP3 (Mishra et al. 2010). Whether this feature is shared by additional ESX-1 substrates is unknown. Potassium efflux is also important for activation of the NLRP3 inflammasome in context of *Mtb* infection (Dorhoi et al. 2012; Wassermann et al. 2015); however, disturbances of other ion fluxes (e.g., Ca^{2+}) through the action of ESX-1 could also potentially contribute to the activation of NLRP3. Recently, Nek7 has been identified as essential molecular component acting upstream of the NLRP3 (Schmid-Burgk et al. 2015; Shi et al. 2016). Nek7 interacts with NLRP3 and is essential for the assembly of the NLRP3 inflammasome. This raises the question whether IL-1β production during *Mtb* infection also relies on Nek7 or whether pathogen-derived proteins or virulence factors from *Mtb* directly engage NLRP3.

The AIM2 inflammasome has also been implicated in the intracellular recognition of *Mtb*. Mouse peritoneal macrophages or THP-1 cells deficient in AIM2 show diminished cleavage of caspase-1 and reduced production of bioactive IL-1β and IL-18 upon infection with virulent bacilli (Saiga et al. 2012; Wassermann et al. 2015). Given its function as intracellular DNA sensor, it is reasonable that similar to cGAS (see below), AIM2 could be directly engaged by mycobacterial genomic DNA. Indeed, co-localization of AIM2 with cytosolic DNA has been observed in macrophages infected with *Mtb* (Saiga et al. 2012) (Fig. 1).

5 *M. tuberculosis* Triggers Inflammasome-Dependent and Inflammasome-Independent IL-1β Production in Various Myeloid Cells

In addition to macrophages, DCs, neutrophils, and monocytes can also contribute to the host defense against *Mtb* by releasing IL-1. DCs infected with *Mtb* display a similar requirement for NLRP3 and ASC for maximal IL-1β secretion. Nevertheless, other pathways must contribute as DCs derived from mice deficient in both caspase-1 and caspase-11 are still capable of producing significant amounts of IL-1β and IL-18 albeit at lower levels than wild-type cells (Abdalla et al. 2012). In line with this, DCs defective in both NLRP3 and AIM2 release considerable amounts of IL-1β in response to various non-tuberculous mycobacteria, including *M. smegmatis*, *M. fortuitum*, and *M. kansaii* (Shah et al. 2013). Whereas a functional ESX-1 system is required for maximal IL-1β responses during *M. smegmatis* infection, ESX-1-deficient bacteria still provoke IL-1β secretion by BMDCs. This indicates that DCs produce bioactive IL-1β irrespective of ESX-1-triggered caspase-1 activation during mycobacterial infection. An intact ESX-1 system and thus an intracellular sensing mode appear to be even more dispensable for IL-1β production by human monocytes. As such, both BCG (lacking a functional ESX-1 system) and virulent *Mtb* induce similar amounts of IL-1β in monocytes (Novikov et al. 2011). It may be that constitutive activation of caspase-1 facilitates release of mature IL-1β by human monocytes in the absence of a bona fide inflammasome activator (Netea et al. 2009). Finally, IL1β production is independent of NLRP3 and caspase-1 in *Mtb*-infected neutrophils, relying perhaps on serine proteases such as proteinase 3, neutrophil elastase, or cathepsins for cleavage of the pre-form of the cytokine (Dorhoi et al. 2012; Netea et al. 2015). Together, these findings highlight that cell-specific mechanisms regulate IL-1β processing during mycobacterial infection, sometimes independently of “classical” inflammasome sensors or effectors (Fig. 2).

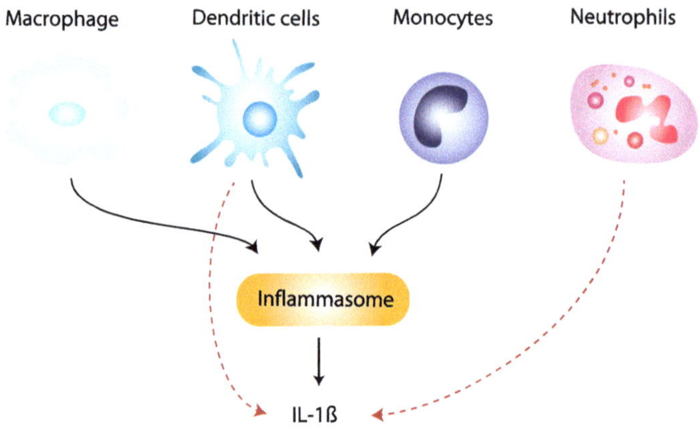

Fig. 2 Differential requirement for IL-1β activation upon *Mtb* infection. Both inflammasome-dependent processing and inflammasome-independent processing are used by immune cells to produce IL-1β during infection with *Mtb* in vivo

6 Activation of Cytosolic "Non-Inflammasome" Sensors

Aside from inflammasome sensors additional cytosolic PRRs mediate detection of mycobacterial components upon their exit from the endosomal compartment. The cGAS-STING DNA sensing pathway is responsible for the induction of type I interferons (IFNs) in macrophages infected with *Mtb* (Collins et al. 2015; Manzanillo et al. 2012; Wassermann et al. 2015; Watson et al. 2015). In addition, engagement of the cGAS-STING pathway is critical for triggering autophagy (Collins et al. 2015; Watson et al. 2012). Mice lacking cGAS display reduced production of type I IFNs and do not control the chronic infection, suggesting that cGAS plays a role in defense against *Mtb*. NOD2, a member of the NLR family of cytoplasmic sensors, recognizes mycobacterial cell wall component mycolylarabinogalactan–peptidoglycan (Gandotra et al. 2007). Whereas macrophages or DCs from NOD2-deficient animals displayed reduced levels of TNF-α, IL-12p40, and RANTES production in vitro (Ferwerda et al. 2005; Gandotra et al. 2007), the contribution of NOD2 for host resistance during aerogenic TB is less clear (Divangahi et al. 2008; Gandotra et al. 2007). Finally, a more recent study reports that AhR acts as a host sensor for *Mtb* by recognizing the pigmented virulence factor phthiocol produced by bacilli (Moura-Alves et al. 2014). AhR restricts mycobacterial growth inside macrophages and modulates abundance of pro-inflammatory cytokine, such as TNF-α, IL-12, and IL-6; as a result, AhR-deficient mice display heightened lethality in mouse models of TB.

7 Regulation of Inflammasome Activity During *M. tuberculosis* Infection

Several cytokine signaling cascades modulate inflammasome activation in a positive or negative manner (Gross et al. 2011). One regulatory pathway particularly relevant to *Mtb* infection is mediated by type I interferons (IFNs). Studies performed in TB patients and mouse models collectively highlight that type I IFNs have a detrimental role in TB (Torrado and Cooper 2013). Type I IFNs negatively regulate IL-1 production in human as well as in murine DCs and macrophages during infection with *Mtb* (Mayer-Barber et al. 2011, 2014; Novikov et al. 2011). Moreover, in vivo expression of IL-1β is increased in DCs from IFNAR1-deficient animals (Mayer-Barber et al. 2011). Transcriptional modulation of IL1B mRNA level is one mechanism by which type I IFNs negatively affect IL-1β in the setting of infection with mycobacteria (Novikov et al. 2011). Downstream inflammasome-mediated maturation of IL-1β is also subject to negative regulation by type I IFNs (Guarda et al. 2011). Notably, during TB a reciprocal regulation of IL-1 and type I IFNs has recently been revealed. IL-1 regulates type I IFN production via the cyclooxygenase-2 product prostaglandin E_2 (PGE_2) (Mayer-Barber et al. 2014). This model wherein IL-1 and type I IFNs form a cross-regulatory network is an attractive target for adjunct immune-modulatory intervention to TB chemotherapy.

IFN-γ also inhibits IL-1 production by monocytes and macrophages infected with *Mtb* (Mayer-Barber et al. 2011). Interestingly, this effect was not observed in DCs highlighting how unique mechanisms of inflammasome-mediated IL-1β processing act in various myeloid cells. Importantly, IFN-γ-mediated suppression of inflammasome activation has been reported in situ during murine TB (Mishra et al. 2013). T-cell-derived IFN-γ activates the inducible nitric oxide synthase (iNOS) and subsequently NO production in macrophages, thereby inhibiting IL-1 processing and dampening tissue damage and immunopathology. IFN-γ-induced reduction in IL-1 secretion was attributed to nitrosylation of NLRP3, rendering NLRP3 incapable of inflammasome assembly and subsequent activation of caspase-1 for cleavage of IL-1β. Modulation of transcriptional levels of NLRP3 through micro-RNAs (e.g., miRNA-223) or regulating of inflammasome complexes by autophagy could also influence inflammasome activation during TB (Dorhoi et al. 2013; Shi et al. 2012).

Another means by which the inflammasome response may be modulated in TB is through *Mtb*-secreted effector molecules. Evidence for such a cross-talk has been provided by the identification ZMP-1, a putative Zn-metalloprotease, which restricts IL-1 production (Master et al. 2008). Virulent *Mtb* may also prevent the activation of the AIM2 by limiting type I IFN production in infected cells (Shah et al. 2013). Although the precise nature of the mycobacterial effector protein has not been identified in this instance, manipulation of the AIM2 pathway requires an intact ESX-1 secretion system (Fig. 3).

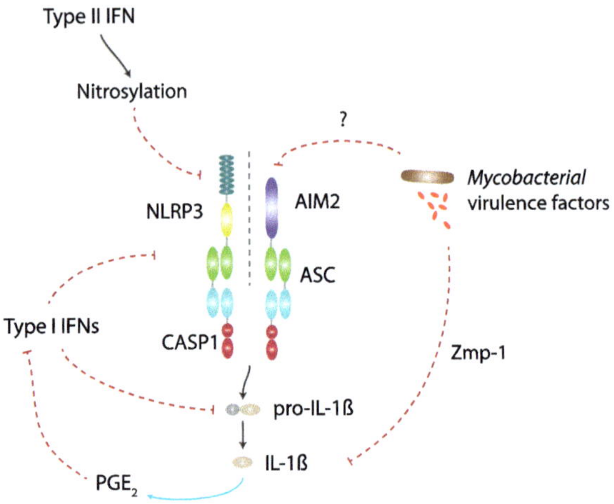

Fig. 3 Regulation of IL-1β in the context of *Mtb* infection. In TB, production of IL-1β is controlled via multiple mechanisms. NLRP3 is negatively regulated by type II IFN involving nitrosylation. Further, type I IFNs interfere with inflammasome activation, but also negatively regulate IL1B mRNA levels. Once released, IL-1 negatively regulates type I IFNs via PGE_2. *Mtb*-derived factors also negatively regulate IL-1β release

8 In Vivo Regulation of IL-1β and the Inflammasome in Host Protection

The balanced production of cytokines by innate immune effectors is a key factor for resistance against *Mtb*. As mentioned earlier, IL-1β is of critical importance for host control of TB, as deficient mice are susceptible to low-dose aerosol infection with *Mtb*, with higher bacterial burden in the lung and BAL and a more severe pathology, characterized by tissue necrosis and large granulomas (Di Paolo et al. 2015; Mayer-Barber et al. 2010). Exploring the regulation of in vivo IL-1β production, however, revealed unexpected findings. In contrast to what has been firmly established in experiments using macrophages and DCs in vitro, both ASC and caspase-1 were found to be dispensable for in vivo IL-1β production and cleavage (Dorhoi et al. 2012; McElvania Tekippe et al. 2010). Moreover, no change in lung bacterial burdens has been observed in ASC-deficient animals and only a moderate elevation of bacterial loads has been reported in caspase-1 deficient animals when compared with wild-type mice (Mayer-Barber et al. 2010; McElvania Tekippe et al. 2010). Whereas caspase-1 and ASC-deficient mice display minor defects in controlling TB, these mutants were much less susceptible to infection than the IL-1β-deficient mice. In addition, deletion of NLRP3 does not affect the outcome of infection with *Mtb* in the low-dose aerosol or dermal infection model of TB (Dorhoi et al. 2012; McElvania Tekippe et al. 2010; Walter et al. 2010). In contrast, AIM2-deficient mice infected intratracheally with *Mtb* displayed significantly reduced survival rates (Saiga et al. 2012). It is to be noted, however, that in this study, high doses of mycobacteria have been employed, which has yet to be tested with ASC and caspase-1 deficient mice. Nevertheless, it seems that the "classical" pathway of inflammasome activation is redundant for host defense during low-dose aerosol infection with *Mtb* and that in the context of pulmonary tuberculosis, IL-1β is processed by concurrent mechanisms. Several scenarios of inflammasome-independent processing of pro-IL-1β have been reported (Netea et al. 2015). As highlighted above, neutrophils are able to process IL-1β through caspase-1-independent mechanisms (Dorhoi et al. 2012; Netea et al. 2015). Another pathway of caspase-1-independent IL-1β production during TB could rely on caspase-8, which is capable of converting pro-IL-1β into its bioactive form (Man and Kanneganti 2016). Indeed, DCs activate IL-1β via caspase-8 in response to *M. leprae* in vitro—a signaling mechanism downstream of Dectin-1 and CARD9-BCL-10-MALT1 complex (Gringhuis et al. 2012). However, this non-canonical cleavage of pro-IL-1β still requires ASC and appears less important for DC-mediated IL-1β production in response to *Mtb*. Macrophages have also been reported to induce secretion of IL-1β via caspase-8 in an ASC- and caspase-1-independent manner through RIPK3 (Vince et al. 2012). These findings offer possible solution toward mechanisms that could back up IL-1β release during *Mtb* infection in the absence of inflammasomes. At this stage, it should be mentioned that in contrast to IL-1β, IL-1α is signaling competent both as cleaved as well as native protein. This could explain a degree of redundancy for the two IL-1 members during infection with *Mtb* in vivo. For example, compared to the heightened susceptibility of the IL-1R mutant mice to TB, deficiency IL-1α or IL-1β alone only recapitulate the IL-1RI phenotype in certain studies (Di Paolo et al. 2015; Mayer-Barber et al. 2011).

9 Concluding Remarks

During the last decade, remarkable advances in the identification and characterization of host factors involved in the innate defense against *Mtb* have been made. However, although the roles of certain cytokines have become clearer, the precise mechanisms governing their production and regulation at the site of infection or within TB granulomas remains partially understood. This is best exemplified by the role of the inflammasome in TB. The involvement of the inflammasome in immune defense against *Mtb* has been largely defined by studies with cultured macrophages. In these experimental settings, production of IL-1β is strictly dependent on the inflammasome components ASC and caspase-1. However, it appears that production of mature IL-1β in vivo during TB occurs partially independently of the canonical inflammasome, suggesting alternative mechanisms for generation of IL-1 are important in TB. Given the dual role of IL-1β in TB, protective by limiting mycobacterial replication and destructive by causing tissue damage identifying cellular sources and mechanism regulating abundance of IL-1 are key issues in the field. TB is characterized by a complex pathology, and thus, diverse inflammatory environments within one host could differentially modulate inflammasome responses. Alternatively, inflammasome-dependent and inflammasome-independent mechanisms could mutually contribute to IL-1β production in vivo. Deciphering the interplay between these pathways represents an important challenge for the future. Once established, such knowledge will be of great value for improving the treatment and the prevention of tuberculosis.

References

Abdalla H, Srinivasan L, Shah S, Mayer-Barber KD, Sher A, Sutterwala FS, Briken V (2012) *Mycobacterium tuberculosis* infection of dendritic cells leads to partially caspase-1/11-independent IL-1beta and IL-18 secretion but not to pyroptosis. PLoS ONE 7: e40722

Abdallah AM, Bestebroer J, Savage ND, de Punder K, van Zon M, Wilson L, Korbee CJ, van der Sar AM, Ottenhoff TH, van der Wel NN et al (2011) Mycobacterial secretion systems ESX-1 and ESX-5 play distinct roles in host cell death and inflammasome activation. J Immunol 187:4744–4753

Abel L, El-Baghdadi J, Bousfiha AA, Casanova JL, Schurr E (2014) Human genetics of tuberculosis: a long and winding road. Philos Trans R Soc Lond B Biol Sci 369:20130428

Ates LS, Houben EN, Bitter W (2016) Type VII secretion: a highly versatile secretion system. Microbiology spectrum 4

Collins AC, Cai H, Li T, Franco LH, Li XD, Nair VR, Scharn CR, Stamm CE, Levine B, Chen ZJ et al (2015) Cyclic GMP-AMP synthase is an innate immune DNA sensor for *Mycobacterium tuberculosis*. Cell Host Microbe 17:820–828

Court N, Vasseur V, Vacher R, Fremond C, Shebzukhov Y, Yeremeev VV, Maillet I, Nedospasov SA, Gordon S, Fallon PG et al (2010) Partial redundancy of the pattern recognition receptors, scavenger receptors, and C-type lectins for the long-term control of *Mycobacterium tuberculosis* infection. J Immunol 184:7057–7070

de Jonge MI, Pehau-Arnaudet G, Fretz MM, Romain F, Bottai D, Brodin P, Honore N, Marchal G, Jiskoot W, England P et al (2007) ESAT-6 from *Mycobacterium tuberculosis* dissociates from its putative chaperone CFP-10 under acidic conditions and exhibits membrane-lysing activity. J Bacteriol 189:6028–6034

Di Paolo NC, Shafiani S, Day T, Papayannoupoulou T, Russell DW, Iwakura Y, Sherman D, Urdahl K, Shayakhmetov DM (2015) Interdependence between Interleukin-1 and tumor necrosis factor regulates TNF-dependent control of *Mycobacterium tuberculosis* infection. Immunity 43:1125–1136

Divangahi M, Mostowy S, Coulombe F, Kozak R, Guillot L, Veyrier F, Kobayashi KS, Flavell RA, Gros P, Behr MA (2008) NOD2-deficient mice have impaired resistance to *Mycobacterium tuberculosis* infection through defective innate and adaptive immunity. J Immunol 181:7157–7165

Dorhoi A, Kaufmann SH (2014) Tumor necrosis factor alpha in mycobacterial infection. Semin Immunol 26:203–209

Dorhoi A, Kaufmann SH (2015) Versatile myeloid cell subsets contribute to tuberculosis-associated inflammation. Eur J Immunol 45:2191–2202

Dorhoi A, Desel C, Yeremeev V, Pradl L, Brinkmann V, Mollenkopf HJ, Hanke K, Gross O, Ruland J, Kaufmann SH (2010) The adaptor molecule CARD9 is essential for tuberculosis control. J Exp Med 207:777–792

Dorhoi A, Nouailles G, Jorg S, Hagens K, Heinemann E, Pradl L, Oberbeck-Muller D, Duque-Correa MA, Reece ST, Ruland J et al (2012) Activation of the NLRP3 inflammasome by *Mycobacterium tuberculosis* is uncoupled from susceptibility to active tuberculosis. Eur J Immunol 42:374–384

Dorhoi A, Iannaccone M, Farinacci M, Fae KC, Schreiber J, Moura-Alves P, Nouailles G, Mollenkopf HJ, Oberbeck-Muller D, Jorg S et al (2013) MicroRNA-223 controls susceptibility to tuberculosis by regulating lung neutrophil recruitment. J Clin Investig 123:4836–4848

Eklund D, Welin A, Andersson H, Verma D, Soderkvist P, Stendahl O, Sarndahl E, Lerm M (2014) Human gene variants linked to enhanced NLRP3 activity limit intramacrophage growth of *Mycobacterium tuberculosis*. J Infect Dis 209:749–753

Ferwerda G, Girardin SE, Kullberg BJ, Le Bourhis L, de Jong DJ, Langenberg DM, van Crevel R, Adema GJ, Ottenhoff TH, Van der Meer JW et al (2005) NOD2 and toll-like receptors are nonredundant recognition systems of *Mycobacterium tuberculosis*. PLoS Pathog 1:279–285

Fremond CM, Yeremeev V, Nicolle DM, Jacobs M, Quesniaux VF, Ryffel B (2004) Fatal *Mycobacterium tuberculosis* infection despite adaptive immune response in the absence of MyD88. J Clin Investig 114:1790–1799

Fremond CM, Togbe D, Doz E, Rose S, Vasseur V, Maillet I, Jacobs M, Ryffel B, Quesniaux VF (2007) IL-1 receptor-mediated signal is an essential component of MyD88-dependent innate response to *Mycobacterium tuberculosis* infection. J Immunol 179:1178–1189

Gandotra S, Jang S, Murray PJ, Salgame P, Ehrt S (2007) Nucleotide-binding oligomerization domain protein 2-deficient mice control infection with *Mycobacterium tuberculosis*. Infect Immun 75:5127–5134

Gringhuis SI, Kaptein TM, Wevers BA, Theelen B, van der Vlist M, Boekhout T, Geijtenbeek TB (2012) Dectin-1 is an extracellular pathogen sensor for the induction and processing of IL-1beta via a noncanonical caspase-8 inflammasome. Nat Immunol 13:246–254

Gross O, Thomas CJ, Guarda G, Tschopp J (2011) The inflammasome: an integrated view. Immunol Rev 243:136–151

Guarda G, Braun M, Staehli F, Tardivel A, Mattmann C, Forster I, Farlik M, Decker T, Du Pasquier RA, Romero P et al (2011) Type I interferon inhibits interleukin-1 production and inflammasome activation. Immunity 34:213–223

Heitmann L, Schoenen H, Ehlers S, Lang R, Holscher C (2013) Mincle is not essential for controlling *Mycobacterium tuberculosis* infection. Immunobiology 218:506–516

Holscher C, Reiling N, Schaible UE, Holscher A, Bathmann C, Korbel D, Lenz I, Sonntag T, Kroger S, Akira S et al (2008) Containment of aerogenic *Mycobacterium tuberculosis* infection in mice does not require MyD88 adaptor function for TLR2, -4 and -9. Eur J Immunol 38:680–694

Houben D, Demangel C, van Ingen J, Perez J, Baldeon L, Abdallah AM, Caleechurn L, Bottai D, van Zon M, de Punder K et al (2012) ESX-1-mediated translocation to the cytosol controls virulence of mycobacteria. Cell Microbiol 14:1287–1298

Hsu T, Hingley-Wilson SM, Chen B, Chen M, Dai AZ, Morin PM, Marks CB, Padiyar J, Goulding C, Gingery M et al (2003) The primary mechanism of attenuation of bacillus Calmette-Guerin is a loss of secreted lytic function required for invasion of lung interstitial tissue. Proc Natl Acad Sci USA 100:12420–12425

Jayaraman P, Sada-Ovalle I, Nishimura T, Anderson AC, Kuchroo VK, Remold HG, Behar SM (2013) IL-1beta promotes antimicrobial immunity in macrophages by regulating TNFR signaling and caspase-3 activation. J Immunol 190:4196–4204

Koo IC, Wang C, Raghavan S, Morisaki JH, Cox JS, Brown EJ (2008) ESX-1-dependent cytolysis in lysosome secretion and inflammasome activation during mycobacterial infection. Cell Microbiol 10:1866–1878

Lowe DM, Redford PS, Wilkinson RJ, O'Garra A, Martineau AR (2012) Neutrophils in tuberculosis: friend or foe? Trends Immunol 33:14–25

Majlessi L, Prados-Rosales R, Casadevall A, Brosch R (2015) Release of mycobacterial antigens. Immunol Rev 264:25–45

Man SM, Kanneganti TD (2016) Converging roles of caspases in inflammasome activation, cell death and innate immunity. Nat Rev Immunol 16:7–21

Manzanillo PS, Shiloh MU, Portnoy DA, Cox JS (2012) *Mycobacterium tuberculosis* activates the DNA-dependent cytosolic surveillance pathway within macrophages. Cell Host Microbe 11:469–480

Master SS, Rampini SK, Davis AS, Keller C, Ehlers S, Springer B, Timmins GS, Sander P, Deretic V (2008) *Mycobacterium tuberculosis* prevents inflammasome activation. Cell Host Microbe 3:224–232

Mayer-Barber KD, Barber DL (2015) Innate and adaptive cellular immune responses to *Mycobacterium tuberculosis* infection. Cold Spring Harbor Perspectives in Medicine 5

Mayer-Barber KD, Barber DL, Shenderov K, White SD, Wilson MS, Cheever A, Kugler D, Hieny S, Caspar P, Nunez G et al (2010) Caspase-1 independent IL-1beta production is critical for host resistance to *Mycobacterium tuberculosis* and does not require TLR signaling in vivo. J Immunol 184:3326–3330

Mayer-Barber KD, Andrade BB, Barber DL, Hieny S, Feng CG, Caspar P, Oland S, Gordon S, Sher A (2011) Innate and adaptive interferons suppress IL-1alpha and IL-1beta production by distinct pulmonary myeloid subsets during *Mycobacterium tuberculosis* infection. Immunity 35:1023–1034

Mayer-Barber KD, Andrade BB, Oland SD, Amaral EP, Barber DL, Gonzales J, Derrick SC, Shi R, Kumar NP, Wei W et al (2014) Host-directed therapy of tuberculosis based on interleukin-1 and Type I interferon crosstalk. Nature 511:99–103

McElvania Tekippe E, Allen IC, Hulseberg PD, Sullivan JT, McCann JR, Sandor M, Braunstein M, Ting JP (2010) Granuloma formation and host defense in chronic *Mycobacterium tuberculosis* infection requires PYCARD/ASC but not NLRP3 or caspase-1. PLoS ONE 5:e12320

Mishra BB, Moura-Alves P, Sonawane A, Hacohen N, Griffiths G, Moita LF, Anes E (2010) *Mycobacterium tuberculosis* protein ESAT-6 is a potent activator of the NLRP3/ASC inflammasome. Cell Microbiol 12:1046–1063

Mishra BB, Rathinam VA, Martens GW, Martinot AJ, Kornfeld H, Fitzgerald KA, Sassetti CM (2013) Nitric oxide controls the immunopathology of tuberculosis by inhibiting NLRP3 inflammasome-dependent processing of IL-1beta. Nat Immunol 14:52–60

Moura-Alves P, Fae K, Houthuys E, Dorhoi A, Kreuchwig A, Furkert J, Barison N, Diehl A, Munder A, Constant P et al (2014) AhR sensing of bacterial pigments regulates antibacterial defence. Nature 512:387–392

Netea MG, Nold-Petry CA, Nold MF, Joosten LA, Opitz B, van der Meer JH, van de Veerdonk FL, Ferwerda G, Heinhuis B, Devesa I et al (2009) Differential requirement for the activation of the inflammasome for processing and release of IL-1beta in monocytes and macrophages. Blood 113:2324–2335

Netea MG, van de Veerdonk FL, van der Meer JW, Dinarello CA, Joosten LA (2015) Inflammasome-independent regulation of IL-1-family cytokines. Annu Rev Immunol 33:49–77

Novikov A, Cardone M, Thompson R, Shenderov K, Kirschman KD, Mayer-Barber KD, Myers TG, Rabin RL, Trinchieri G, Sher A et al (2011) *Mycobacterium tuberculosis* triggers host type I IFN signaling to regulate IL-1beta production in human macrophages. J Immunol 187:2540–2547

O'Garra A, Redford PS, McNab FW, Bloom CI, Wilkinson RJ, Berry MP (2013) The immune response in tuberculosis. Annu Rev Immunol 31:475–527

Philips JA, Ernst JD (2012) Tuberculosis pathogenesis and immunity. Ann Rev Pathol 7:353–384

Pym AS, Brodin P, Brosch R, Huerre M, Cole ST (2002) Loss of RD1 contributed to the attenuation of the live tuberculosis vaccines *Mycobacterium bovis* BCG and *Mycobacterium microti*. Mol Microbiol 46:709–717

Renshaw PS, Panagiotidou P, Whelan A, Gordon SV, Hewinson RG, Williamson RA, Carr MD (2002) Conclusive evidence that the major T-cell antigens of the *Mycobacterium tuberculosis* complex ESAT-6 and CFP-10 form a tight, 1:1 complex and characterization of the structural properties of ESAT-6, CFP-10, and the ESAT-6*CFP-10 complex. Implications for pathogenesis and virulence. J Biol Chem 277:21598–21603

Repasy T, Lee J, Marino S, Martinez N, Kirschner DE, Hendricks G, Baker S, Wilson AA, Kotton DN, Kornfeld H (2013) Intracellular bacillary burden reflects a burst size for *Mycobacterium tuberculosis* in vivo. PLoS Pathog 9:e1003190

Saiga H, Kitada S, Shimada Y, Kamiyama N, Okuyama M, Makino M, Yamamoto M, Takeda K (2012) Critical role of AIM2 in *Mycobacterium tuberculosis* infection. Int Immunol 24:637–644

Schmid-Burgk JL, Chauhan D, Schmidt T, Ebert TS, Reinhardt J, Endl E, Hornung V (2015) A genome-wide CRISPR screen identifies NEK7 as an essential component of NLRP3 inflammasome activation. J Biol Chem

Shah S, Bohsali A, Ahlbrand SE, Srinivasan L, Rathinam VA, Vogel SN, Fitzgerald KA, Sutterwala FS, Briken V (2013) Cutting edge: *Mycobacterium tuberculosis* but not nonvirulent mycobacteria inhibits IFN-beta and AIM2 inflammasome-dependent IL-1beta production via its ESX-1 secretion system. J Immunol 191:3514–3518

Shah S, Cannon JR, Fenselau C, Briken V (2015) A Duplicated ESAT-6 Region of ESX-5 Is involved in protein export and virulence of Mycobacteria. Infect Immun 83:4349–4361

Shi CS, Shenderov K, Huang NN, Kabat J, Abu-Asab M, Fitzgerald KA, Sher A, Kehrl JH (2012) Activation of autophagy by inflammatory signals limits IL-1beta production by targeting ubiquitinated inflammasomes for destruction. Nat Immunol 13:255–263

Shi H, Wang Y, Li X, Zhan X, Tang M, Fina M, Su L, Pratt D, Bu CH, Hildebrand S et al (2016) NLRP3 activation and mitosis are mutually exclusive events coordinated by NEK7, a new inflammasome component. Nat Immunol 17:250

Simeone R, Bottai D, Brosch R (2009) ESX/type VII secretion systems and their role in host-pathogen interaction. Curr Opin Microbiol 12:4–10

Simeone R, Bobard A, Lippmann J, Bitter W, Majlessi L, Brosch R, Enninga J (2012) Phagosomal rupture by *Mycobacterium tuberculosis* results in toxicity and host cell death. PLoS Pathog 8: e1002507

Srivastava S, Ernst JD, Desvignes L (2014) Beyond macrophages: the diversity of mononuclear cells in tuberculosis. Immunol Rev 262:179–192

Stamm CE, Collins AC, Shiloh MU (2015) Sensing of *Mycobacterium tuberculosis* and consequences to both host and bacillus. Immunol Rev 264:204–219

Stoop EJ, Bitter W, van der Sar AM (2012) Tubercle bacilli rely on a type VII army for pathogenicity. Trends Microbiol 20:477–484

Torrado E, Cooper AM (2013) Cytokines in the balance of protection and pathology during mycobacterial infections. Adv Exp Med Biol 783:121–140

van der Wel N, Hava D, Houben D, Fluitsma D, van Zon M, Pierson J, Brenner M, Peters PJ (2007) *M. tuberculosis* and *M. leprae* translocate from the phagolysosome to the cytosol in myeloid cells. Cell 129:1287–1298

Vince JE, Wong WW, Gentle I, Lawlor KE, Allam R, O'Reilly L, Mason K, Gross O, Ma S, Guarda G et al (2012) Inhibitor of apoptosis proteins limit RIP3 kinase-dependent interleukin-1 activation. Immunity 36:215–227

Walter K, Holscher C, Tschopp J, Ehlers S (2010) NALP3 is not necessary for early protection against experimental tuberculosis. Immunobiology 215:804–811

Wassermann R, Gulen MF, Sala C, Perin SG, Lou Y, Rybniker J, Schmid-Burgk JL, Schmidt T, Hornung V, Cole ST et al (2015) *Mycobacterium tuberculosis* differentially activates cGAS- and inflammasome-dependent intracellular immune responses through ESX-1. Cell Host Microbe 17:799–810

Watson RO, Manzanillo PS, Cox JS (2012) Extracellular *M. tuberculosis* DNA targets bacteria for autophagy by activating the host DNA-sensing pathway. Cell 150:803–815

Watson RO, Bell SL, MacDuff DA, Kimmey JM, Diner EJ, Olivas J, Vance RE, Stallings CL, Virgin HW, Cox JS (2015) The cytosolic sensor cGAS detects *Mycobacterium tuberculosis* DNA to induce type I interferons and activate autophagy. Cell Host Microbe 17:811–819

Role of Canonical and Non-canonical Inflammasomes During *Burkholderia* Infection

Manoranjan Sahoo, Louis Lantier and Fabio Re

Abstract *Burkholderia pseudomallei* is a Gram-negative flagellate bacterium that causes melioidosis, a disease endemic to Southeast Asia and other tropical regions. Following infection of macrophages and other non-phagocytic cell types, *B. pseudomallei* or *B. thailandensis* (a related species that causes disease in mice but not humans) are able to escape the phagosome and replicate in the host cell cytoplasm. Resistance to infection with *Burkholderia* is dependent on the Nlrp3 and Nlrc4 inflammasomes and the non-canonical caspase-11 inflammasome. Nlrc4 mediates protection through induction of pyroptosis in the early phase of infection. As the infection progresses and as IL-18-dependent IFNγ production increases, caspase-11-dependent pyroptosis acquires a preponderant protective role. Production of IL-1β and IL-18 during infection is primarily mediated by Nlrp3. IL-18 is essential for survival because of its ability to induce IFNγ production, which in turn activates macrophage microbicidal functions and primes for caspase-11 expression. In contrast, during melioidosis, IL-1β has deleterious effects due to excessive recruitment of neutrophils to the lung and consequent tissue damage.

Contents

M. Sahoo · L. Lantier · F. Re (✉)
Department of Microbiology & Immunology, Rosalind Franklin University of Medicine and Science, 3333 Green Bay Road, North Chicago, IL 60064, USA
e-mail: fabio.re@rosalindfranklin.edu

S. Backert (ed.), *Inflammasome Signaling and Bacterial Infections*,
Current Topics in Microbiology and Immunology 397,
DOI 10.1007/978-3-319-41171-2_10

1 Introduction

Burkholderia pseudomallei is a Gram-negative, rod-shaped, motile, flagellated saprophyte bacterium that causes melioidosis, a serious and often fatal disease endemic to Southeast Asia and Northern Australia (Lazar Adler et al. 2009; Wiersinga et al. 2006). *B. pseudomallei* is often found in moist tropical soils and stagnant water where it infects protozoa. In humans, infection can be contracted through ingestion, inhalation, or subcutaneous inoculation, and can lead to a broad spectrum of disease forms including septicemia, organ abscesses, and pneumonia, which is the most lethal form of this disease. Because melioidosis carries a high fatality rate, even after medical interventions, *B. pseudomallei* is classified as a category B potential bioterrorism agent by the CDC and NIAID. While any healthy individual traveling to or living in the endemic area can contract melioidosis, underlying conditions such as diabetes mellitus, renal disease, or alcohol abuse are known to act as predisposing factors (Currie et al. 2004). Although melioidosis is an acute disease, *B. pseudomallei* can establish an asymptomatic persistent infection with a latency period that can last several years and lead to relapsing disease (Cheng and Currie 2005; Gan 2005). *Burkholderia thailandensis* is a closely related species that occupies a similar environmental niche as *B. pseudomallei,* but is unable to cause disease in humans. The reason for the attenuation of *B. thailandensis* appears to be the presence of a functional arabinose biosynthesis operon that is missing in *B. pseudomallei* (Chaiyaroj et al. 1999; Moore et al. 2004). However, in mice, both *B. pseudomallei* and *B. thailandensis* induce a similar disease, and for this reason *B. thailandensis* is often used as a mouse model of melioidosis (Haraga et al. 2008; Yu et al. 2006; West et al. 2008b).

B. pseudomallei is a facultative, intracellular pathogen that is able to infect and multiply within the cytoplasm of non-phagocytic epithelial cells and phagocytic cells such as macrophages, monocytes, and neutrophils (Pruksachartvuthi et al. 1990; Jones et al. 1996). Cell-to-cell spread and actin-mediated motility have been reported (Kespichayawattana et al. 2000). *B. pseudomallei* escape from the phagosome is dependent on one of the three type-III secretion systems (T3SS) expressed by this bacterium (Stevens et al. 2002). Bacteria that carry mutations within the T3SS are severely attenuated, indicating that ability to lyse the vacuole is an important virulence determinant (Burtnick et al. 2008; Muangsombut et al. 2008). In fact, escape from the phagosome allows bacteria to evade killing mediated by ROS and RNI. While the respiratory burst is a protective effector mechanism against melioidosis, whether nitric oxide and neutrophils significantly contribute to protection from *B. pseudomallei* infection remains controversial (Utaisincharoen et al. 2001; Miyagi et al. 1997; Breitbach et al. 2006; Chanchamroen et al. 2009; Easton et al. 2007).

2 Innate Immune Response to *Burkholderia* Infection

Activation of the complement cascade by *B. pseudomallei* occurs through the alternative pathway, but is ineffective due to the presence of a polysaccharide capsule that protects the bacterium from deposition of the membrane attack complex (Egan and Gordon 1996; DeShazer et al. 1998; Reckseidler-Zenteno et al. 2005).

Like most Gram-negative bacteria, *B. pseudomallei* activates TLR2, TLR4, and TLR5 leading to production of a variety of inflammatory cytokines. *B. pseudomallei* LPS was reported to have weak TLR4 agonist activity (Matsuura et al. 1996; West et al. 2008a), and in fact TLR4-deficient mice were as resistant to *B. pseudomallei* infection as wild-type mice (Wiersinga et al. 2007a). Recognition of *B. pseudomallei* flagellin by TLR5 appears to have a physiologic role as shown by the higher susceptibility of $Tlr5^{-/-}$ mice (West et al. 2014). Surprisingly, mice that lack TLR2 or CD14 were shown to be more resistant than wild-type mice to infection (Wiersinga et al. 2007a, 2008a), suggesting that the inflammatory response against *B. pseudomallei*, which is primarily mediated by TLR2, may be deleterious. In contrast, mice deficient in MyD88, an adaptor molecule critical for signaling through most TLRs, but also through the IL-1 and IL-18 receptors, had increased susceptibility (Wiersinga et al. 2008b). Our discovery that IL-1β plays a deleterious role in melioidosis, while IL-18 is protective (Ceballos-Olvera et al. 2011; Sahoo et al. 2014), may provide an explanation to these apparently puzzling results.

During the acute phase of melioidosis, the level of several proinflammatory cytokines (including IFNγ, IL-12, IL-15, IL-18, TNFα) is elevated and poor prognosis is associated with excessive serum levels of IL-6 and IL-18 (Lauw et al. 1999). Mouse studies have shown that IFNγ plays a critically protective role during infection with *B. pseudomallei*. This IFNγ was produced by NK cells, $CD8^+$ T cells, and NKT cells, and the administration of anti-IL-12 and anti-IL-18 mAb inhibited its production (Lertmemongkolchai et al. 2001; Haque et al. 2006). In agreement with IL-18s function as an IFNγ-inducing factor, $Il\text{-}18^{-/-}$ mice were found to be highly susceptible to intranasal *B. pseudomallei* infection, showing high organ bacterial burden and bacteremia (Wiersinga et al. 2007b). Expression of other proinflammatory cytokines such as IL-1β, IL-6, and TNFα is also increased in the liver of mice infected with *B. pseudomallei* (Ulett et al. 2000). IL-1β mRNA and protein levels were also elevated in PBMC and purified monocytes infected with *B. pseudomallei* or *B. thailandensis* (Pongcharoen et al. 2008).

3 Role of Caspase-1

The first study to investigate the role of caspase-1 during *B. pseudomallei* infection comes from Sun et al. (2005). The authors demonstrated that *B. pseudomallei* induced rapid cell death in THP1 cells, primary human monocytes, macrophages, dendritic cells, and mouse peritoneal macrophages. Bacteria internalization and T3SS-dependent pore formation were responsible for induction of cell death and IL-1β and IL-18 release. Caspase-1-deficient mice were shown to be highly susceptible to intranasal infection with *B. pseudomallei* and have higher organ bacterial burden (Breitbach et al. 2009). These mice also showed reduced IL-18 and IFNγ serum levels. Bone marrow-derived macrophages (BMMs) from caspase-1-deficient mice infected with *B. pseudomallei* showed significantly decreased pyroptotic cell death and IL-1β secretion, and exhibited impaired bactericidal activity compared to wild-type BMMs. Activation of caspase-9 and caspase-7 and induction of apoptosis have been shown in the early phase of infection of macrophages with *B. pseudomallei* (Bast et al. 2014), although the physiologic role of these responses in vivo remains unclear.

4 Role of NLRC4 Inflammasome

The demonstration of caspase-1-dependent cell death and IL-1β and IL-18 secretion in response to *B. pseudomallei* infection suggested the involvement of the inflammasome. The first indication of which inflammasome may be activated by *B. pseudomallei* came from the works of Ed Miao who demonstrated that the *B. pseudomallei* protein BsaK, the T3SS basal body rod component, activates the Nlrc4 inflammasome (Miao et al. 2010b). In vivo, caspase-1-dependent pyroptosis was important to restrict the intracellular bacterial growth independently of IL-1β and IL-18 (Miao et al. 2010a). More recently, another group showed that the NLR molecule Naip2 binds *B. pseudomallei* BsaK upstream of Nlrc4, leading to its activation (Zhao et al. 2011). Surprisingly, the same work also found that Naip5, which participates with Nlrc4 in recognition of various bacterial flagellins, is not involved in recognition of *B. pseudomallei* flagellin. However, a *B. pseudomallei* FliC mutant failed to induce Nlrc4/caspase-1-dependent cell death suggesting the existence of a Nlrc4-dependent mechanism to detect *B. pseudomallei* flagellin (Bast et al. 2014). Finally, Naip1 (and human NAIP, the only member of the NAIP family expressed in humans) was shown to recognize the *B. pseudomallei* BsaL-needled protein (another structural constituent of T3SS) and to trigger Nlrc4-dependent caspase-1 activation (Yang et al. 2013). Thus, in mice, the Nlrc4 inflammasome mediates innate immunity against at least two structural proteins of *Burkholderia* species by pairing with different NAIP molecules. Humans appear to recognize only BsaL through NAIP.

Our laboratory has analyzed the role of the Nlrc4 inflammasome during infection with *B. pseudomallei* or *B. thailandensis* in vitro and in vivo (Ceballos-Olvera et al. 2011; Sahoo et al. 2014). In agreement with previous work, we demonstrated that induction of pyroptosis in BMM or bone marrow-derived dendritic cells (BMDC) was dependent on Nlrc4 and caspase-1 and efficiently restricted bacteria replication. The adaptor ASC was not required for this response, as shown during infection with other bacteria (Suzuki et al. 2007; Case et al. 2009). Using a mouse model of pneumonic melioidosis, Nlrc4/caspase-1-dependent pyroptosis was also shown to be very efficient in controlling the bacteria replication during in vivo infection. Consequently, *Nlrc4*$^{-/-}$ mice were very susceptible to intranasal *B. pseudomallei* or *B. thailandensis* infection compared to their wild-type counterparts and showed higher bacterial burden in their organs. Surprisingly, the level of IL-1β or IL-18 in the BALF (or serum IL-18) of *Nlrc4*$^{-/-}$ mice was significantly higher than that in wild-type mice suggesting that activation of the Nlrc4 inflammasome is not required for IL-1β or IL-18 production during pneumonic melioidosis. The increased level of IL-1β and IL-18 in Nlrc4-deficient mice is likely a result of the increased bacteria burden in these mice as well as the aborted pyroptotic response of infected macrophages/DC, which are then allowed to continue secreting these cytokines. Although one study has shown that *Nlrc4*$^{-/-}$ mice were as resistant to infection with *B. thailandensis* as wild-type mice (Aachoui et al. 2013), in a follow-up study, the same group reported that, in fact, Nlrc4-deficient mice had increased bacteria burden compared to wild-type mice (Aachoui et al. 2015). The reason for this discrepancy is unknown, but it is possibly due to the fact that in the earlier study the route of infection used was intraperitoneal, rather than intranasal as in the other studies, or that the infectious dose used to infect animals was too low to detect an increased susceptibility of the knockout strain. The fact that a mouse-passaged *B. thailandensis* strain was used for the second study is a further source of variability.

5 Role of NLRP3 Inflammasome

The ability of *B. pseudomallei* to lyse the phagosome membrane and invade the cytoplasm suggested that other inflammasomes could also be involved. Our studies, in fact, showed that both the Nlrp3 and Nlrc4 inflammasomes participate in the innate immune response to intranasal infection with *B. pseudomallei* or *B. thailandensis* (Ceballos-Olvera et al. 2011; Sahoo et al. 2014). Importantly, we showed that activation of the Nlrp3 and Nlrc4 inflammasome plays non-redundant roles in melioidosis. In contrast to *Nlrc4*$^{-/-}$ mice, the levels of IL-1β and IL-18 were drastically reduced in the BALF of *Nlrp3*$^{-/-}$, *Asc*$^{-/-}$, or *Casp1*$^{-/-}$ mice despite high bacteria replication. Mice deficient in Nlrp3 or ASC were also more susceptible than wild type to intranasal infection with *B. pseudomallei* or *B. thailandensis*, though not as susceptible as mice lacking Nlrc4, an observation that may suggest that pyroptosis has a more important protective role than IL-1β or IL-18 in this type of infection. A recent study has confirmed that *Nlrc4*$^{-/-}$ mice have a more severe phenotype than *Nlrp3*$^{-/-}$ mice (Aachoui et al. 2015). Caspase-1-deficient mice, lacking both the NLRC4-dependent pyroptosis and the NLRP3-dependent production of IL-1β and IL-18, were significantly more susceptible to *B. pseudomallei* infection than *Nlrp3*$^{-/-}$, *Asc*$^{-/-}$, or *Nlrc4*$^{-/-}$ mice, confirming the non-redundant role of the Nlrp3 and the Nlrc4 inflammasome in melioidosis. Activation of the Nlrp3 and Nlrc4 inflammasome during *B. thailandensis* infection has been recently confirmed by other groups (Bast et al. 2014; West et al. 2014). Notably, it has been shown that *Nlrc4*$^{-/-}$*Asc*$^{-/-}$ mice had higher bacteria burden than *Nlrc4*$^{-/-}$*Nlrp3*$^{-/-}$ mice suggesting the existence of an additional ASC-dependent protective mechanism, possibly another inflammasome (Aachoui et al. 2015). Analysis of *Nlrc4*$^{-/-}$*Nlrp3*$^{-/-}$ mice also showed that the residual induction of IL-18 observed in *Nlrp3*$^{-/-}$ mice was completely eliminated by the concomitant deficiency in *Nlrc4*$^{-/-}$, suggesting that the Nlrc4 inflammasome also contributes to IL-18 production, but in a much smaller proportion than Nlrp3 (Aachoui et al. 2015).

Whereas we have a relatively good understanding of the *B. pseudomallei* molecules that activate the Nlrc4 inflammasome, it is still unknown what mechanism *B. pseudomallei* uses to trigger the Nlrp3 inflammasome. Hemolysins released by various bacteria, including *Staphylococcus aureus* and *Streptococcus pneumoniae*, have been shown to activate the NLRP3 inflammasome (Munoz-Planillo et al. 2009; Craven et al. 2009; Costa et al. 2012). Interestingly, *B. pseudomallei* expresses at least two different hemolysins (Ashdown and Koehler 1990; Haussler et al. 1998) and possesses ATP-binding cassette transport systems that are predicted to be able to export these toxins (Harland et al. 2007). It is unclear whether these toxins are required for phagosome escape, but they are likely candidate as NLRP3 activators. An additional mechanism through which *Burkholderia* can activate the NLRP3 inflammasome may be represented by caspase-11, which is activated by *Burkholderia* species and has been proposed to regulate activation of the NLRP3 inflammasome (Rathinam et al. 2012, Broz et al. 2012). Interferon-inducible GTPases (GBP) have been shown to promote lysis of pathogen-containing vacuole allowing escape of bacteria into the cytosol and activation of caspase-11 (Meunier et al. 2014; Pilla et al. 2014). However, GBP-deficient mice appear to be as susceptible as wild-type mice to *B. thailandensis* infection (Aachoui et al. 2015), suggesting that phagosome escape and Nlrp3 activation during infection with *B. thailandensis* occur independently of GBP.

6 Opposing Function of IL-1β and IL-18

The increased susceptibility to melioidosis of *Nlrp3*$^{-/-}$ or *Asc*$^{-/-}$ mice was associated with absence of IL-1β or IL-18 (Ceballos-Olvera et al. 2011), two cytokines that have been shown to be protective in many infection models (Garlanda et al. 2013). To further understand the role of these cytokines during *B. pseudomallei* infection, we analyzed mice deficient in IL-1RI or IL-18. Predictably, *Il-18*$^{-/-}$ mice were found to be more susceptible to *B. pseudomallei* or *B. thailandensis* infection and were unable to restrict bacteria replication and systemic dissemination (as previously shown (Wiersinga et al. 2007b) and subsequently confirmed (Aachoui et al. 2015)). In agreement with its role as an interferon gamma-inducing factor, absence of IL-18 translated into drastically reduced level of IFNγ, suggesting that the increased susceptibility of *Il-18*$^{-/-}$ mice was due to inability to produce IFNγ. Confirming this hypothesis, exogenous administration of IFNγ to *Il-18*$^{-/-}$ mice reverted their phenotype. A previous work has shown that IFNγ production by NK or T cells is critical for resistance to melioidosis (Santanirand et al. 1997).

To our surprise, mice lacking IL-1RI appeared to be more resistant than wild-type mice to *B. thailandensis* or *B. pseudomallei* infection suggesting that IL-1 may play a deleterious role in melioidosis. While both IL-1α and IL-1β can signal through IL1RI and both are produced during infection, only IL-1β was found to be deleterious (Sahoo et al. 2014). The realization of the contrasting roles played by IL-1β and IL-18 provided an elegant explanation to the previously reported increased resistance of *Tlr2*$^{-/-}$ mice as opposed to the higher susceptibility of *Myd88*$^{-/-}$ mice (Wiersinga et al. 2007a, 2008b). Because of *B. pseudomallei*'s weak LPS, production of inflammatory cytokines (including IL-1β) during infection is mostly dependent on TLR2. Thus, *Tlr2*$^{-/-}$ mice are protected from the deleterious action of IL-1β but still can benefit from the protective action of IL-18 (whose mRNA is constitutively expressed and not dependent on TLR stimulation). In contrast, *Myd88*$^{-/-}$ mice are deprived of the negative influence of IL-1β but also of the more critical and beneficial action of IL-18. The observation that *Il-1r1*$^{-/-}$ mice were more resistant than wild type to *B. pseudomallei* infection is quite surprising, considering that this cytokine has been shown to be protective in several bacterial, viral, and fungal infection models (Dinarello 2009). Studies in humans have also shown that inhibition of the function of IL-1 using the IL-1R antagonist IL-1ra (Kineret) is associated with increased susceptibility to bacterial infection. Infected *Il-1r1*$^{-/-}$ mice had lower BALF levels of proinflammatory cytokines as well as reduced neutrophil influx into the lungs, bacterial burdens, and lung pathology. Consistent with a deleterious role of IL-1β in melioidosis, administration of recombinant IL-1β drastically increased mortality, inflammation, pathology, and bacteria burdens while administration of IL-1ra (Kineret) rescued the survival of wild-type mice infected with a lethal dose of *B. pseudomallei*. Interestingly, a clinical study has shown reduced mortality among melioidosis patients receiving glyburide, a hypoglycemic agent used to treat diabetes mellitus that also inhibits the NLRP3 inflammasome. This suggests that also in human melioidosis IL-1β may be deleterious (Koh et al. 2011).

The reason for the detrimental effect of IL-1β during melioidosis is unclear, and it is likely that several factors determine this outcome. Our recent work (Sahoo et al. 2014) indicates that IL-1β drives excessive neutrophil recruitment to the infected lung and that these cells can be detrimental in melioidosis. It has been shown that neutrophils are not very effective at killing *B. pseudomallei* infection (Egan and Gordon 1996) and, in fact, may foster its spread despite their strong microbicidal activities, including production of ROS, which is effective against *B. pseudomallei*

(Breitbach et al. 2006). This notion is supported by our observation that human or mouse neutrophils infected with *B. pseudomallei* failed to undergo pyroptosis, consistent with the finding that neutrophils do not express NLRC4 (Miao et al. 2010a; Ceballos-Olvera et al. 2011). At the same time, activated neutrophils were found to release elastase. This is the main neutrophil protease, and it was found to be detrimental in melioidosis through at least two distinct mechanisms: direct damage to lung tissue and augmentation of vascular permeability through the bradykinin pathway (Sahoo et al. 2014).

7 Role of the Caspase-11 Non-canonical Inflammasome

Although caspase-1-deficient mice have been used in hundred of studies, they were known to be also deficient of caspase-11, a fact that most of the scientific community has only recently acknowledged. This realization prompted a re-examination of the role of these caspases during bacterial infections, a task that is rendered more difficult by the absence of "bona fide" caspase-1-deficient mice. In a seminal paper, the Dixit laboratory showed that caspase-11 is part of a non-canonical inflammasome responsive to cholera toxin B and certain bacteria such as *E. coli*, *C. rodentium*, and *V. cholera* (Kayagaki et al. 2011). Subsequently, it was shown that caspase-11 protects against Gram-negative bacteria that invade the cytosol, including *B. pseudomallei* and *B. thailandensis*, by inducing pyroptosis (Aachoui et al. 2013; Casson et al. 2013). Caspase-11 was shown to recognize LPS localized in the cytosol (Hagar et al. 2013; Kayagaki et al. 2013), possibly by direct binding through the CARD domain (Shi et al. 2014). Caspase-11 expression in unstimulated macrophages is low and inducible by type-I IFN, IFNγ, or TLR agonists and, therefore, investigation of its function requires a priming step (Broz et al. 2012; Gurung et al. 2012; Rathinam et al. 2012; Wang et al. 1996).

Aachoui et al. (2015) analyzed the role of caspase-1 and caspase-11 in a model of intraperitoneal *B. thailandensis* infection. Due to the fact that "pure" caspase-1-deficient mice are not available, the authors used $Nlrc4^{-/-}Asc^{-/-}$ mice as surrogate for caspase-1 deficiency (these mice are predicted to be deficient for all known canonical inflammasomes that trigger caspase-1 activation). Whether this assumption is completely valid awaits confirmation with the analysis of pure $Casp1^{-/-}$ mice. In that study, $Casp1^{-/-}Casp11^{-/-}$, $Casp11^{-/-}$, and $Nlrc4^{-/-}Asc^{-/-}$ mice were highly susceptible to infection with high-dose *B. thailandensis*. However, when mice were infected with medium/low doses, $Casp1^{-/-}Casp11^{-/-}$, and $Casp11^{-/-}$ mice were equally susceptible whereas $Nlrc4^{-/-}Asc^{-/-}$ mice were resistant. This suggested to the authors that under this infection condition (low inoculum), the protection conferred by caspase-11-mediated pyroptosis is more important than that conferred by activation of caspase-1. In preliminary experiments, we have observed that $Casp1^{-/-}Casp11^{-/-}$ and $Casp11^{-/-}$ mice intranasally infected with *B. thailandensis* at high dose are very susceptible and indistinguishable. However, by lowering the infectious dose, we could show that $Casp11^{-/-}$ mice were moderately, but significantly, less susceptible than *Casp1*$^{-/-}$*Casp11*$^{-/-}$ mice and had lower bacteria burden in organs (manuscript in preparation). Thus, it seems that even in mice infected with low doses, caspase-1 confers a measurable protection, though the one mediated by caspase-11 may be more decisive.

8 Different Contribution of the Canonical and Non-canonical Inflammasomes

Taken together, these results indicate that the Nlrp3 and Nlrc4 and non-canonical caspase-11 inflammasomes are activated at different time points during the infection cycle, and to a different extent, reflecting different levels of protection (see Fig. 1). The highest level of protection appears to be conferred by caspase-11-dependent pyroptosis followed by NLRC4-dependent pyroptosis and finally NLRP3-dependent IL-18 secretion. This may reflect the fact that caspase-1-mediated pyroptosis depends on activation of NLRC4 through deployment of the T3SS and release of relatively small quantities of BasK or BsaL, which occurs early and only transiently in the infection cycle, before the bacterium has escaped the vacuole. In contrast, once the bacterium starts to replicate in the cytosol, LPS would be present in higher

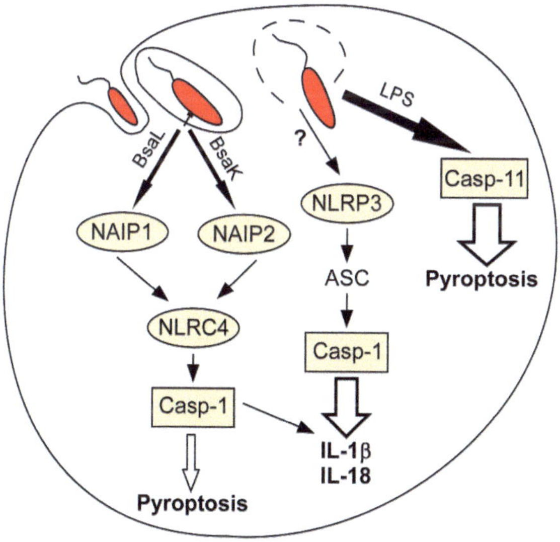

Fig. 1 Different inflammasomes are activated during infection with *B. pseudomallei*. Deployment of T3SS components Bsak and BsaL activates the NLRC4 inflammasome and pyroptosis in the early phase of infection before cytosol invasion. Phagosome escape activates the NLRP3 inflammasome and triggers production of IL-1β and IL-18. Presence of LPS in the cytosol activates the caspase-11 inflammasome and results in pyroptosis. The thickness of the arrows is meant to reflect the extent to which each pathway is activated and its protective effect

amounts and would lead to a more sustained activation of caspase-11. This delayed activation of caspase-11 is also due to the fact that its expression depends on priming of macrophages by IFNγ or TLR agonists. The protection conferred by activation of the Nlrp3 inflammasome appears less decisive, compared to the other inflammasomes, and is primarily mediated by production of IL-18. However, as noted above, residual production of IL-18 in *Nlrp3*$^{-/-}$ mice is mediated by the Nlrc4 inflammasome (Aachoui et al. 2015), which can explain why *Il-18*$^{-/-}$ mice are considerably more susceptible than *Nlrp3*$^{-/-}$ (Ceballos-Olvera et al. 2011). It should also be noted that the IL-18-mediated protective role of the Nlrp3 inflammasome in melioidosis is mitigated by the simultaneous and deleterious action of IL-1β, as shown by the fact that *Il-18*$^{-/-}$ mice are more susceptible than *Il-1r1*$^{-/-}$/*Il-18*$^{-/-}$ (Ceballos-Olvera et al. 2011; Sahoo et al. 2014). Figure 2 illustrates the opposite roles of IL-1β and IL-18 in melioidosis.

Based on the fact that caspase-11 requires priming by IFNγ, whose production is severely reduced in *B. pseudomallei*- or *B. thailandensis*-infected *Il-18*$^{-/-}$ mice, it has been proposed that the main role of the canonical inflammasome in melioidosis is to activate the IL-18-IFNγ axis that then primes cells for caspase-11-mediated pyroptosis (Aachoui et al. 2015). Our preliminary experiments support this model, but also show that the protective role of IFNγ in melioidosis is also mediated by activation of the NADPH oxidase and production of ROS. Thus, treatment with IFNγ reduced bacteria burden and improved survival much more efficiently in *Casp11*$^{-/-}$ or *Casp1*$^{-/-}$*Casp11*$^{-/-}$ mice than in *gp91*$^{phox-/-}$ mice intranasally

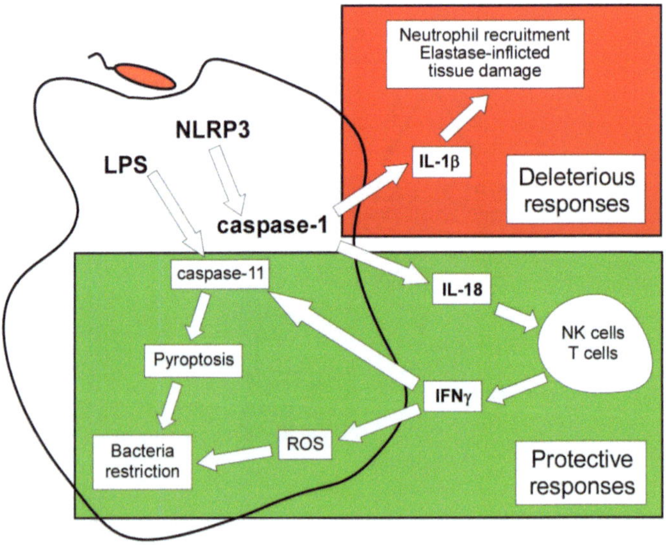

Fig. 2 IL-1β and IL-18 play opposite roles in melioidosis. IL-18 activates protective responses through induction of IFNγ production by NK cells and T cells. IFNγ stimulates generation of reactive oxygen species and induces caspase-11 and pyroptosis. IL-1β stimulates deleterious responses through excessive neutrophil recruitment and consequent lung tissue damage

infected with *B. thailandensis* (manuscript in preparation), suggesting that ROS-mediated killing rather than caspase-11-dependent pyroptosis is the microbicidal mechanism primarily induced by IFNγ.

9 Role of the Human Non-canonical Inflammasome

The caspase-11 gene appears duplicated in the human genome as caspase-4 and caspase-5. Recent studies suggest that both caspase-4 and caspase-5 recognize LPS and induce pyroptosis similar to caspase-11 (Shi et al. 2014; Kajiwara et al. 2014; Casson et al. 2015). Transgenic expression of caspase-4 in *Casp1*$^{-/-}$*Casp11*$^{-/-}$ mice or *Casp11*$^{-/-}$ mice was shown to complement their defective pyroptosis and mortality induced by infection with *B. thailandensis* (Aachoui et al. 2015). In contrast to caspase-11, caspase-4 expression is not IFNγ-inducible (Lin et al. 2000), a fact that may explain the higher susceptibility of humans, compared to mice, to endotoxic shock, which in mice is dependent on caspase-11 (Hagar et al. 2013; Kayagaki et al. 2013). This observation may suggest that humans may be better equipped than mice to fight infection with *B. pseudomallei* or other bacteria that are exquisitely sensitive to pyroptosis.

10 Concluding Remarks

A relatively small number of laboratories have examined the role of the inflammasome in melioidosis and, reassuringly, have come to similar general conclusions. However, some discrepancies among studies have been noted. One possible source of variability in the in vivo studies is represented by the route of infection used. For example, the deleterious role played by IL-1β during pneumonic melioidosis is not observed when mice are infected through the intraperitoneal route (Aachoui et al. 2015, and our unpublished results). An additional confounding factor is the fact that, although most laboratories use the *B. thailandensis* E264 or the *B. pseudomallei* 1026b reference strains, it is known that high rate of recombination, horizontal gene transfer, and mutation render the *Burkholderia* genome extraordinarily dynamic. It has been reported that up to 14 % of the *B. pseudomallei* genome differs among a panel of strains recovered from human cases of melioidosis (Sim et al. 2008). This portion of the genome is commonly referred to as the "accessory genome" and contains several genomic islands that contribute to pathogenesis and antibiotic resistance. Within-host evolution and niche adaptation have been documented during establishment of persistence, but also during the shorter time course of an acute infection in humans (Price et al. 2010). This high genome instability may explain the great variability in the LD_{50} for the *B. pseudomallei* 1026b strain reported in different studies (Gelhaus et al. 2013; Jeddeloh et al. 2003). It has been recently reported that it is possible to obtain a more consistent virulent strain by passaging *B. thailandensis* E264 into *Casp1*$^{-/-}$ *Casp11*$^{-/-}$ mice (Aachoui et al. 2015). We have also observed variability in the virulence of different batches of *B. thailandensis*/*B. pseudomallei* despite the fact that they were grown from the same glycerol stocks. This is an important consideration as it is clear from the above studies that the inoculum dose affects the outcome of the infection in different mouse strains and may lead to incorrect conclusions.

It is interesting to note that while several pathogens appear to have evolved mechanisms to evade or inhibit inflammasome activation, such strategies do not seem to have been adopted by *Burkholderia* species. This may explain why inflammasome deficiency seems to affect survival of *Burkholderia*-infected mice to a much greater degree than infection with other bacteria species.

References

Aachoui Y, Leaf IA, Hagar JA, Fontana MF, Campos CG, Zak DE, Tan MH, Cotter PA, Vance RE, Aderem A et al (2013) Caspase-11 protects against bacteria that escape the vacuole. Science 339:975–978

Aachoui Y, Kajiwara Y, Leaf IA, Mao D, Ting JP, Coers J, Aderem A, Buxbaum JD, Miao EA (2015) Canonical inflammasomes drive IFN-gamma to prime caspase-11 in defense against a cytosol-invasive bacterium. Cell Host Microbe 18:320–332

Ashdown LR, Koehler JM (1990) Production of hemolysin and other extracellular enzymes by clinical isolates of pseudomonas pseudomallei. J Clin Microbiol 28:2331–2334

Bast A, Krause K, Schmidt IH, Pudla M, Brakopp S, Hopf V, Breitbach K, Steinmetz I (2014) Caspase-1-dependent and -independent cell death pathways in Burkholderia pseudomallei infection of macrophages. PLoS Pathog 10:e1003986

Breitbach K, Klocke S, Tschernig T, van Rooijen N, Baumann U, Steinmetz I (2006) Role of inducible nitric oxide synthase and NADPH oxidase in early control of Burkholderia pseudomallei infection in mice. Infect Immun 74:6300–6309

Breitbach K, Sun GW, Kohler J, Eske K, Wongprompitak P, Tan G, Liu Y, Gan YH, Steinmetz I (2009) Caspase-1 mediates resistance in murine melioidosis. Infect Immun 77:1589–1595

Broz P, Ruby T, Belhocine K, Bouley DM, Kayagaki N, Dixit VM, Monack DM (2012) Caspase-11 increases susceptibility to Salmonella infection in the absence of caspase-1. Nature 490:288–291

Burtnick MN, Brett PJ, Nair V, Warawa JM, Woods DE, Gherardini FC (2008) Burkholderia pseudomallei type III secretion system mutants exhibit delayed vacuolar escape phenotypes in RAW 264.7 murine macrophages. Infect Immun 76:2991–3000

Case CL, Shin S, Roy CR (2009) Asc and Ipaf Inflammasomes direct distinct pathways for caspase-1 activation in response to Legionella pneumophila. Infect Immun 77:1981–1991

Casson CN, Copenhaver AM, Zwack EE, Nguyen HT, Strowig T, Javdan B, Bradley WP, Fung TC, Flavell RA, Brodsky IE et al (2013) Caspase-11 activation in response to bacterial secretion systems that access the host cytosol. PLoS Pathog 9:e1003400

Casson CN, Yu J, Reyes VM, Taschuk FO, Yadav A, Copenhaver AM, Nguyen HT, Collman RG, Shin S (2015) Human caspase-4 mediates noncanonical inflammasome activation against gram-negative bacterial pathogens. Proc Natl Acad Sci USA 112:6688–6693

Ceballos-Olvera I, Sahoo M, Miller MA, Del Barrio L, Re F (2011) Inflammasome-dependent pyroptosis and IL-18 protect against Burkholderia pseudomallei lung infection while IL-1beta is deleterious. PLoS Pathog 7:e1002452

Chaiyaroj SC, Kotrnon K, Koonpaew S, Anantagool N, White NJ, Sirisinha S (1999) Differences in genomic macrorestriction patterns of arabinose-positive (Burkholderia thailandensis) and arabinose-negative Burkholderia pseudomallei. Microbiol Immunol 43:625–630

Chanchamroen S, Kewcharoenwong C, Susaengrat W, Ato M, Lertmemongkolchai G (2009) Human polymorphonuclear neutrophil responses to Burkholderia pseudomallei in healthy and diabetic subjects. Infect Immun 77:456–463

Cheng AC, Currie BJ (2005) Melioidosis: epidemiology, pathophysiology, and management. Clin Microbiol Rev 18:383–416

Costa A, Gupta R, Signorino G, Malara A, Cardile F, Biondo C, Midiri A, Galbo R, Trieu-Cuot P, Papasergi S et al (2012) Activation of the NLRP3 inflammasome by group B streptococci. J Immunol 188:1953–1960

Craven RR, Gao X, Allen IC, Gris D, Bubeck Wardenburg J, McElvania-Tekippe E, Ting JP, Duncan JA (2009) Staphylococcus aureus alpha-hemolysin activates the NLRP3-inflammasome in human and mouse monocytic cells. PLoS ONE 4:e7446

Currie BJ, Jacups SP, Cheng AC, Fisher DA, Anstey NM, Huffam SE, Krause VL (2004) Melioidosis epidemiology and risk factors from a prospective whole-population study in northern Australia. Trop Med Int Health 9:1167–1174

DeShazer D, Brett PJ, Woods DE (1998) The type II O-antigenic polysaccharide moiety of Burkholderia pseudomallei lipopolysaccharide is required for serum resistance and virulence. Mol Microbiol 30:1081–1100

Dinarello CA (2009) Immunological and inflammatory functions of the interleukin-1 family. Annu Rev Immunol 27:519–550

Easton A, Haque A, Chu K, Lukaszewski R, Bancroft GJ (2007) A critical role for neutrophils in resistance to experimental infection with Burkholderia pseudomallei. J Infect Dis 195:99–107

Egan AM, Gordon DL (1996) Burkholderia pseudomallei activates complement and is ingested but not killed by polymorphonuclear leukocytes. Infect Immun 64:4952–4959

Gan YH (2005) Interaction between Burkholderia pseudomallei and the host immune response: sleeping with the enemy? J Infect Dis 192:1845–1850

Garlanda C, Dinarello CA, Mantovani A (2013) The interleukin-1 family: back to the future. Immunity 39:1003–1018

Gelhaus HC, Anderson MS, Fisher DA, Flavin MT, Xu ZQ, Sanford DC (2013) Efficacy of post exposure administration of doxycycline in a murine model of inhalational melioidosis. Sci Rep 3:1146

Gurung P, Malireddi RK, Anand PK, Demon D, Vande Walle L, Liu Z, Vogel P, Lamkanfi M, Kanneganti TD (2012) Toll or interleukin-1 receptor (TIR) domain-containing adaptor inducing interferon-beta (TRIF)-mediated caspase-11 protease production integrates Toll-like receptor 4 (TLR4) protein- and Nlrp3 inflammasome-mediated host defense against enteropathogens. J Biol Chem 287:34474–34483

Hagar JA, Powell DA, Aachoui Y, Ernst RK, Miao EA (2013) Cytoplasmic LPS activates caspase-11: implications in TLR4-independent endotoxic shock. Science 341:1250–1253

Haque A, Easton A, Smith D, O'Garra A, Van Rooijen N, Lertmemongkolchai G, Titball RW, Bancroft GJ (2006) Role of T cells in innate and adaptive immunity against murine Burkholderia pseudomallei infection. J Infect Dis 193:370–379

Haraga A, West TE, Brittnacher MJ, Skerrett SJ, Miller SI (2008) Burkholderia thailandensis as a model system for the study of the virulence-associated type III secretion system of Burkholderia pseudomallei. Infect Immun 76:5402–5411

Harland DN, Dassa E, Titball RW, Brown KA, Atkins HS (2007) ATP-binding cassette systems in Burkholderia pseudomallei and Burkholderia mallei. BMC Genom 8:83

Haussler S, Nimtz M, Domke T, Wray V, Steinmetz I (1998) Purification and characterization of a cytotoxic exolipid of Burkholderia pseudomallei. Infect Immun 66:1588–1593

Jeddeloh JA, Fritz DL, Waag DM, Hartings JM, Andrews GP (2003) Biodefense-driven murine model of pneumonic melioidosis. Infect Immun 71:584–587

Jones AL, Beveridge TJ, Woods DE (1996) Intracellular survival of Burkholderia pseudomallei. Infect Immun 64:782–790

Kajiwara Y, Schiff T, Voloudakis G, Gama Sosa MA, Elder G, Bozdagi O, Buxbaum JD (2014) A critical role for human caspase-4 in endotoxin sensitivity. J Immunol 193:335–343
Kayagaki N, Warming S, Lamkanfi M, Vande Walle L, Louie S, Dong J, Newton K, Qu Y, Liu J, Heldens S et al (2011) Non-canonical inflammasome activation targets caspase-11. Nature 479:117–121
Kayagaki N, Wong MT, Stowe IB, Ramani SR, Gonzalez LC, Akashi-Takamura S, Miyake K, Zhang J, Lee WP, Muszynski A et al (2013) Noncanonical inflammasome activation by intracellular LPS independent of TLR4. Science 341:1246–1249
Kespichayawattana W, Rattanachetkul S, Wanun T, Utaisincharoen P, Sirisinha S (2000) Burkholderia pseudomallei induces cell fusion and actin-associated membrane protrusion: a possible mechanism for cell-to-cell spreading. Infect Immun 68:5377–5384
Koh GC, Maude RR, Schreiber MF, Limmathurotsakul D, Wiersinga WJ, Wuthiekanun V, Lee SJ, Mahavanakul W, Chaowagul W, Chierakul W et al (2011) Glyburide is anti-inflammatory and associated with reduced mortality in melioidosis. Clin Infect Dis 52:717–725
Lauw FN, Simpson AJ, Prins JM, Smith MD, Kurimoto M, van Deventer SJ, Speelman P, Chaowagul W, White NJ, van der Poll T (1999) Elevated plasma concentrations of interferon (IFN)-gamma and the IFN-gamma-inducing cytokines interleukin (IL)-18, IL-12, and IL-15 in severe melioidosis. J Infect Dis 180:1878–1885
Lazar Adler NR, Govan B, Cullinane M, Harper M, Adler B, Boyce JD (2009) The molecular and cellular basis of pathogenesis in melioidosis: how does Burkholderia pseudomallei cause disease? FEMS Microbiol Rev 33:1079–1099
Lertmemongkolchai G, Cai G, Hunter CA, Bancroft GJ (2001) Bystander activation of CD8+ T cells contributes to the rapid production of IFN-gamma in response to bacterial pathogens. J Immunol 166:1097–1105
Lin XY, Choi MS, Porter AG (2000) Expression analysis of the human caspase-1 subfamily reveals specific regulation of the CASP5 gene by lipopolysaccharide and interferon-gamma. J Biol Chem 275:39920–39926
Matsuura M, Kawahara K, Ezaki T, Nakano M (1996) Biological activities of lipopolysaccharide of Burkholderia (pseudomonas) pseudomallei. FEMS Microbiol Lett 137:79–83
Meunier E, Dick MS, Dreier RF, Schurmann N, Kenzelmann Broz D, Warming S, Roose-Girma M, Bumann D, Kayagaki N, Takeda K et al (2014) Caspase-11 activation requires lysis of pathogen-containing vacuoles by IFN-induced GTPases. Nature 509:366–370
Miao EA, Leaf IA, Treuting PM, Mao DP, Dors M, Sarkar A, Warren SE, Wewers MD, Aderem A (2010a) Caspase-1-induced pyroptosis is an innate immune effector mechanism against intracellular bacteria. Nat Immunol 11:1136–1142
Miao EA, Mao DP, Yudkovsky N, Bonneau R, Lorang CG, Warren SE, Leaf IA, Aderem A (2010b) Innate immune detection of the type III secretion apparatus through the NLRC4 inflammasome. Proc Natl Acad Sci USA 107:3076–3080
Miyagi K, Kawakami K, Saito A (1997) Role of reactive nitrogen and oxygen intermediates in gamma interferon-stimulated murine macrophage bactericidal activity against Burkholderia pseudomallei. Infect Immun 65:4108–4113
Moore RA, Reckseidler-Zenteno S, Kim H, Nierman W, Yu Y, Tuanyok A, Warawa J, DeShazer D, Woods DE (2004) Contribution of gene loss to the pathogenic evolution of Burkholderia pseudomallei and Burkholderia mallei. Infect Immun 72:4172–4187
Muangsombut V, Suparak S, Pumirat P, Damnin S, Vattanaviboon P, Thongboonkerd V, Korbsrisate S (2008) Inactivation of Burkholderia pseudomallei bsaQ results in decreased invasion efficiency and delayed escape of bacteria from endocytic vesicles. Arch Microbiol 190:623–631
Munoz-Planillo R, Franchi L, Miller LS, Nunez G (2009) A critical role for hemolysins and bacterial lipoproteins in Staphylococcus aureus-induced activation of the Nlrp3 inflammasome. J Immunol 183:3942–3948
Pilla DM, Hagar JA, Haldar AK, Mason AK, Degrandi D, Pfeffer K, Ernst RK, Yamamoto M, Miao EA, Coers J (2014) Guanylate binding proteins promote caspase-11-dependent pyroptosis in response to cytoplasmic LPS. Proc Natl Acad Sci USA 111:6046–6051

Pongcharoen S, Niumsup PR, Butkhamchot P (2008) Comparative study of interleukin-1beta expression by peripheral blood mononuclear cells and purified monocytes experimentally infected with Burkholderia pseudomallei and Burkholderia thailandensis. Immunol Invest 37:704–713

Price EP, Hornstra HM, Limmathurotsakul D, Max TL, Sarovich DS, Vogler AJ, Dale JL, Ginther JL, Leadem B, Colman RE et al (2010) Within-host evolution of Burkholderia pseudomallei in four cases of acute melioidosis. PLoS Pathog 6:e1000725

Pruksachartvuthi S, Aswapokee N, Thankerngpol K (1990) Survival of pseudomonas pseudomallei in human phagocytes. J Med Microbiol 31:109–114

Rathinam VA, Vanaja SK, Waggoner L, Sokolovska A, Becker C, Stuart LM, Leong JM, Fitzgerald KA (2012) TRIF licenses caspase-11-dependent NLRP3 inflammasome activation by gram-negative bacteria. Cell 150:606–619

Reckseidler-Zenteno SL, DeVinney R, Woods DE (2005) The capsular polysaccharide of Burkholderia pseudomallei contributes to survival in serum by reducing complement factor C3b deposition. Infect Immun 73:1106–1115

Sahoo M, Del Barrio L, Miller MA, Re F (2014) Neutrophil elastase causes tissue damage that decreases host tolerance to lung infection with burkholderia species. PLoS Pathog 10: e1004327

Santanirand P, Harley VS, Dance DA, Raynes JG, Drasar BS, Bancroft GJ (1997) Interferon-gamma mediates host resistance in a murine model of melioidosis. Biochem Soc Trans 25:287S

Shi J, Zhao Y, Wang Y, Gao W, Ding J, Li P, Hu L, Shao F (2014) Inflammatory caspases are innate immune receptors for intracellular LPS. Nature 514:187–192

Sim SH, Yu Y, Lin CH, Karuturi RK, Wuthiekanun V, Tuanyok A, Chua HH, Ong C, Paramalingam SS, Tan G et al (2008) The core and accessory genomes of Burkholderia pseudomallei: implications for human melioidosis. PLoS Pathog 4:e1000178

Stevens MP, Wood MW, Taylor LA, Monaghan P, Hawes P, Jones PW, Wallis TS, Galyov EE (2002) An Inv/Mxi-Spa-like type III protein secretion system in Burkholderia pseudomallei modulates intracellular behaviour of the pathogen. Mol Microbiol 46:649–659

Sun GW, Lu J, Pervaiz S, Cao WP, Gan YH (2005) Caspase-1 dependent macrophage death induced by Burkholderia pseudomallei. Cell Microbiol 7:1447–1458

Suzuki T, Franchi L, Toma C, Ashida H, Ogawa M, Yoshikawa Y, Mimuro H, Inohara N, Sasakawa C, Nunez G (2007) Differential regulation of caspase-1 activation, pyroptosis, and autophagy via Ipaf and ASC in Shigella-infected macrophages. PLoS Pathog 3:e111

Ulett GC, Ketheesan N, Hirst RG (2000) Proinflammatory cytokine mRNA responses in experimental Burkholderia pseudomallei infection in mice. Acta Trop 74:229–234

Utaisincharoen P, Tangthawornchaikul N, Kespichayawattana W, Chaisuriya P, Sirisinha S (2001) Burkholderia pseudomallei interferes with inducible nitric oxide synthase (iNOS) production: a possible mechanism of evading macrophage killing. Microbiol Immunol 45:307–313

Wang S, Miura M, Jung Y, Zhu H, Gagliardini V, Shi L, Greenberg AH, Yuan J (1996) Identification and characterization of Ich-3, a member of the interleukin-1beta converting enzyme (ICE)/Ced-3 family and an upstream regulator of ICE. J Biol Chem 271:20580–20587

West TE, Ernst RK, Jansson-Hutson MJ, Skerrett SJ (2008a) Activation of Toll-like receptors by Burkholderia pseudomallei. BMC Immunol 9:46

West TE, Frevert CW, Liggitt HD, Skerrett SJ (2008b) Inhalation of Burkholderia thailandensis results in lethal necrotizing pneumonia in mice: a surrogate model for pneumonic melioidosis. Trans R Soc Trop Med Hyg 102(Suppl 1):S119–S126

West TE, Myers ND, Chantratita N, Chierakul W, Limmathurotsakul D, Wuthiekanun V, Miao EA, Hajjar AM, Peacock SJ, Liggitt HD et al (2014) NLRC4 and TLR5 each contribute to host defense in respiratory melioidosis. PLoS Negl Trop Dis 8:e3178

Wiersinga WJ, van der Poll T, White NJ, Day NP, Peacock SJ (2006) Melioidosis: insights into the pathogenicity of Burkholderia pseudomallei. Nat Rev Microbiol 4:272–282

Wiersinga WJ, Wieland CW, Dessing MC, Chantratita N, Cheng AC, Limmathurotsakul D, Chierakul W, Leendertse M, Florquin S, de Vos AF et al (2007a) Toll-like receptor 2 impairs

host defense in gram-negative sepsis caused by Burkholderia pseudomallei (melioidosis). PLoS Med 4:e248
Wiersinga WJ, Wieland CW, van der Windt GJ, de Boer A, Florquin S, Dondorp A, Day NP, Peacock SJ, van der Poll T (2007b) Endogenous interleukin-18 improves the early antimicrobial host response in severe melioidosis. Infect Immun 75:3739–3746
Wiersinga WJ, de Vos AF, Wieland CW, Leendertse M, Roelofs JJ, van der Poll T (2008a) CD14 impairs host defense against gram-negative sepsis caused by Burkholderia pseudomallei in mice. J Infect Dis 198:1388–1397
Wiersinga WJ, Wieland CW, Roelofs JJ, van der Poll T (2008b) MyD88 dependent signaling contributes to protective host defense against Burkholderia pseudomallei. PLoS ONE 3:e3494
Yang J, Zhao Y, Shi J, Shao F (2013) Human NAIP and mouse NAIP1 recognize bacterial type III secretion needle protein for inflammasome activation. Proc Natl Acad Sci USA 110: 14408–14413
Yu Y, Kim HS, Chua HH, Lin CH, Sim SH, Lin D, Derr A, Engels R, DeShazer D, Birren B et al (2006) Genomic patterns of pathogen evolution revealed by comparison of Burkholderia pseudomallei, the causative agent of melioidosis, to avirulent Burkholderia thailandensis. BMC Microbiol 6:46
Zhao Y, Yang J, Shi J, Gong YN, Lu Q, Xu H, Liu L, Shao F (2011) The NLRC4 inflammasome receptors for bacterial flagellin and type III secretion apparatus. Nature 477:596–600

Inflammasomes in Pneumococcal Infection: Innate Immune Sensing and Bacterial Evasion Strategies

Anne Rabes, Norbert Suttorp and Bastian Opitz

Abstract *Streptococcus pneumoniae* frequently colonizes the upper respiratory tract of healthy individuals, but also commonly causes severe invasive infections such as community-acquired pneumonia and meningitis. One of the key virulence factors of pneumococci is the pore-forming toxin pneumolysin which stimulates cell death and is involved in the evasion of some defense mechanisms. The immune system, however, employs different inflammasomes to sense pneumolysin-induced pore formation, cellular membrane damage, and/or subsequent leakage of bacterial nucleic acid into the host cell cytosol. Canonical inflammasomes are cytosolic multiprotein complexes consisting of a receptor molecule such as NLRP3 or AIM2, the adapter ASC, and caspase-1. NLRP3 and AIM2 inflammasomes mediate cell death and production of important IL-1 family cytokines to recruit leukocytes and defend against *S. pneumoniae*. Here, we review recent evidence that highlights inflammasomes as critical sensors of *S. pneumoniae*-induced cellular perturbations, summarize their role in pneumococcal infections, and discuss potential evasion strategies of some emerging pneumococcal strains.

Contents

A. Rabes · N. Suttorp · B. Opitz (✉)
Department of Internal Medicine/Infectious Diseases and Pulmonary Medicine, Charité Universitätsmedizin Berlin, Augustenburger Platz 1, 13353 Berlin, Germany
e-mail: bastian.opitz@charite.de

S. Backert (ed.), *Inflammasome Signaling and Bacterial Infections*, Current Topics in Microbiology and Immunology 397,
DOI 10.1007/978-3-319-41171-2_11

1 Introduction

The family Streptococcaceae comprises several clinically important species of which *Streptococcus pneumoniae* (the pneumococcus) is of particular importance. Similar to group A and B streptococci, *S. pneumoniae* can be classified as pathobiont, as it colonizes the oro—and nasopharynx of up to 60 % of healthy small children and 10 % of adults (Henriques-Normark and Tuomanen 2013). This colonization is in most cases temporary and asymptomatic; however, it is also a prerequisite of local and life-threatening invasive infections as well as transmission to new susceptible hosts (Siegel and Weiser 2015). Although the mechanism how the pathogen causes disease starting from an asymptomatic colonization is poorly understood, strain-specific virulence traits, chronic underlying diseases, a poor immune status of the host, and preceding viral infections can promote progression to invasive pneumococcal infection. These invasive infections include community-acquired pneumonia and meningitis, which cause high morbidity and mortality (van der Poll and Opal 2009; Henriques-Normark and Tuomanen 2013). Estimations suggest that *S. pneumoniae* is responsible for over 1 million infant deaths every year worldwide and probably even more in elderly people as well as immunocompromised individuals (van der Poll and Opal 2009; Krone et al. 2014).

Currently, 93 capsular serotypes of *S. pneumoniae* have been described, which vary markedly in their ability to cause invasive infection. For example, serotypes 3, 6B, 9N, and others are associated with nasopharyngeal colonization and serious infections in patients with underlying diseases. In contrast, dominating clones of serotypes 1 and 8 are frequently found in invasive diseases that, however, show lower case fatality rates (Weinberger et al. 2010). Among serotype 1 pneumococci, the recently emerging MLST306 clone now dominates by over 80 % in many parts of the world (Henriques Normark et al. 2001; Kirkham et al. 2006; Le Hello et al. 2010; Staples et al. 2015; Lamb et al. 2014; Marimon et al. 2009), although the molecular basis for its "success" is incompletely understood.

Important virulence factors of *S. pneumoniae* include the polysaccharide capsule, surface adhesins, extracellular matrix-degrading enzymes, and the exotoxin pneumolysin (PLY) (Henriques-Normark and Tuomanen 2013; Kadioglu et al. 2008; Voss et al. 2012). The capsule inhibits phagocytosis, complement factor binding, and entrapment by neutrophil extracellular traps (Abeyta et al. 2003; Hyams et al. 2010; Wartha et al. 2007). PLY is a member of the cholesterol-dependent cytolysins expressed by various gram-positive bacteria. PLY binds to cholesterol-containing membranes, forms pores upon oligomerization, and thereby typically causes cell lysis (Mitchell and Mitchell 2010). Moreover, the toxin inhibits the classical complement pathway and is involved in biofilm formation (Shak et al. 2013; Rubins et al. 1996; Mitchell et al. 1991). The critical role of PLY in causing disease has been demonstrated in various models of pneumococcal pneumonia (Kadioglu et al. 2002; Rubins et al. 1995), lung injury (Maus et al. 2004; Witzenrath et al. 2006), and meningitis (Hirst et al. 2008; Wellmer et al. 2002). Interestingly, PLY exists in at least 16 different protein variants with variable hemolytic activity. For example, allele 5 PLY expressed by serotype 1 MLST306 and some serotype 8 pneumococci is non-hemolytic in contrast to the PLY variants of most other pneumococcal strains (Jefferies et al. 2007; Kirkham et al. 2006; Lock et al. 1996). It is very likely that the expression of PLY variants as well as the degree of encapsulation and the production of other virulence factors affect the pneumococci's ability to colonize and to cause invasive diseases by modulating the interaction with the host's immune system (Weinberger et al. 2010; Jefferies et al. 2007).

2 Brief Overview of the Innate Immune Response to *S. Pneumoniae* Infection

An efficient antibacterial defense depends on the detection of the microbe by the innate immune system and the subsequent production of inflammatory cytokines and chemokines. These mediators stimulate the recruitment of antibacterial neutrophils and inflammatory macrophages, activate the acute-phase response, and help to initiate an adaptive immune response. *S. pneumoniae,* as well as virtually any microbe, is sensed by pattern recognition receptors (PRRs) belonging to different protein families. PRRs detect microbial components which are typically called pathogen-associated molecular patterns (PAMPs) although also commensal symbionts and pathobionts produce many of these molecules (Opitz et al. 2010). Pneumococcal cell wall lipopeptides and peptidoglycan, for example, are recognized by Toll-like receptor (TLR)-2 at the cellular surface and the cytosolic NOD-like receptor (NLR) NOD2, respectively (Davis et al. 2011; Knapp et al. 2004; Opitz et al. 2004) (Fig. 1). Moreover, pneumococcal DNA is detected by TLR9 in endosomes, and by an unidentified cytosolic sensor that signals through the adapter molecule STING (Albiger et al. 2007; Koppe et al. 2012a). Interestingly, recognition of pneumococcal peptidoglycan and DNA by the NOD2 and STING pathways, respectively, depends on the expression of PLY, as this toxin mediates leakage of these PAMPs into the cytosol following bacterial phagocytosis and phagosomal degradation (Davis et al. 2011; Koppe et al. 2012b). Upon activation, these TLRs, NOD2, and the cytosolic DNA sensor regulate the production of inflammatory mediators, including TNFα, IL-1β, IL-6, different chemokines, and type I interferons by activation of the transcription factors NF-κB and/or IRF3/7 (Koppe et al. 2012b). Whereas most of these mediators are immediately secreted following their transcription and translation, production of some very potent cytokines including IL-1β is additionally regulated on a posttranslational level by inflammasomes.

3 Sensing of *S. Pneumoniae* by Inflammasomes

Canonical inflammasomes have been first described by Tschopp and colleagues as cytosolic protein complexes that serve as platforms for recruitment and activation of caspase-1 (Martinon et al. 2002). In addition to caspase-1, these complexes usually

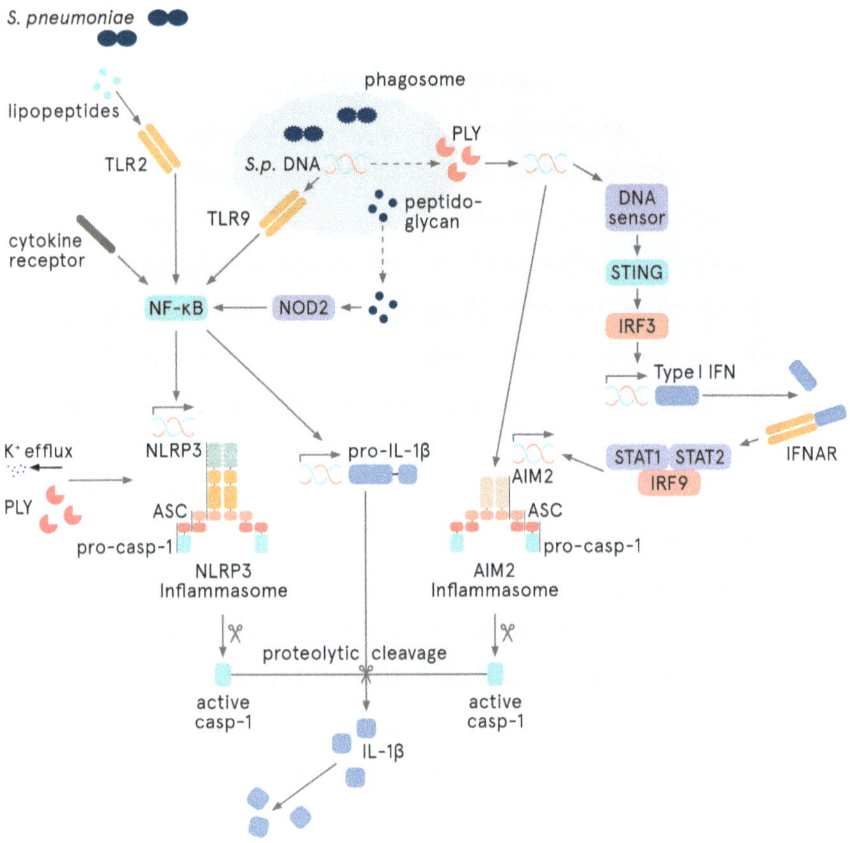

Fig. 1 Overview of the inflammasomes and other PRRs involved in recognition of *S. pneumoniae*. Cell wall lipopeptides are recognized by TLR2. Moreover, *S. pneumoniae* (*S.p.*) is internalized by phagocytic cells and subsequently degraded in phagosomes leading to the release of bacterial peptidoglycan and nucleic acids. While unmethylated CpG-containing DNA is sensed by TLR9 within the endosomes, different bacterial components get also access to the cytosol dependent on PLY-mediated membrane disruption. In the cytosol, pneumococcal peptidoglycan fragments are detected by NOD2, and pneumococcal DNA is detected by a still-not-identified cytosolic PRR that signals through STING and IRF3. The TLRs, NOD2, and cytokine receptors subsequently stimulate the production of NF-κB-dependent genes including pro-IL-1β and NLRP3. The STING-dependent pathway primarily activates production of type I IFNs. Type I IFNs induce the expression of hundreds of IFN-stimulated genes, such as, e.g., AIM2. NLRP3 forms together with ASC and caspase-1 (casp-1) the NLRP3 inflammasomes. This complex is activated by *S. pneumoniae* expressing hemolytic PLY, a process that requires K^+ efflux. The AIM2 inflammasome, consisting of AIM2, ASC, and caspase-1, is activated by pneumococcal DNA in the cytosol. Both inflammasomes mediate caspase-1 activation to cleave pro-IL-1β into mature IL-1β. In addition, inflammasomes also regulate IL-18 production as well as pyroptosis (not depicted)

consist of the adapter molecule ASC and a receptor which in most cases belongs to the NLR or PYHIN protein families (Schroder and Tschopp 2010; Chaput et al. 2013). Prominent inflammasomes are, for example, the NLRP3 and AIM2 inflammasomes that are formed by the respective NLR and PYHIN family member. The activation of inflammasomes is usually a two-step process that involves a NF-κB-dependent transcriptional upregulation of pro-IL-1β and NLRP3 and/or an upregulation of AIM2 mediated by type I IFNs (Fang et al. 2014; Bauernfeind et al. 2009). The second signal is generated by the inflammasome stimulus itself. Whereas the NLRP3 inflammasome responds to cellular disturbance induced by a wide variety of microbial and endogenous molecules (Martinon et al. 2006; Eisenbarth et al. 2008; Duewell et al. 2010; Kanneganti et al. 2006; Mariathasan et al. 2006; Halle et al. 2008; Hornung et al. 2008), AIM2 directly senses different kinds of dsDNA (Burckstummer et al. 2009; Fernandes-Alnemri et al. 2009; Hornung et al. 2009; Roberts et al. 2009). Inflammasomes regulate production of IL-1β and IL-18 on a posttranslational level by cleaving their zymogenic pro-forms pro-IL-1β and pro-IL-18 into functional mature cytokines. In addition, activation of these protein complexes induces a caspase-1-dependent inflammatory cell death that is called pyroptosis (Schroder and Tschopp 2010).

Although PLY is involved in the evasion of some host defense mechanisms (Mitchell and Mitchell 2010), the immune system is also able to sense the presence of active PLY through inflammasomes. It has been known that host cells respond to hemolytic PLY and other pore-forming toxins by activating caspase-1 and producing IL-1β/IL-18 (Koedel et al. 2002; Shoma et al. 2008; Tsukada et al. 1992; Mariathasan et al. 2006). Subsequent studies from our and other groups have demonstrated that the caspase-1-dependent responses to *S. pneumoniae* are mediated by the NLRP3 (McNeela et al. 2010; Witzenrath et al. 2011; Fang et al. 2011; Hoegen et al. 2011) and AIM2 inflammasomes (Fang et al. 2011; Koppe et al. 2012a). Infection of macrophages, dendritic cells (DCs), and neutrophils, for example, showed that caspase-1 activation as well as IL-1β and IL-18 production require the expression of active PLY by the pneumococcus and of ASC by the host cell (Fang et al. 2011; Hoegen et al. 2011; McNeela et al. 2010; Witzenrath et al. 2011; Karmakar et al. 2015). Moreover, NLRP3- or AIM2-deficient cells produced diminished but still measureable amounts of these cytokines, indicating that NLRP3 and AIM2 inflammasomes play partly redundant roles in responding to *S. pneumoniae* infection (Fang et al. 2011).

The exact mechanism of NLRP3 activation by PLY (and other stimuli) is unknown. It is most likely indirect and involves, or is modulated by, potassium efflux (Witzenrath et al. 2011; Karmakar et al. 2015; McNeela et al. 2010). Given that microbial RNA was recently identified as a direct NLRP3 activator in *S. pyogenes* infections (Gupta et al. 2014), it appears reasonable to speculate that NLRP3 is stimulated by pneumococcal RNA as well. In this hypothetical scenario, PLY might be required for lysosomal rupture and leakage of bacterial RNA (as well as other endogenous or bacterial molecules) into the cytosol, as it has been indicated for β-hemolysin in *S. pyogenes* infection (Gupta et al. 2014). Similarly, PLY appears to be required for release of pneumococcal DNA from the phagolysosome into the cytosol and activation of the AIM2 inflammasome (Fang et al. 2011; Koppe et al. 2012a). In summary, the immune system has evolved different mechanisms to indirectly sense the presence of *S. pneumoniae* and other virulent bacteria expressing pore-forming toxins.

4 Role of Inflammasomes in Pneumococcal Infections

Several recent studies have examined the function of inflammasomes in different infection models. ASC-deficient animals, which lack functional NLRP3 and AIM2 inflammasomes, for example, have been shown to be highly susceptible toward pneumococcal pneumonia, whereas *Nlrp3*$^{-/-}$ mice demonstrated only a slightly decreased resistance compared to controls (Witzenrath et al. 2011; Fang et al. 2011; McNeela et al. 2010; van Lieshout et al. 2014). *Asc*$^{-/-}$ mice, as well as in some studies—and to a lesser extent also *Nlrp3*$^{-/-}$ animals—showed reduced IL-1β and IL-18 levels and enhanced bacterial dissemination and mortality compared to wild-type controls. Mice lacking IL-1β or IL-18 were also more susceptible toward pneumococcal infection as compared to wild-type animals (Kafka et al. 2008; Lauw et al. 2002). Moreover, IL-1 production was crucial for bacterial restriction in a model of pneumococcal upper respiratory tract colonization (Lemon et al. 2015). The protective effects of the inflammasome/IL-1β pathway in pneumococcal infections might at least partly depend on neutrophil and/or macrophage recruitment and the antibacterial activities of these cell types (Lemon et al. 2015; Marriott et al. 2012).

In parallel to most other immune pathways, excessive activation of inflammasomes can also have detrimental effects for the host by exacerbating inflammation and causing extensive tissue damage. In models of pneumococcal meningitis, for example, NLRP3 inflammasomes appear to contribute to pathology rather than to host protection. It was shown that NLRP3- and ASC-deficient mice or animals treated with IL-1 or IL-18 antagonists exhibited less inflammation, tissue damage, and improved clinical scores compared to wild-type animals (Hoegen et al. 2011). These detrimental effects have been attributed to the production of IFNγ which is induced by inflammasome-dependent production of IL-18 (Mitchell et al. 2012; Zwijnenburg et al. 2003). Moreover, cerebrospinal fluid levels of inflammasome-related cytokines were related to complications and adverse disease outcome in patients with bacterial meningitis (Geldhoff et al. 2013). Together, inflammasomes play an important function in defending the host from pneumococcal infection. Depending on the location, magnitude, and the context of their activation, however, inflammasomes can also have detrimental consequences for the host.

5 Evasion of Inflammasome-Dependent Sensing by Emerging Strains

As discussed above, PLY is a key virulence factor of *S. pneumoniae* involved in tissue damage, inhibition of the classical complement pathway, and in biofilm formation (Shak et al. 2013; Rubins et al. 1996; Mitchell et al. 1991); however, its activation at the same time triggers strong antibacterial immune pathways. Interestingly, some clinically important pneumococcal strains, such as the emerging serotype 1 strain MLST306, express a different type of PLY. This allele 5 PLY differs from other toxin variants by only a few amino acids (Jefferies et al. 2007), maintains its ability to bind to cholesterol-containing membranes and to oligomerize, but it cannot form pores (Kirkham et al. 2006). Studies comparing bacterial mutants expressing hemolytic allele 1 PLY, non-hemolytic allele 5 PLY, or no toxin on the same genetic background suggest that non-hemolytic PLY still significantly contributes to virulence (Harvey et al. 2014). Although these studies used an intraperitoneal infection model that poorly reflects the natural route of pneumococcal infections, the results nonetheless suggest that pore formation may not be essential for the pneumococci to cause disease or that lack of pore formation by, e.g., MLST306 is partly "compensated" by other virulence strategies such as immune evasion.

We recently showed that indeed MLST306 and other pneumococci expressing allele 5 PLY evade recognition by inflammasomes (Fig. 2) (Fatykhova et al. 2015; Witzenrath et al. 2011). Experiments with murine and human cells as well as human

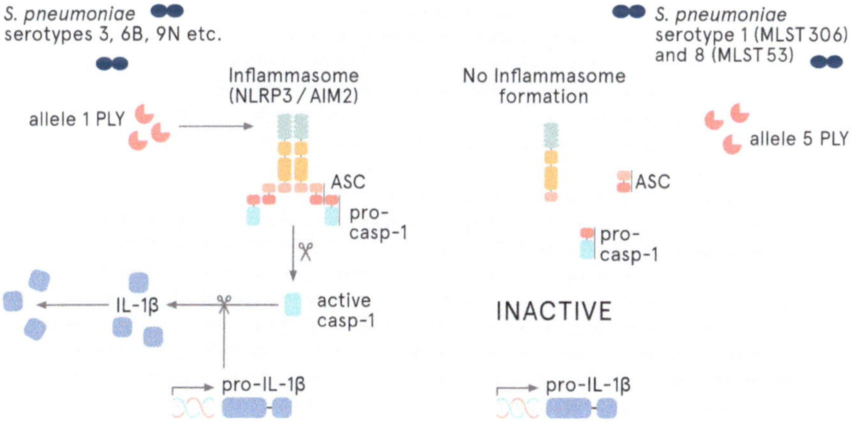

Fig. 2 Differential sensing of pneumococcal strains expressing allele 1 or allele 5 PLY. Most pneumococcal strains including serotype 3, 6B, and 9N bacteria express a hemolytic allele 1 PLY, which is sensed by inflammasomes. In contrast, serotype 1 MLST306, serotype 8 MLST53, and some other strains produce a non-hemolytic allele 5 PLY, which poorly activates inflammasome responses

lung tissue demonstrated that MLST306 or the serotype 8 strain MLST53 induced only little IL-1β production compared to the much stronger response induced by other pneumococcal strains. Moreover, the non-hemolytic allele 5 PLY which is expressed by MLST306 and MLST53 poorly activated inflammasomes, whereas hemolytic allele 1 PLY stimulated strong IL-1β responses (Harvey et al. 2014; Fatykhova et al. 2015; Witzenrath et al. 2011). We speculate that the evasion of inflammasome-dependent innate immune responses by MLST306 and other strains might contribute to their remarkable ability to enter sterile body sites and cause invasive infections in humans. Moreover, the avoidance of other cytosolic innate immune sensors whose activation also relies on PLY activities (see above) might additionally play a role. On the other hand, the weaker inflammatory response initiated by, e.g., MLST306 might also explain the reduced severity associated with serotype 1 disease (Weinberger et al. 2010).

6 Concluding Remarks

Inflammasomes are key players of the immune system to sense potentially pathogenic microbes. Accordingly, the NLRP3 and AIM2 inflammasomes are activated by infections with most pneumococcal strains, which are characterized by the expression of a hemolytic PLY as major virulence factor. Some emerging strains of *S. pneumoniae*, however, have evolved to express a PLY variant to evade inflammasome-dependent immune detection but apparently retaining virulence. A better understanding of the exact mechanisms of inflammasome activation by *S. pneumoniae* and the evasion strategies employed by some strains, as well as of the function of inflammasomes during different types of infections, will be critical for a comprehensive understanding of the bacterial pathogenesis and the host response. This knowledge might additionally lead to the development of novel strategies to treat invasive pneumococcal diseases.

Acknowledgments The work of the authors is supported by the Deutsche Forschungsgemeinschaft (GRK1673/B5 to A.R. and B.O., OP 86/7-2 and OP 86/10-1 to B.O, and SFB/TR84 to N.S. and B.O.).

References

Abeyta M, Hardy GG, Yother J (2003) Genetic alteration of capsule type but not PspA type affects accessibility of surface-bound complement and surface antigens of Streptococcus pneumoniae. Infect Immun 71(1):218–225

Albiger B, Dahlberg S, Sandgren A, Wartha F, Beiter K, Katsuragi H, Akira S, Normark S, Henriques-Normark B (2007) Toll-like receptor 9 acts at an early stage in host defence against pneumococcal infection. Cell Microbiol 9(3):633–644. doi:10.1111/j.1462-5822.2006.00814.x

Bauernfeind FG, Horvath G, Stutz A, Alnemri ES, MacDonald K, Speert D, Fernandes-Alnemri T, Wu J, Monks BG, Fitzgerald KA, Hornung V, Latz E (2009) Cutting edge: NF-kappaB activating pattern recognition and cytokine receptors license NLRP3 inflammasome activation by regulating NLRP3 expression. J Immunol 183(2):787–791. doi:10.4049/jimmunol.0901363

Burckstummer T, Baumann C, Bluml S, Dixit E, Durnberger G, Jahn H, Planyavsky M, Bilban M, Colinge J, Bennett KL, Superti-Furga G (2009) An orthogonal proteomic-genomic screen identifies AIM2 as a cytoplasmic DNA sensor for the inflammasome. Nat Immunol 10 (3):266–272. doi:10.1038/ni.1702

Chaput C, Sander LE, Suttorp N, Opitz B (2013) NOD-like receptors in lung diseases. Front Immunol 4:393. doi:10.3389/fimmu.2013.00393

Davis KM, Nakamura S, Weiser JN (2011) Nod2 sensing of lysozyme-digested peptidoglycan promotes macrophage recruitment and clearance of *S. pneumoniae* colonization in mice. J Clin Investig 121(9):3666–3676. doi:10.1172/JCI57761

Duewell P, Kono H, Rayner KJ, Sirois CM, Vladimer G, Bauernfeind FG, Abela GS, Franchi L, Nunez G, Schnurr M, Espevik T, Lien E, Fitzgerald KA, Rock KL, Moore KJ, Wright SD, Hornung V, Latz E (2010) NLRP3 inflammasomes are required for atherogenesis and activated by cholesterol crystals. Nature 464(7293):1357–1361. doi:10.1038/nature08938

Eisenbarth SC, Colegio OR, O'Connor W, Sutterwala FS, Flavell RA (2008) Crucial role for the Nalp3 inflammasome in the immunostimulatory properties of aluminium adjuvants. Nature 453 (7198):1122–1126. doi:10.1038/nature06939

Fang R, Tsuchiya K, Kawamura I, Shen Y, Hara H, Sakai S, Yamamoto T, Fernandes-Alnemri T, Yang R, Hernandez-Cuellar E, Dewamitta SR, Xu Y, Qu H, Alnemri ES, Mitsuyama M (2011) Critical roles of ASC inflammasomes in caspase-1 activation and host innate resistance to Streptococcus pneumoniae infection. J Immunol 187(9):4890–4899. doi:10.4049/jimmunol.1100381

Fang R, Hara H, Sakai S, Hernandez-Cuellar E, Mitsuyama M, Kawamura I, Tsuchiya K (2014) Type I interferon signaling regulates activation of the absent in melanoma 2 inflammasome during Streptococcus pneumoniae infection. Infect Immun 82(6):2310–2317. doi:10.1128/IAI.01572-14

Fatykhova D, Rabes A, Machnik C, Guruprasad K, Pache F, Berg J, Toennies M, Bauer TT, Schneider P, Schimek M, Eggeling S, Mitchell TJ, Mitchell AM, Hilker R, Hain T, Suttorp N, Hippenstiel S, Hocke AC, Opitz B (2015) Serotype 1 and 8 pneumococci evade sensing by inflammasomes in human lung tissue. PLoS ONE 10(8):e0137108. doi:10.1371/journal.pone.0137108

Fernandes-Alnemri T, Yu JW, Datta P, Wu J, Alnemri ES (2009) AIM2 activates the inflammasome and cell death in response to cytoplasmic DNA. Nature 458(7237):509–513. doi:10.1038/nature07710

Geldhoff M, Mook-Kanamori BB, Brouwer MC, Troost D, Leemans JC, Flavell RA, Van der Ende A, Van der Poll T, Van de Beek D (2013) Inflammasome activation mediates inflammation and outcome in humans and mice with pneumococcal meningitis. BMC Infect Dis 13:358. doi:10.1186/1471-2334-13-358

Gupta R, Ghosh S, Monks B, DeOliveira RB, Tzeng TC, Kalantari P, Nandy A, Bhattacharjee B, Chan J, Ferreira F, Rathinam V, Sharma S, Lien E, Silverman N, Fitzgerald K, Firon A, Trieu-Cuot P, Henneke P, Golenbock DT (2014) RNA and beta-hemolysin of group B Streptococcus induce interleukin-1beta (IL-1beta) by activating NLRP3 inflammasomes in mouse macrophages. J Biol Chem 289(20):13701–13705. doi:10.1074/jbc.C114.548982

Halle A, Hornung V, Petzold GC, Stewart CR, Monks BG, Reinheckel T, Fitzgerald KA, Latz E, Moore KJ, Golenbock DT (2008) The NALP3 inflammasome is involved in the innate immune response to amyloid-beta. Nat Immunol 9(8):857–865. doi:10.1038/ni.1636

Harvey RM, Hughes CE, Paton AW, Trappetti C, Tweten RK, Paton JC (2014) The impact of pneumolysin on the macrophage response to Streptococcus pneumoniae is strain-dependent. PLoS ONE 9(8):e103625. doi:10.1371/journal.pone.0103625

Henriques Normark B, Kalin M, Ortqvist A, Akerlund T, Liljequist BO, Hedlund J, Svenson SB, Zhou J, Spratt BG, Normark S, Kallenius G (2001) Dynamics of penicillin-susceptible clones in invasive pneumococcal disease. J Infect Dis 184(7):861–869. doi:10.1086/323339

Henriques-Normark B, Tuomanen EI (2013) The pneumococcus: epidemiology, microbiology, and pathogenesis. Cold Spring Harb Perspect Med 3(7). doi:10.1101/cshperspect.a010215

Hirst RA, Gosai B, Rutman A, Guerin CJ, Nicotera P, Andrew PW, O'Callaghan C (2008) Streptococcus pneumoniae deficient in pneumolysin or autolysin has reduced virulence in meningitis. J Infect Dis 197(5):744–751. doi:10.1086/527322

Hoegen T, Tremel N, Klein M, Angele B, Wagner H, Kirschning C, Pfister HW, Fontana A, Hammerschmidt S, Koedel U (2011) The NLRP3 inflammasome contributes to brain injury in pneumococcal meningitis and is activated through ATP-dependent lysosomal cathepsin B release. J Immunol 187(10):5440–5451. doi:10.4049/jimmunol.1100790

Hornung V, Bauernfeind F, Halle A, Samstad EO, Kono H, Rock KL, Fitzgerald KA, Latz E (2008) Silica crystals and aluminum salts activate the NALP3 inflammasome through phagosomal destabilization. Nat Immunol 9(8):847–856. doi:10.1038/ni.1631

Hornung V, Ablasser A, Charrel-Dennis M, Bauernfeind F, Horvath G, Caffrey DR, Latz E, Fitzgerald KA (2009) AIM2 recognizes cytosolic dsDNA and forms a caspase-1-activating inflammasome with ASC. Nature 458(7237):514–518. doi:10.1038/nature07725

Hyams C, Camberlein E, Cohen JM, Bax K, Brown JS (2010) The Streptococcus pneumoniae capsule inhibits complement activity and neutrophil phagocytosis by multiple mechanisms. Infect Immun 78(2):704–715. doi:10.1128/IAI.00881-09

Jefferies JM, Johnston CH, Kirkham LA, Cowan GJ, Ross KS, Smith A, Clarke SC, Brueggemann AB, George RC, Pichon B, Pluschke G, Pfluger V, Mitchell TJ (2007) Presence of nonhemolytic pneumolysin in serotypes of Streptococcus pneumoniae associated with disease outbreaks. J Infect Dis 196(6):936–944. doi:10.1086/520091

Kadioglu A, Taylor S, Iannelli F, Pozzi G, Mitchell TJ, Andrew PW (2002) Upper and lower respiratory tract infection by Streptococcus pneumoniae is affected by pneumolysin deficiency and differences in capsule type. Infect Immun 70(6):2886–2890

Kadioglu A, Weiser JN, Paton JC, Andrew PW (2008) The role of Streptococcus pneumoniae virulence factors in host respiratory colonization and disease. Nat Rev Microbiol 6(4):288–301. doi:10.1038/nrmicro1871

Kafka D, Ling E, Feldman G, Benharroch D, Voronov E, Givon-Lavi N, Iwakura Y, Dagan R, Apte RN, Mizrachi-Nebenzahl Y (2008) Contribution of IL-1 to resistance to Streptococcus pneumoniae infection. Int Immunol 20(9):1139–1146. doi:10.1093/intimm/dxn071

Kanneganti TD, Ozoren N, Body-Malapel M, Amer A, Park JH, Franchi L, Whitfield J, Barchet W, Colonna M, Vandenabeele P, Bertin J, Coyle A, Grant EP, Akira S, Nunez G (2006) Bacterial RNA and small antiviral compounds activate caspase-1 through cryopyrin/Nalp3. Nature 440(7081):233–236. doi:10.1038/nature04517

Karmakar M, Katsnelson M, Malak HA, Greene NG, Howell SJ, Hise AG, Camilli A, Kadioglu A, Dubyak GR, Pearlman E (2015) Neutrophil IL-1beta processing induced by pneumolysin is mediated by the NLRP3/ASC inflammasome and caspase-1 activation and is dependent on K+ efflux. J Immunol 194(4):1763–1775. doi:10.4049/jimmunol.1401624

Kirkham LA, Jefferies JM, Kerr AR, Jing Y, Clarke SC, Smith A, Mitchell TJ (2006) Identification of invasive serotype 1 pneumococcal isolates that express nonhemolytic pneumolysin. J Clin Microbiol 44(1):151–159. doi:10.1128/JCM.44.1.151-159.2006

Knapp S, Wieland CW, van't Veer C, Takeuchi O, Akira S, Florquin S, van der Poll T (2004) Toll-like receptor 2 plays a role in the early inflammatory response to murine pneumococcal pneumonia but does not contribute to antibacterial defense. J Immunol 172 5):3132–3138

Koedel U, Winkler F, Angele B, Fontana A, Flavell RA, Pfister HW (2002) Role of Caspase-1 in experimental pneumococcal meningitis: evidence from pharmacologic Caspase inhibition and Caspase-1-deficient mice. Ann Neurol 51(3):319–329

Koppe U, Hogner K, Doehn JM, Muller HC, Witzenrath M, Gutbier B, Bauer S, Pribyl T, Hammerschmidt S, Lohmeyer J, Suttorp N, Herold S, Opitz B (2012a) Streptococcus pneumoniae stimulates a STING- and IFN regulatory factor 3-dependent type I IFN production

in macrophages, which regulates RANTES production in macrophages, cocultured alveolar epithelial cells, and mouse lungs. J Immunol 188(2):811–817. doi:10.4049/jimmunol.1004143
Koppe U, Suttorp N, Opitz B (2012b) Recognition of Streptococcus pneumoniae by the innate immune system. Cell Microbiol 14(4):460–466. doi:10.1111/j.1462-5822.2011.01746.x
Krone CL, van de Groep K, Trzcinski K, Sanders EA, Bogaert D (2014) Immunosenescence and pneumococcal disease: an imbalance in host-pathogen interactions. Lancet Respir Med 2 (2):141–153. doi:10.1016/S2213-2600(13)70165-6
Lamb KE, Flasche S, Diggle M, Inverarity D, Greenhalgh D, Jefferies JM, Smith A, Edwards GF, Denham B, McMenamin J, McDonald E, Mitchell TJ, Clarke SC, Robertson C (2014) Trends in serotypes and sequence types among cases of invasive pneumococcal disease in Scotland, 1999–2010. Vaccine 32(34):4356–4363. doi:10.1016/j.vaccine.2013.05.079
Lauw FN, Branger J, Florquin S, Speelman P, van Deventer SJ, Akira S, van der Poll T (2002) IL-18 improves the early antimicrobial host response to pneumococcal pneumonia. J Immunol 168(1):372–378
Le Hello S, Watson M, Levy M, Marcon S, Brown M, Yvon JF, Missotte I, Garin B (2010) Invasive serotype 1 Streptococcus pneumoniae outbreaks in the South Pacific from 2000 to 2007. J Clin Microbiol 48(8):2968–2971. doi:10.1128/JCM.01615-09
Lemon JK, Miller MR, Weiser JN (2015) Sensing of interleukin-1 cytokines during Streptococcus pneumoniae colonization contributes to macrophage recruitment and bacterial clearance. Infect Immun 83(8):3204–3212. doi:10.1128/IAI.00224-15
Lock RA, Zhang QY, Berry AM, Paton JC (1996) Sequence variation in the Streptococcus pneumoniae pneumolysin gene affecting haemolytic activity and electrophoretic mobility of the toxin. Microb Pathog 21(2):71–83. doi:10.1006/mpat.1996.0044
Mariathasan S, Weiss DS, Newton K, McBride J, O'Rourke K, Roose-Girma M, Lee WP, Weinrauch Y, Monack DM, Dixit VM (2006) Cryopyrin activates the inflammasome in response to toxins and ATP. Nature 440(7081):228–232. doi:10.1038/nature04515
Marimon JM, Ercibengoa M, Alonso M, Zubizarreta M, Perez-Trallero E (2009) Clonal structure and 21-year evolution of Streptococcus pneumoniae serotype 1 isolates in northern Spain. Clin Microbiol Infect Official Publ Eur Soc Clin Microbiol Infect Dis 15(9):875–877. doi:10.1111/j.1469-0691.2009.02883.x
Marriott HM, Gascoyne KA, Gowda R, Geary I, Nicklin MJ, Iannelli F, Pozzi G, Mitchell TJ, Whyte MK, Sabroe I, Dockrell DH (2012) Interleukin-1beta regulates CXCL8 release and influences disease outcome in response to Streptococcus pneumoniae, defining intercellular cooperation between pulmonary epithelial cells and macrophages. Infect Immun 80(3):1140–1149. doi:10.1128/IAI.05697-11
Martinon F, Burns K, Tschopp J (2002) The inflammasome: a molecular platform triggering activation of inflammatory caspases and processing of proIL-beta. Mol Cell 10(2):417–426
Martinon F, Petrilli V, Mayor A, Tardivel A, Tschopp J (2006) Gout-associated uric acid crystals activate the NALP3 inflammasome. Nature 440(7081):237–241. doi:10.1038/nature04516
Maus UA, Srivastava M, Paton JC, Mack M, Everhart MB, Blackwell TS, Christman JW, Schlondorff D, Seeger W, Lohmeyer J (2004) Pneumolysin-induced lung injury is independent of leukocyte trafficking into the alveolar space. Journal of immunology 173(2):1307–1312
McNeela EA, Burke A, Neill DR, Baxter C, Fernandes VE, Ferreira D, Smeaton S, El-Rachkidy R, McLoughlin RM, Mori A, Moran B, Fitzgerald KA, Tschopp J, Petrilli V, Andrew PW, Kadioglu A, Lavelle EC (2010) Pneumolysin activates the NLRP3 inflammasome and promotes proinflammatory cytokines independently of TLR4. PLoS Pathog 6(11):e1001191. doi:10.1371/journal.ppat.1001191
Mitchell AM, Mitchell TJ (2010) Streptococcus pneumoniae: virulence factors and variation. Clin Microbiol Infect 16(5):411–418. doi:10.1111/j.1469-0691.2010.03183.x
Mitchell TJ, Andrew PW, Saunders FK, Smith AN, Boulnois GJ (1991) Complement activation and antibody binding by pneumolysin via a region of the toxin homologous to a human acute-phase protein. Mol Microbiol 5(8):1883–1888
Mitchell AJ, Yau B, McQuillan JA, Ball HJ, Too LK, Abtin A, Hertzog P, Leib SL, Jones CA, Gerega SK, Weninger W, Hunt NH (2012) Inflammasome-dependent IFN-gamma drives

pathogenesis in Streptococcus pneumoniae meningitis. J Immunol 189(10):4970–4980. doi:10.4049/jimmunol.1201687
Opitz B, Puschel A, Schmeck B, Hocke AC, Rosseau S, Hammerschmidt S, Schumann RR, Suttorp N, Hippenstiel S (2004) Nucleotide-binding oligomerization domain proteins are innate immune receptors for internalized Streptococcus pneumoniae. J Biol Chem 279(35):36426–36432. doi:10.1074/jbc.M403861200
Opitz B, van Laak V, Eitel J, Suttorp N (2010) Innate immune recognition in infectious and noninfectious diseases of the lung. Am J Respir Crit Care Med 181(12):1294–1309. doi:10.1164/rccm.200909-1427SO
Roberts TL, Idris A, Dunn JA, Kelly GM, Burnton CM, Hodgson S, Hardy LL, Garceau V, Sweet MJ, Ross IL, Hume DA, Stacey KJ (2009) HIN-200 proteins regulate caspase activation in response to foreign cytoplasmic DNA. Science 323(5917):1057–1060. doi:10.1126/science.1169841
Rubins JB, Charboneau D, Paton JC, Mitchell TJ, Andrew PW, Janoff EN (1995) Dual function of pneumolysin in the early pathogenesis of murine pneumococcal pneumonia. J Clin Investig 95 (1):142–150. doi:10.1172/JCI117631
Rubins JB, Charboneau D, Fasching C, Berry AM, Paton JC, Alexander JE, Andrew PW, Mitchell TJ, Janoff EN (1996) Distinct roles for pneumolysin's cytotoxic and complement activities in the pathogenesis of pneumococcal pneumonia. Am J Respir Crit Care Med 153(4 Pt 1):1339–1346. doi:10.1164/ajrccm.153.4.8616564
Schroder K, Tschopp J (2010) The inflammasomes. Cell 140(6):821–832. doi:10.1016/j.cell.2010.01.040
Shak JR, Ludewick HP, Howery KE, Sakai F, Yi H, Harvey RM, Paton JC, Klugman KP, Vidal JE (2013) Novel role for the Streptococcus pneumoniae toxin pneumolysin in the assembly of biofilms. MBio 4(5):e00655–00613. doi:10.1128/mBio.00655-13
Shoma S, Tsuchiya K, Kawamura I, Nomura T, Hara H, Uchiyama R, Daim S, Mitsuyama M (2008) Critical involvement of pneumolysin in production of interleukin-1alpha and caspase-1-dependent cytokines in infection with Streptococcus pneumoniae in vitro: a novel function of pneumolysin in caspase-1 activation. Infect Immun 76(4):1547–1557. doi:10.1128/IAI.01269-07
Siegel SJ, Weiser JN (2015) Mechanisms of bacterial colonization of the respiratory tract. Annu Rev Microbiol 69:425–444. doi:10.1146/annurev-micro-091014-104209
Staples M, Graham RM, Jennison AV, Ariotti L, Hicks V, Cook H, Krause V, Giele C, Smith HV (2015) Molecular characterization of an Australian serotype 1 Streptococcus pneumoniae outbreak. Epidemiol Infect 143(2):325–333. doi:10.1017/S0950268814000648
Tsukada H, Kawamura I, Fujimura T, Igarashi K, Arakawa M, Mitsuyama M (1992) Induction of macrophage interleukin-1 production by Listeria monocytogenes hemolysin. Cell Immunol 140(1):21–30
van der Poll T, Opal SM (2009) Pathogenesis, treatment, and prevention of pneumococcal pneumonia. Lancet 374(9700):1543–1556. doi:10.1016/S0140-6736(09)61114-4
van Lieshout MH, Scicluna BP, Florquin S, van der Poll T (2014) NLRP3 and ASC differentially affect the lung transcriptome during pneumococcal pneumonia. Am J Respir Cell Mol Biol 50 (4):699–712. doi:10.1165/rcmb.2013-0015OC
Voss S, Gamez G, Hammerschmidt S (2012) Impact of pneumococcal microbial surface components recognizing adhesive matrix molecules on colonization. Mol Oral Microbiol 27 (4):246–256. doi:10.1111/j.2041-1014.2012.00654.x
Wartha F, Beiter K, Albiger B, Fernebro J, Zychlinsky A, Normark S, Henriques-Normark B (2007) Capsule and D-alanylated lipoteichoic acids protect Streptococcus pneumoniae against neutrophil extracellular traps. Cell Microbiol 9(5):1162–1171. doi:10.1111/j.1462-5822.2006.00857.x
Weinberger DM, Harboe ZB, Sanders EA, Ndiritu M, Klugman KP, Ruckinger S, Dagan R, Adegbola R, Cutts F, Johnson HL, O'Brien KL, Scott JA, Lipsitch M (2010) Association of serotype with risk of death due to pneumococcal pneumonia: a meta-analysis. Clin Infect Dis Official Publ Infect Dis Soc Am 51(6):692–699. doi:10.1086/655828

Wellmer A, Zysk G, Gerber J, Kunst T, Von Mering M, Bunkowski S, Eiffert H, Nau R (2002) Decreased virulence of a pneumolysin-deficient strain of Streptococcus pneumoniae in murine meningitis. Infect Immun 70(11):6504–6508

Witzenrath M, Gutbier B, Hocke AC, Schmeck B, Hippenstiel S, Berger K, Mitchell TJ, de los Toyos JR, Rosseau S, Suttorp N, Schutte H (2006) Role of pneumolysin for the development of acute lung injury in pneumococcal pneumonia. Crit Care Med 34(7):1947–1954. doi:10.1097/01.CCM.0000220496.48295.A9

Witzenrath M, Pache F, Lorenz D, Koppe U, Gutbier B, Tabeling C, Reppe K, Meixenberger K, Dorhoi A, Ma J, Holmes A, Trendelenburg G, Heimesaat MM, Bereswill S, van der Linden M, Tschopp J, Mitchell TJ, Suttorp N, Opitz B (2011) The NLRP3 inflammasome is differentially activated by pneumolysin variants and contributes to host defense in pneumococcal pneumonia. J Immunol 187(1):434–440. doi:10.4049/jimmunol.1003143

Zwijnenburg PJ, van der Poll T, Florquin S, Akira S, Takeda K, Roord JJ, van Furth AM (2003) Interleukin-18 gene-deficient mice show enhanced defense and reduced inflammation during pneumococcal meningitis. J Neuroimmunol 138(1–2):31–37

Francisella Inflammasomes: Integrated Responses to a Cytosolic Stealth Bacterium

Pierre Wallet, Brice Lagrange and Thomas Henry

Abstract *Francisella tularensis* is a facultative intracellular bacterium causing tularemia, a zoonotic disease. *Francisella* replicates in the macrophage cytosol and eventually triggers cytosolic immune responses. In murine macrophages, *Francisella novicida* and *Francisella tularensis* live vaccine strain lyse in the host cytosol and activate the cytosolic DNA receptor Aim2. Here, we review the mechanisms leading or contributing to Aim2 inflammasome activation, including the role of TLRs and of IFN signaling and the implication of the guanylate-binding proteins 2 and 5 in triggering cytosolic bacteriolysis. Furthermore, we present how this cytosolic Gram-negative bacterium escapes recognition by caspase-11 but can trigger a non-canonical caspase-8 inflammasome. In addition, we highlight the differences in inflammasome activation in murine and human cells with pyrin, NLRP3, and AIM2 involved in sensing *Francisella* in human phagocytes. From a bacterial prospective, we describe the hiding strategy of *Francisella* to escape recognition by innate sensors and to resist to bacteriolysis in the host cytosol. Finally, we discuss the inability of the inflammasome sensors to detect *F. tularensis* subspecies *tularensis* strains, making them highly pathogenic stealth microbes.

P. Wallet · B. Lagrange · T. Henry
Centre International de Recherche en Infectiologie, Lyon, France

P. Wallet · B. Lagrange · T. Henry
Inserm U1111, CNRS UMR5308, Lyon, France

P. Wallet · B. Lagrange · T. Henry (✉)
ENS Lyon, Université Claude Bernard Lyon 1, Lyon, France
e-mail: thomas.henry@inserm.fr

S. Backert (ed.), *Inflammasome Signaling and Bacterial Infections*,
Current Topics in Microbiology and Immunology 397,
DOI 10.1007/978-3-319-41171-2_12

Contents

1 Introduction

1.1 The Francisella *Genus*

Francisella is the only genus within the *Francisellaceae* family (McLendon et al. 2006). These gamma-proteobacteria from the *Thiotrichales* order are small and pleiomorphic Gram-negative bacteria. The *Francisella* genus is divided into several species including *Francisella tularensis* and *Francisella novicida* (Table 1). *Francisella tularensis* is the most pathogenic species for humans. Yet, the virulence of *F. tularensis* strains is highly dependent on the subspecies they belong to. *F. tularensis* subspecies *tularensis* (also named type A strains) and its representative strain SCHU S4 are highly infectious and highly virulent with a lethal dose estimated to be lower than 10 bacteria for humans. *F. tularensis* subspecies *holarctica* (type B strains, hereinafter referred to as *F. holarctica*), although responsible for numerous infections of immunocompetent humans, is slightly less pathogenic. The live vaccine strain (LVS) is a strain derived from a virulent *F. holarctica* strain through sequential, in vitro passages. Although not licensed for vaccination, *F. holarctica* LVS is commonly used to study *Francisella* virulence factors and host immune responses. *F. novicida* is a species closely related to *F. tularensis* with greater than 97 % nucleotide identities between the two species (Larsson et al. 2009). *F. novicida* strains have been isolated from a few human

Table 1 Summary of the nomenclature, the virulence in mice and men, and the ability to activate the inflammasome complex of the most studied *Francisella* strains

Name	Representative strain	Alternative strain name	Virulence in mice	Virulence in humans	Inflammasome activation
F. novicida	U112	–	++	–	++
F. tularensis subsp. *holarctica*	LVS	Type B	+[a]	–[a]	+[a]
F. tularensis subsp. *tularensis*	SCHU S4	Type A	+++	+++	–

[a]Data are shown for *F. holarctica* LVS. Clinical strains from *F. holarctica* are virulent in mice and humans but, to our knowledge, have not been characterized in regard to inflammasome activation

patients, most of them with immune deficits or underlying medical conditions (Kingry and Petersen 2014). Due to its high virulence in mouse models of tularemia and an intracellular life cycle very similar to the one of virulent *F. tularensis* strains, *F. novicida* strain U112 is used to study tularemia.

In this review, we will present the modalities of the inflammasome activation upon *F. novicida* infections and we will discuss the similarities and differences observed between *F. novicida*, *F. holarctica* LVS, and the highly virulent *F. tularensis* type A strains.

1.2 **Francisella tularensis,** *the Agent of Tularemia*

Tularemia is a zoonotic disease with no transmission from human to human. Several forms of tularemia are observed depending on *F. tularensis* route of entry into the human host. Tularemia is associated with flu-like symptoms with high fever and lymphadenopathies. The most common form is the ulceroglandular tularemia following entry of the bacterium through the skin either through the bite of an infected arthropod or after direct contact with an infected animal (Sjostedt 2007). Ulceroglandular tularemia is characterized by a necrotic ulcer at the site of infection and a swollen draining lymph node. Consumption of contaminated food or freshwater leads to oropharyngeal tularemia (Gurcan 2014) with the rapid development of cervical lymphadenopathy. Finally, the most life-threatening disease is respiratory tularemia, which is triggered upon inhalation of *F. tularensis*. The severity of this infection coupled to high infectivity of *F. tularensis* has led the Center for Diseases Control to classify *F. tularensis* as a bioterrorism class A agent.

1.3 Intracellular Life Cycle

F. tularensis is a facultative intracellular pathogen found in vivo mostly in phagocytic cells including macrophages, neutrophils, dendritic cells, and monocytes (Hall et al. 2008). In contrast to other intracellular bacteria such as *Salmonella* or *Shigella*, *F. tularensis* is unable to direct its entry into non-phagocytic cells. *F. tularensis* thus relies on host phagocytic receptors to invade host cells. *F. tularensis* uptake by macrophages proceeds through an atypical phagocytosis termed "looping phagocytosis." Bacterial opsonization and several host receptors are involved in facilitating *F. tularensis* entry (Clemens et al. 2005; Clemens and Horwitz 2007). The engagement of a specific entry receptor modulates the subsequent pro-inflammatory response including IL-1β release (Dai et al. 2013).

Following uptake, *F. tularensis* rapidly lyses the phagosome (Golovliov et al. 2003; Clemens et al. 2004; Checroun et al. 2006) to reach the host cytosol. Escape into the host cytosol is dependent on an atypical type VI secretion system (T6SS) encoded in the *Francisella Pathogenicity Island* (FPI) (Nano et al. 2004; Barker et al. 2009; Broms et al. 2010). While the T6SS is thought to secrete lytic effectors into the host cytosol, the mechanisms underlying vacuolar escape remain to be identified (Broms et al. 2010). Upon reaching the host cytosol, *F. tularensis* replicates rapidly thanks to a metabolic adaptation allowing bacterial growth in this cellular compartment (Barel and Charbit 2013). Due to the cytosolic localization of *Francisella* strains, these bacteria are used as a model to study the cytosolic innate immune responses including the inflammasome complex. The vacuolar mutants deleted of the *FPI* locus, of key *FPI* genes, or mutated in the master regulator gene *mglA* (Baron and Nano 1998) are often used in comparison with wild-type strains to discriminate the immune responses elicited in the vacuole/phagosome from the cytosol-specific responses.

2 Overview of the Innate Immune Responses to *Francisella* Infection

The inflammasome is highly connected with other innate immune pathways (Henry et al. 2007; Cole et al. 2008; Jones and Weiss 2011; Man et al. 2015; Meunier et al. 2015). We will thus briefly review the roles of the Toll-like receptors (TLRs) and of the pathways leading to type I IFN secretion during *Francisella* infection.

2.1 TLRs and NF-κB Activation

Bacteria from the *Francisellaceae* family are Gram-negative bacteria harboring lipopolysaccharide (LPS) on the external leaflet of their outer membrane. It was noted early on that, in contrast to the typical *Enterobacteriaceae* LPS, *Francisella* LPS does not have any endotoxin properties and is unable to induce IL-1 in human monocytes (Sandstrom et al. 1992). Indeed, *Francisella* LPS is not recognized by TLR4 (Hajjar et al. 2006). This lack of recognition is due to three peculiar features of *Francisella* lipid A. First, while most Gram-negative bacteria have hexa-acylated

LPS, *Francisella* lipid A is tetra-acylated. Second, the four fatty acyl chains are made of a chain of 16–18 carbons in length as opposed to 12–14 for *Escherichia coli* lipid A (Gunn and Ernst 2007; Kanistanon et al. 2008). Third, the carbon in position 4′ of the glucosamine is lacking the typical phosphate group observed in most Gram-negative lipid A (Wang et al. 2007).

TLR2 is thus the main pattern recognition receptor (PRR) responsible for transcription factor NF-κB activation during *Francisella* infection (Katz et al. 2006; Li et al. 2006). TLR2 activation occurs through the recognition of bacterial lipoproteins (Thakran et al. 2008) and is observed both at the plasma membrane and in the phagosome before *Francisella* escapes into the host cytosol (Cole et al. 2010).

2.2 Signaling Pathways Leading to Type I IFN Production

In addition to the pro-inflammatory response triggered at the plasma membrane or in the phagosome, escape of *Francisella* into the host cytosol triggers the "cytosolic innate immune responses." Henry et al. studied these responses by comparing the transcriptional responses of bone marrow-derived macrophages (BMDMs) infected with either the cytosolic wild-type *F. novicida* strain (U112) or a vacuolar mutant (ΔFPI). They demonstrated that macrophages detecting *Francisella* specifically in the cytosol secrete type I interferon (IFN-α and IFN-β) (Henry et al. 2007). Cole et al. confirmed this finding using *F. holarctica* LVS (Cole et al. 2008). In agreement with the cytosolic localization required to trigger IFN-β secretion, the type I IFN response upon *F. novicida* infection is independent of TLR2 and of the TLR adaptors MyD88 and TRIF (Henry et al. 2007; Jones and Weiss 2011). Type I IFN induction during *Francisella* infection is dependent on STING (Jones et al. 2010; Jin et al. 2011) and could be recapitulated by transfecting into the host cytosol of a proteinase K-resistant, DNase1-sensitive factor contained in a *Francisella* extract (Jones et al. 2010). The cytosolic DNA sensor cyclic GMP-AMP synthase (cGAS) is required for type I IFN production in response to *F. novicida* infection (Storek et al. 2015; Man et al. 2015). In addition to cGAS, Storek et al. demonstrated that in infected BMDMs, the DNA sensor Ifi204 (the murine homologue of human IFI16) is required to induce IFN-β production. The respective roles of the cGAS and Ifi204 remain unknown (Storek et al. 2015). Altogether, the results demonstrate that detection of *F. novicida* DNA in the host cytosol by the synergistic action of two DNA sensors triggers type I IFN secretion. Downstream of the DNA sensors, STING contributes to activate the transcription factors IRF1 and 3 (Henry et al. 2007; Jin et al. 2011; Man et al. 2015) to induce IFN-β and expression of numerous interferon-stimulated genes (ISGs) (Fig. 1).

The pathway leading to type I IFN secretion upon *F. holarctica* LVS infection might be slightly different. Indeed, type I IFN secretion is neither observed in *Sting*$^{-/-}$ BMDMs (Jin et al. 2011) nor in *Tlr2*$^{-/-}$ BMDMs (Cole et al. 2007). At the

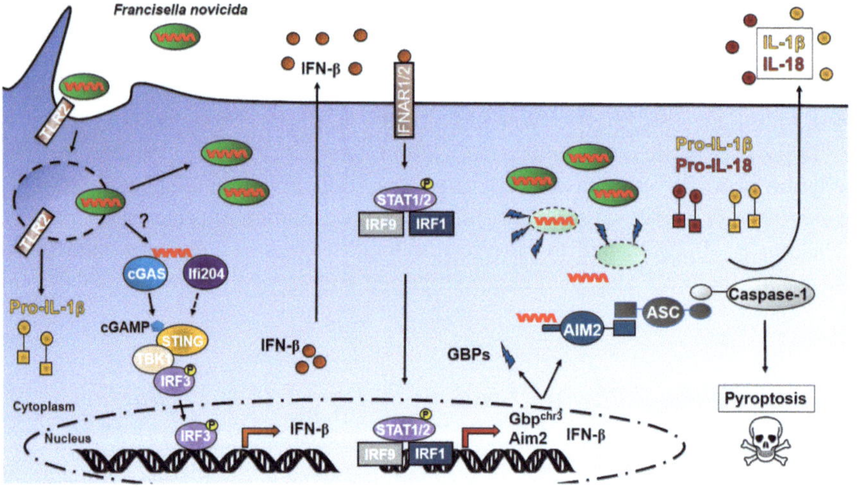

Fig. 1 Innate immune pathways leading to the Aim2 inflammasome activation during *F. novicida* infection of BMDMs (see text for details)

present time, it remains unclear whether TLR2 is directly involved in type I IFN response upon *F. holarctica* LVS infection, possibly through the MyD88-IRF1-IRF7 pathways elicited in the endosomal compartment (Dietrich et al. 2010) or whether TLR2 is required upstream of the release of LVS DNA into the host cytosol to elicit the cGAS/Ifi204-STING pathway.

3 *Francisella* Activates the Aim2 Inflammasome in Murine Phagocytes

Most of the published work on the modalities of inflammasome activation during *Francisella* infection has been performed in murine macrophages using *F. novicida* or *F. holarctica* LVS. First, we will describe activation of the inflammasome in mice and in murine phagocytes upon *F. novicida* and *F. holarctica* LVS infections (Fig. 1). In a second time, we will highlight the similarities and differences between human and murine cells and between *F. novicida*, *F. holarctica* LVS, and the highly pathogenic *F. tularensis* subsp. *tularensis* strains.

3.1 Aim2 Inflammasome and the IFN Requirement

3.1.1 In Vitro and In Vivo Role of the Inflammasome

The role of the inflammasome was demonstrated early on by Monack and colleagues (Mariathasan et al. 2005). In 2005, Mariathasan et al. described a cell death response that was dependent on both ASC and caspase-1 as well as on the cytosolic localization of *F. novicida* in BMDMs. This cell death is associated with caspase-1 processing, IL-1β, and IL-18 release. As expected, TNF-α release (which levels are regulated by NF-κB) is independent of both ASC and caspase-1. The in vitro results were validated in a mouse model of tularemia. Mariathasan et al. demonstrated that ASC and caspase-1 are required for the survival of the mice to *F. novicida* infection while Nlrc4 (known at the time as IPAF) is not. Indeed, when 1.5×10^5 *F. novicida* colony-forming units (CFU) were injected subcutaneously, 100 % of *Asc*$^{-/-}$ and *Casp1*$^{-/-}$ mice died within 3–5 days. In contrast, with the same inoculum, 25 % of the wild-type mice survived the infection (Mariathasan et al. 2005). Cells expressing active caspase-1 are detected in the liver, in the spleen of LVS-infected mice (Wickstrum et al. 2009), and in the lung of *F. novicida*-infected mice (Mares et al. 2008). In agreement with the key role of the inflammasome pathway in these organs, *F. novicida* burden at 2 days post-inoculation is 100- to 1000-fold higher in the spleen, liver, and lung of inflammasome-deficient mice than in the corresponding organs of wild-type mice (Mariathasan et al. 2005).

Caspase-1-dependent cytokines orchestrate the innate immune response against *Francisella*. Indeed, IL-18 is key to induce IFN-γ production by NK cells in *F. novicida*-infected mice (Pierini et al. 2013; del Barrio et al. 2015). Interestingly, the survival of *Il-18*$^{-/-}$ mice to LVS infection can be rescued by IFN-γ administration. In contrast, *Il-1r1*$^{-/-}$ mice remained highly susceptible to LVS infection even upon IFN-γ administration (del Barrio et al. 2015). Re and colleagues identified that IL-1β is required early during infection to produce anti-LPS IgM (del Barrio et al. 2015). Anti LVS-IglM was observed in the serum of infected wild-type mice 7 days post-infection but were greatly reduced in *Asc*$^{-/-}$ and *Il-1b*$^{-/-}$ mice. B1a B cells are responsible for this specific IgM response, and their numbers were reduced in the spleen of infected *Il-1b*^{KO} mice. This IglM response was demonstrated to be protective likely through opsonization and increased phagocytosis (del Barrio et al. 2015). Finally, caspase-1 inhibition reduces IL-23p19 level (an IL-1β-regulated cytokine controlling IL-17 levels) in LVS-infected macrophages (Skyberg et al. 2013).

Using IL-1β- and IL-18-neutralizing antibodies injections in wild-type mice, Monack and colleagues demonstrated that mature IL-1β and IL-18 are required to fight *Francisella* infection. However, the susceptibility of caspase-1-deficient mice is due only in part to the lack of these two active cytokines. These results were confirmed by comparing mice doubly deficient for IL-1β and IL-18 (*Il-1β/Il-18*^{DKO}) to wild-type and *Casp1*$^{-/-}$ mice. *Il-1β/Il-18*^{DKO} mice display an intermediate phenotype in terms of bacterial burden in the spleen, liver, and lung between wild-type and *Casp1*$^{-/-}$ mice (Henry and Monack 2007). These results led the authors to conclude that the cell death defect in caspase-1-deficient mice was also participating in the strong susceptibility of caspase-1-deficient mice (Mariathasan et al. 2005; Henry and Monack 2007). Similarly to what have been reported for *Salmonella typhimurium*, *Legionella pneumophila*, and *Burkholderia thailandensis* (Miao et al. 2010), pyroptosis is thus a potent innate immune mechanism against

Francisella by removing its replicative niche and contributing to the clearance of this pathogen.

Of note, a Fas-mediated apoptotic pathway has been described for macrophages infected with *F. novicida* (Rajaram et al. 2009) but this pathway was characterized in RAW 264.7 cells, which are deficient for the inflammasome adaptor ASC (Pelegrin et al. 2008). While both apoptotic and pyroptotic pathways may occur during LVS or *F. tularensis* infections, in primary macrophages infected with *F. novicida*, the cell death is clearly inflammasome dependent.

3.1.2 Type I IFN Signaling Is Required for *Francisella*-mediated Inflammasome Activation

This inflammasome response in infected BMDMs was rapidly characterized as Nlrp3 independent (known at the time as cryopyrin) (Mariathasan et al. 2006). The first clue on the potential receptor-sensing *Francisella* in the cytosol came in 2007 with the demonstration that macrophages deficient in the type I IFN receptor (IFNAR1) were specifically deficient for inflammasome activation in response to *Francisella* infection (Henry et al. 2007). While the induction of pro-IL-1β protein is similar in wild-type and in *Ifnar1*$^{-/-}$ macrophages, secretion of mature IL-1β, IL-18, and caspase-1 processing and cell death are fully deficient in the latter cells. The same phenotype is also observed in *Irf3*$^{-/-}$ macrophages, which are deficient for IFN-β production during *F. novicida* infection. However, addition of rIFN-β to *Irf3*$^{-/-}$ macrophages restores inflammasome activation in these cells (Henry et al. 2007). These results were strengthened by the study of cGAS- and STING-deficient BMDMs, which are deficient both for type I IFN induction and inflammasome activation in response to *F. novicida* infection (Jones et al. 2010; Storek et al. 2015). Altogether, these studies indicate that type I IFN production and signaling are required to trigger inflammasome activation in response to *F. novicida* infection. In 2007, this finding led Henry et al. (2007) to suggest that the inflammasome receptor-sensing *Francisella* in the cytosol could be a member of the interferon-stimulated genes (ISGs) family, which includes proteins with pyrin domain.

3.1.3 Aim2 Is the Inflammasome Receptor Detecting *Francisella* in the Host Cytosol

Absent in melanoma 2 (AIM2) was identified in 2009 as an inflammasome receptor-sensing cytoplasmic DNA (Roberts et al. 2009; Burckstummer et al. 2009; Fernandes-Alnemri et al. 2009; Hornung et al. 2009). *Aim2* is an ISG coding a protein with a N-terminal pyrin domain and a C-terminal HIN domain. Through its HIN domain, AIM2, a cytosolic protein, binds double-stranded (ds)DNA. dsDNA is recognized by AIM2 in a sequence-independent manner as the HIN domain binds the dsDNA sugar-phosphate backbone (Jin et al. 2012). A DNA strand of at least

80 bp in length is required upon transfection to trigger IL-1β release. dsDNA in the cytosol may thus act as a scaffold to oligomerize AIM2. The resulting complex in turn recruits ASC oligomers through homotypic interactions of AIM2 and ASC pyrin domains leading to the formation of the active inflammasome complex (Jin et al. 2012).

The role of Aim2 as the *Francisella* inflammasome receptor was demonstrated independently in 2010 by three groups (the Fitzgerald, Alnemri, and Monack teams) who had generated *Aim2*$^{-/-}$ mice (Fernandes-Alnemri et al. 2010; Rathinam et al. 2010; Jones et al. 2010). Among several Aim2-activating pathogens, Fitzgerald and colleagues demonstrated that *Aim2*$^{-/-}$ macrophages were deficient for caspase-1 processing and IL-1β release in response to LVS infection (Rathinam et al. 2010). The Alnemri and Monack groups used *F. novicida* to demonstrate that, even at high multiplicity of infection, Aim2 was required for ASC oligomerization, caspase-1 processing, pro-IL-1β processing, IL-1β release, and macrophage death (Fernandes-Alnemri et al. 2010; Jones et al. 2010). The requirement for Aim2 is highly specific since neither *Nlrp3*$^{-/-}$ nor *Mefv*$^{-/-}$ BMDMs (the latter being deficient in the pyrin receptor) display any deficiency in response to *F. novicida* infection (Fernandes-Alnemri et al. 2010). In agreement, with the cytosolic localization of AIM2 and with previous studies (Mariathasan et al. 2005; Gavrilin et al. 2006), the vacuole-restricted mutant (Δ*FPI*) is unable to trigger Aim2 activation. The direct recognition of *Francisella* DNA by Aim2 was demonstrated by immunofluorescence. Macrophages were infected with *F. Novicida* pre-labeled with Hoechst, a DNA intercalating dye. Following infection, Aim2 specks or clusters were observed co-localizing with Hoechst-labeled material in close proximity to bacteria. In addition, Aim2 specks co-localize with ASC specks likely corresponding to the mature inflammasome complex. Importantly, Aim2 specks are also observed in ASC-deficient macrophages further demonstrating the role of Aim2 as a receptor upstream of ASC (Fernandes-Alnemri et al. 2010; Jones et al. 2010). Of note, Aim2 is also required in bone marrow-derived dendritic cells (BMDC) to trigger inflammasome activation in response to *F. novicida* infection (Belhocine and Monack 2012).

Aim2 is also critical for the immune response in vivo in a mouse model of tularemia. As previously reported for *Casp1*$^{-/-}$ and *Asc*$^{-/-}$, *Aim2*$^{-/-}$ mice are highly susceptible to *F. novicida* infection with 100 % of the mice dying in 5 days as opposed to 30 % of wild-type mice (Fernandes-Alnemri et al. 2010). The fast death is associated with a 100- to 1000-fold increase in the bacterial burden in the spleen, the lung, and the liver and a much lower concentration of IL-18 in the serum of *Aim2*$^{-/-}$ mice compared to wild-type mice (Fernandes-Alnemri et al. 2010; Jones et al. 2010). The Aim2 inflammasome is thus a key receptor for the innate immune defense against *Francisella*.

The discovery of Aim2 as the receptor-sensing *Francisella* in the host cytosol gave an opportunity to question the role of type I IFN signaling in Aim2 induction and Aim2 inflammasome activation. While Aim2 expression was strongly induced upon *Francisella* infection in a STING- and type I IFN receptor-dependent manner in experiments performed by Jones and colleagues, Aim2 was not induced during similar experiments performed by Fernandes-Alnemri et al. Interestingly, while several groups confirmed a role for the type I IFN pathway [IRF3/IRF7 (Rathinam et al. 2010); IRF3, IFNAR1 (Fernandes-Alnemri et al. 2010); STING, IFNAR1 (Jones et al. 2010)] in potentiating the Aim2 inflammasome, the type I IFN pathway is not required for Aim2 activation following transfection with synthetic DNA (p (dA:dT)) or following transfection with *F. novicida* genomic DNA (Fernandes-Alnemri et al. 2010; Rathinam et al. 2010; Meunier et al. 2015). This result indicates that the basal level of Aim2 is sufficient to detect cytosolic DNA and suggests that type I IFN and ISGs are required specifically during *Francisella* infection upstream of the Aim2 inflammasome.

3.2 Gbp-mediated Bacteriolysis Is Required to Trigger Aim2 Inflammasome Activation

One of the key questions regarding the activation of the cGAS/Ifi204 DNA sensor or of the Aim2 inflammasome during *F. novicida* infection is the source of cytosolic DNA. Indeed, bacterial DNA is normally enclosed within the bacterium and not accessible to cytosolic DNA sensors. During immunostaining experiments, Jones and colleagues had observed that Aim2 specks were localized in proximity to irregular-shaped bacterial remnants (Jones et al. 2010) suggesting that bacteriolysis may be a prerequisite for the release of genomic bacterial DNA into the host cytosol. This observation coupled to the lack of requirement for type I IFN signaling to activate the Aim2 inflammasome in response to synthetic DNA led Meunier et al. (2015) to hypothesize that type I IFN signaling may be involved in the bacteriolysis mechanism. As type I IFN signaling induces hundreds of ISGs in *F. novicida*-infected macrophages (Henry et al. 2007), they performed a RNA interference screening on 483 ISGs in *F. novicida*-infected BMDMs. They identified that the knockdown of the genes encoding two guanylate-binding proteins (Gbps): Gbp2 and Gbp5 strongly reduced Aim2 inflammasome activation upon *F. novicida* infection (Meunier et al. 2015). At the meantime, Kanneganti and colleagues demonstrated that in addition to IFNAR1, IFNAR2, Stat1, and IRF9, the IFN-inducible transcription factor IRF1 was required for *F. novicida*-mediated Aim2 inflammasome activation (Man et al. 2015). IRF1 regulates IFN-β induction during *F. novicida* infection. However, Man et al. demonstrated that the addition of IFN-β in *Irf1*$^{-/-}$ BMDM does not rescue Aim2 inflammasome activation during *F. novicida* infection. This result suggested that IRF1 action is required downstream of the IFNAR receptor to induce one of several specific ISGs. Using a gene expression profiling approach, Man and colleagues identified Gbps as a family of protein strongly up-regulated by IRF1 and IFNAR1 in *F. novicida*-infected macrophages. A subsequent siRNA screen on the Gbp family confirmed Gbp2 and Gbp5 as two specific Gbps required for Aim2 inflammasome activation in response to *F. novicida* infection (Man et al. 2015).

Gbps are interferon-inducible, dynamin-like GTPases involved in cell-autonomous immunity. They may act as scaffolding proteins to target antimicrobial proteins such as the NADPH oxidase, autophagic proteins, and the IFN-γ-inducible GTPase p47 to membrane-bound intracellular pathogens (Meunier and Broz 2016). In addition, Gbp5 directly binds NLRP3 and may promote NLRP3 inflammasome assembly in response to pathogenic bacteria and to ATP and Nigericin (Shenoy et al. 2012) although the latter observation is controversial (Meunier et al. 2014). In addition, Gbps contribute to caspase-11 activation following infection with *Salmonella typhimurium* (Meunier et al. 2014) and *Legionella pneumophila* (Pilla et al. 2014). The mechanisms controlling the enhanced activation of this non-canonical inflammasome are still unclear with three mechanisms proposed. (1) Gbps could favor the disruption of pathogen-containing vacuole, thus enhancing the release of the bacteria into the host cytosol (Meunier et al. 2014). (2) Gbps could act downstream of the release of the bacteria into the host cytosol (Pilla et al. 2014) by targeting cytosolic bacteria and triggering the shedding of LPS into the cytosol. (3) Gbps could directly act onto the non-canonical caspase-11 complex to promote its multimerization (Finethy et al. 2015). Eleven *Gbp* genes (and two *Gbp* pseudogenes) are present in two clusters on mouse chromosomes 3 and 5. Several of the chromosome 3-encoded Gbps control bacterial (Kim et al. 2011) and parasitic replication (Yamamoto et al. 2012; Degrandi et al. 2013). Using BMDMs from $Gbp5^{-/-}$ *and* $Gbp2^{-/-}$ mice and from mice deficient for the cluster of *Gbps* (*Gbp1, Gbp2, Gbp3, Gbp5, and Gbp7*) present on the chromosome 3 (Gbp^{chr3}), Meunier et al. and Man et al. confirmed that Gbps and particularly Gbp2 and Gbp5 are required for Aim2 inflammasome activation during *F. novicida* infection. Indeed, Gbp^{chr3} macrophages secrete lower levels of IL-1β and IL-18 than wild-type macrophages during the infection. Similarly, Gbp^{chr3} macrophages are largely resistant to *Francisella*-induced pyroptosis. Since Gbps are known to target pathogen-containing phagosomes/vacuoles and destabilize the integrity of these compartments (Yamamoto et al. 2012; Meunier et al. 2014), Meunier et al. and Man et al. hypothesized that Gbps might attack *Francisella*-containing phagosomes leading to the release of bacteria into the host cytosol. Yet, there was no difference in the *F. novicida* escape rate between wild-type and Gbp^{chr3}-deficient macrophages. This result indicates that Gbps act after *Francisella* escape into the host cytosol strengthening a role of Gbps in triggering cytosolic bacteriolysis.

By confocal microscopy, both Gbp2 and Gbp5 were observed surrounding the bacteria. Interestingly, using GFP-expressing bacteria, Man et al. noticed that bacteria decorated with Gbp5 tend to lose their GFP expression suggesting a loss of viability. Accordingly, a greater number of cytosolic bacteria are permeable to propidium iodide in wild-type macrophages compared to Gbp^{chr3}-deficient macrophages indicating a Gbps-dependent loss of bacterial membrane integrity (Meunier et al. 2015). In addition, *Francisella* replication is impaired in wild-type macrophages compared to Gbp^{chr3}-deficient and $Irf1^{-/-}$ BMDMs as shown by CFU assay and by quantification of bacterial loads in single cells using confocal microscopy or high-resolution microscopy in flow. How Gbps trigger bacteriolysis following recruitment onto/on close proximity to cytosolic *Francisella* is still unclear. Bosio and colleagues have shown that the anti-oxidant N-acetyl cysteine and the mitochondrial-specific reactive oxygen species (ROS) scavenger, mitoTEMPO inhibit *F. novicida*-mediated inflammasome activation (Crane et al. 2014). ROS can target bacterial lipids and outer membrane proteins (Fang 2004). Mitochondrial ROS may thus participate in the killing of *F. novicida* inside the host cytosol leading to the release of DNA and Aim2 activation. Yet, the precise role of mitochondrial ROS and their connection with Gbps remain to be deciphered.

The roles of IRF1 and Gbps in inflammasome activation were validated in vivo in a mouse model of tularemia (Man et al. 2015; Meunier et al. 2015). Indeed, after *F. novicida* intradermal inoculation, both $Irf1^{-/-}$ and Gbp^{chr3}-deficient mice showed reduced IL-18 concentration in the serum. Furthermore, $Irf1^{-/-}$, $Gbp2^{-/-}$, and Gbp^{chr3}-deficient mice had a higher bacterial burden in the liver and the spleen than wild-type mice. Finally, in a survival experiment, all $Irf1^{-/-}$, $Gbp2^{-/-}$, and Gbp^{chr3}-deficient mice died within 4 days of infection, while most wild-type mice survived until the end of the experiment (day 8–10).

Importantly, while Gbps expression is induced by type I IFN, Gbps are also strongly induced by IFN-γ in a manner requiring the transcription factors Stat1 and IRF1 (Briken et al. 1995). In vitro, BMDMs do not produce IFN-γ and the IFN signaling is only due to autocrine and paracrine type I IFN signaling following engagement of the cytosolic DNA sensors. However, in vivo, IFN-γ produced early on by NK cells, T cells, and several different phagocyte populations (Bokhari et al. 2008; De Pascalis et al. 2008; Pierini et al. 2013) is likely to control Gbps induction. Indeed, while IFN-γ signaling is critical in vivo to fight *Francisella* infection (Anthony et al. 1989; Elkins et al. 1996), type I IFN signaling is deleterious in vivo during infections with several intracellular bacteria including *Francisella* (Auerbuch et al. 2004; Stanley et al. 2007; Henry et al. 2010).

3.3 TLR2 Controls pro-IL-1β Level and the Kinetics of AIM2 Inflammasome Activation

As previously mentioned, TLR2 is the main TLR responsible for NF-κB activation during *Francisella* infection (Katz et al. 2006). As such, induction of pro-IL-1β is fully abolished in *F. holarctica* LVS-infected $Tlr2^{-/-}$ BMDMs and BMDCs (Li et al. 2006; Cole et al. 2007; Dotson et al. 2013). Upon *F. holarctica* LVS infection, IFN-β induction is also fully dependent on TLR2 (Cole et al. 2007). During *F. novicida* infection, TLR2 controls pro-IL-1β level but does not play a significant role in IFN-β induction (Jones and Weiss 2011). In addition, during *F. novicida* infection, TLR2 signaling modulates inflammasome activation kinetics. Indeed, wild-type macrophages display more ASC specks, secrete more mature caspase-1 and IL-18, and die faster than $Tlr2^{-/-}$ macrophages (Jones and Weiss 2011). Importantly, the requirement for TLR2 signaling is also observed in vivo since (i) immunostaining with an antibody against active caspase-1 p20 subunit reveals more caspase-1 specks in the spleen of infected wild-type mice than in the spleen of

Tlr2$^{-/-}$ mice; (ii) TLR2-deficient mice present less IL-18 in their serum than wild-type mice do (Jones and Weiss 2011). While differences exist between LVS- and *F. novicida*-infected macrophages, these studies clearly demonstrate that TLR2 is required for full inflammasome activation during *Francisella* infection. Besides the role of TLR2 on IFN-β induction observed during LVS infection, the contributions of TLR2 to the induction of mitochondrial ROS (West et al. 2011), to Gbps-induced bacteriolysis, or to a possible priming of the Aim2 inflammasome remain to be evaluated.

4 *Francisella* and the Non-canonical Inflammasomes

4.1 Francisella *LPS Escapes Caspase-11 Recognition*

The presence of LPS inside the cytosol is detected by caspase-11 (Kayagaki et al. 2013; Hagar et al. 2013; Shi et al. 2014) leading to activation of this inflammatory caspase in a complex termed a non-canonical inflammasome (Kayagaki et al. 2013). The caspase-11 inflammasome is thus a molecular sentinel monitoring the cytosol for the presence of Gram-negative bacteria (Aachoui et al. 2013). As a Gram-negative bacterium replicating in the cytosol, *Francisella* was thus a likely pathogen to activate caspase-11. Although the expression of caspase-11, an IFN-inducible gene, is induced upon *F. novicida* infection (Akhter et al. 2012), *F. novicida* does not activate caspase-11 (Hagar et al. 2013). Miao and colleagues identified that *F. novicida*-mediated evasion from caspase-11 recognition is associated with its specific lipid A structure (Hagar et al. 2013). Indeed, as previously described, *Francisella* LPS is tetra-acylated while LPS from most Gram-negative bacteria are hexa-acylated (Okan and Kasper 2013). LpxF is a periplasmic phosphatase, which specifically removes the 4′ monophosphate group in the lipid A. Raetz and collaborators have observed that the presence of the 4′ monophosphate group in the lipid A of the *lpxF* mutant impairs the cleavage of the 3′ acyl chain. The lack of the LpxF enzyme thus results in the generation of a penta-acylated lipid A (Wang et al. 2007). Strikingly, while purified tetra-acylated LPS from *F. novicida* does not activate caspase-11, purified penta-acylated LPS from *lpxF* mutant does so (Hagar et al. 2013). This result highlights LpxF as a key enzyme to escape caspase-11 recognition. However, the *lpxF* mutant is also severely impaired for growth in axenic medium and for resistance to antimicrobial peptides indicating that this enzyme is also required for membrane integrity and the general fitness of the bacterium (Wang et al. 2007). Of note, LPS from the *lpxF* mutant does not activate TLR4 (Wang et al. 2007) while it does activate caspase-11 in the cytosol indicating that TLR4 and caspase-11 pathways have different structural requirements (Hagar et al. 2013).

4.2 ASC-dependent Caspase-1-independent Pathways

While the canonical inflammasome is defined by caspase-1 activation, Mariathasan et al. (2005) observed early on that, upon *F. novicida* infection, the kinetics of cell death differed between ASC- and caspase-1-deficient macrophages. This observation suggested the presence of an ASC-dependent caspase-1-independent cell death pathway. This non-canonical pathway was characterized by Pierini et al. (2012). They identified that while wild-type macrophages die rapidly by pyroptosis, caspase-1-deficient macrophages die 4–6 h later with features typical of apoptosis such as condensed nuclei and a ladder-like DNA pattern indicative of internucleosomal fragmentation. Importantly, neither Aim2-deficient nor ASC-deficient macrophages display any apoptotic nor necrotic features at the same time point. While in wild-type macrophages, the Aim2/ASC speck co-localizes with active caspase-1 (Jones et al. 2010), in caspase-1-deficient macrophages, the Aim2/ASC complex recruits and activates caspase-8 (Pierini et al. 2012). The Aim2/ASC-dependent, caspase-1-independent apoptosis is inhibited by caspase-8 or caspase-9 inhibitors or by ectopic expression of Bcl-2 or Bcl-X_L, two anti-apoptotic proteins blocking the mitochondrial intrinsic pathway. Altogether, these data led to the conclusion that upon *F. novicida* infection, in the absence of caspase-1, Aim2/ASC complex recruits and activates caspase-8, which in turn activates the mitochondrial intrinsic pathway and caspase-9. The combined action of both caspase-8 and caspase-9 then triggers activation of the executioner caspase-3, leading to apoptosis of the infected macrophages (Pierini et al. 2012). Importantly, the ability of ASC platforms to activate caspase-8 is not restricted to *Francisella* infection or to Aim2 activation since Nlrp3, Aim2 and Nlrc4 complexes can recruit and activate caspase-8 to trigger apoptosis in response to a variety of stimuli (Sagulenko et al. 2013; Man et al. 2014). In addition to promoting apoptotic cell death, the ASC-dependent caspase-1-independent pathway also regulates the IL-18/IFN-γ cascade in a mouse model of tularemia (Pierini et al. 2013). While the level of IL-18 and IFN-γ in the serum of *F. novicida*-infected mice are much lower in caspase-1-deficient mice than in wild-type mice, the levels of the two cytokines are much higher in $Casp1^{-/-}$ mice than in $Asc^{-/-}$ mice. Neutralization of IL-18 in infected $Casp1^{-/-}$ mice reduces IFN-γ levels in the serum demonstrating that bioactive IL-18 can be generated in a caspase-1-independent manner. In vitro, caspase-8, expressed in the presence of Aim2, ASC, and pro-IL-18, triggers specifically the release of IL-18. Altogether, these results demonstrate that, as presented above for the apoptotic cell death, in vivo, in the absence of caspase-1, caspase-8 can be activated in the Aim2/ASC complex leading to pro-IL-18 processing, release of bioactive IL-18, and activation of NK cells to trigger IFN-γ production (Pierini et al. 2013). In a caspase-1-proficient context, the relevance of this caspase-8 pathway remains to be understood. Indeed, caspase-8 is inducing an apoptotic cell death outcompeted by the rapid pyroptosis and caspase-8 is triggering the IL-18/IFN-γ cascade in a manner less efficient than the canonical caspase-1-inflammasome. Such a fail-safe mechanism could be important in humans bearing defective caspase-1 variants (Luksch et al. 2012) or during infections with pathogens blocking specifically caspase-1 activity.

5 Inflammasome Activation in Human Cells

BMDMs have been the main cellular model to study inflammasome activation upon *Francisella* infection. Yet, there is clear evidence that the inflammasome complexes assembled in human phagocytes contain receptors that differ from the ones sensing *Francisella* in murine macrophages (Gavrilin et al. 2009; Atianand et al. 2011).

Inflammasome activation in human cells was first studied in monocytes (Fig. 2). Gavrilin et al. demonstrated that the cytosolic localization of *F. novicida* is required in human monocytes to trigger IL-1β release (Gavrilin et al. 2006). Indeed, various compounds blocking *F. novicida* escape into the cytosol abolish IL-1β secretion while pro-IL-1β level in the cell lysates is unaffected. Similarly, a *mglA* mutant, which remains in the vacuole, is unable to trigger IL-1β release in human monocytes (Gavrilin et al. 2006). Finally, while heat-killed *Francisella* readily induces pro-IL-1β expression, live bacteria are required to activate caspase-1 and IL-1β release in primary human dendritic cells (Li et al. 2006).

Inflammasome activation by *Francisella* in human cells (Fig. 2) is associated with three intracellular protein sensors, AIM2, NLRP3, and pyrin (encoded by the *Mefv* gene) (Gavrilin et al. 2009; Atianand et al. 2011). As described above, several groups have observed that murine *Nlrp3*$^{-/-}$ and *Mefv*$^{-/-}$ BMDMs have a normal inflammasome response upon *F. novicida* infection indicating the key requirement for Aim2 in this experimental system (Mariathasan et al. 2006; Fernandes-Alnemri et al. 2010; Ulland et al. 2010; Atianand et al. 2011). Surprisingly, human macrophages rely on both NLRP3 and AIM2 to activate the inflammasome upon

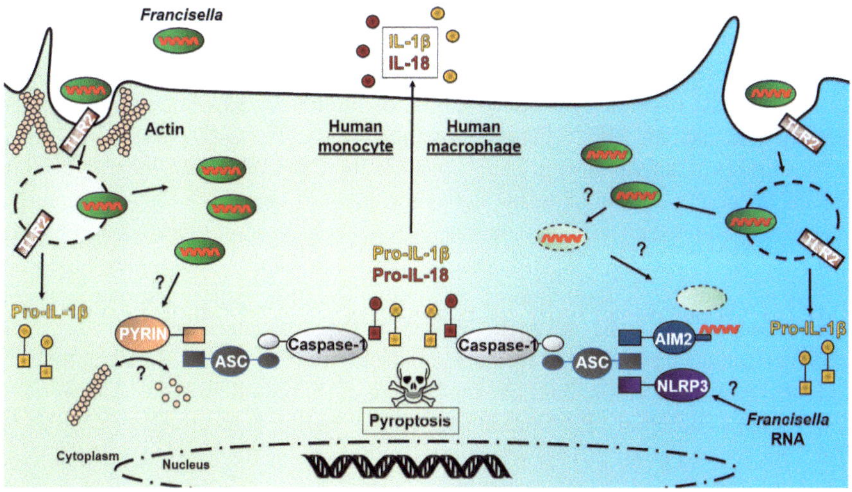

Fig. 2 Models for the different inflammasomes activated upon *Francisella* infection of human monocytes (*left panel*) and human macrophages (*right panel*) (see text for details)

F. novicida infection (Atianand et al. 2011). Indeed, knockdown of either *Nlrp3* or *Aim2* reduced IL-1β release in PMA-differentiated THP-1 macrophages. A slightly higher IL-1β inhibition was observed upon combined knockdown of both *Nlrp3* and *Aim2* suggesting that the two PRRs are involved in sensing *F. novicida* in the human macrophage cytosol (Atianand et al. 2011). The role of NLRP3 was further confirmed in an inflammasome reconstitution system in 293T cells. In this system, NLRP3 but not NLRP12 co-localizes with ASC specks upon *F. novicida* infection and expression of NLRP3 strongly increases IL-1β release upon LVS or *F. novicida* infections. While this study by Harton and colleagues (Bedoya et al. 2007) clearly demonstrates the role of NLRP3 in human cells, the reasons explaining the species specificity of the NLRP3 response are unknown. The nucleotide-binding domain, leucine-rich repeat region of human NLRP3 differs from murine Nlrp3 in the distribution of potential serine/threonine phosphorylation sites, oxidant-sensitive cysteine residues, and lysines that may serve as ubiquitination sites to regulate NLRP3 activation (Atianand et al. 2011; Py et al. 2013). Furthermore, human NLRP3 may recognize a broader panel of ligands than its murine counterpart. Indeed, human NLRP3 senses multiple types of RNAs (bacterial mRNA, rRNA, tRNA; synthetic ssRNA) whereas murine Nlrp3 preferentially recognizes only bacterial mRNA (Sha et al. 2014). The exact nature of the PAMPs or of the danger signal leading to NLRP3 activation in human cells during *Francisella* infection is still elusive.

Pyrin is expressed in primary monocytes and in THP-1 monocytes but its expression is greatly reduced in primary macrophages cultured in the absence of M-CSF or in PMA-differentiated THP-1 cells. In primary human monocytes and in THP-1 monocytes, *Mefv* knockdown reduces IL-1β release upon *F. novicida* infection. Conversely, ectopic expression of pyrin in PMA-differentiated THP-1 cells increases IL-1β release upon *F. novicida* infection (Gavrilin et al. 2009). These results indicate that Pyrin is required in human monocytes to activate the inflammasome complex in response to *F. novicida*. Recently, pyrin was identified as a sensor of the modification of Rho GTPases activity. It is still unclear whether pyrin directly senses Rho GTPases activity or the consequences of Rho dysfunction on the actin cytoskeleton (Xu et al. 2014). Numerous bacteria target Rho GTPases and the actin cytoskeleton (Lemichez and Aktories 2013; Welch and Way 2013). Yet, in contrast to other cytosolic bacteria, *F. tularensis* does not use actin-based motility and has no Rho GTPases-targeting toxins. The molecular mechanisms governing pyrin activation during *F. novicida* infection of human monocytes thus remain mysterious. Inflammasome activation has not been investigated in murine monocytes. However, pyrin is expressed at functional level in murine macrophages (Xu et al. 2014) and *Mefv*$^{-/-}$ BMDMs, in contrast to *Aim2*$^{-/-}$ macrophages, do not display any defect in inflammasome activation. The discrepancy observed between human and murine cells might be linked to the differences between the pyrin proteins in the two species. Murine pyrin lacks the C-terminal B30.2 domain, a domain known in certain immune proteins (e.g., butyrophilin 3A1, TRIM5α) to bind PAMPs (bacterial phospho-antigen, HIV-1 capsid, respectively) (Yang et al. 2012; Sandstrom et al. 2014). While the identity of the PAMP (cytosolic gDNA) and the inflammasome receptor (Aim2) is well established in murine macrophages, further work is needed in human cells to fully understand the nature of the detected PAMP/DAMP and the interaction of the different inflammasome receptors.

6 Lessons from the Bacterial Side: Study of Hypercytotoxic Mutants

Numerous *F. novicida* and *F. holarctica* LVS mutants with a hypercytotoxic phenotypes have been identified either through genetic screens (Jones 2010; Lai et al. 2010) or through phenotypic characterization of defined mutants of interest (Weiss et al. 2007; Platz et al. 2010; Ulland et al. 2010; Huang et al. 2010; Jayakar et al. 2011; Lindemann et al. 2011; Peng et al. 2011; Mahawar et al. 2012; Dotson et al. 2013; Doyle et al. 2014). As both *F. novicida* and *F. holarctica* LVS have the ability to replicate extensively in the macrophage cytosol before being detected by Aim2, the hypercytotoxic mutants were thought to be mutated in specific virulence factors inhibiting either the recognition by innate immune receptors or directly innate immune pathways. Deletion of the gene encoding an OmpA-like protein (FTL_0325/FTT_0831c) results in an *F. holarctica* LVS mutant hyperactivating the inflammasome pathway (Mahawar et al. 2012). Upon ectopic expression in HEK293T cells, FTL_0325/FTT_0831c inhibited NF-κB activation suggesting that the hyperinflammasome phenotype of the corresponding mutant is due to an inability of the mutant to specifically block this pathway (Mahawar et al. 2012). However, Norgard and colleagues identified that the OmpA-like protein FTL_0325 contributes to cell division, maintenance of cell shape, and structural integrity suggesting that the hyperinflammasome phenotype was more likely to result from enhanced bacteriolysis in the host cytosol (Robertson et al. 2014). TolC is an outer membrane protein involved in type I secretion pathway. The hypercytotoxic phenotype of the *tolC* mutant led Platz et al. to suggest that TolC was implicated in the secretion of a toxin inhibiting host cell death (Platz et al. 2010). While this hypothesis cannot be formerly excluded, TolC is well known in *E. coli* as a member of the Tol-Pal system required for membrane integrity (Lloubes et al. 2001) suggesting that this mutant might be more sensitive to bacteriolysis. Infection with a LVS mutant in *ripA*, a gene coding an inner membrane protein, leads to a stronger MAPK response and a faster inflammasome response than upon wild-type LVS infection (Huang et al. 2010). RipA was later on identified as a protein-modulating lipid A synthesis, suggesting that the ability of *F. tularensis* strains to resist to host antimicrobial factors and/or to remain undetected may be associated with membrane remodeling (Miller et al. 2014). This idea was strengthened by work from Weiss and colleagues, who identified that *F. novicida* used the CRISPR-cas9 endonuclease (FTN_0757) to down-regulate the expression level of a bacterial lipoprotein (FTN_1103) during infection. This down-regulation enhances envelope integrity, decreases TLR2 recognition, and reduces inflammasome activation (Jones et al. 2012; Sampson et al. 2014). In addition to the *ripA* mutant, several mutants in LPS/capsule biosynthesis genes are hypercytotoxic (Lai et al. 2010; Jayakar et al. 2011). Δ*kdtA*, Δ*lpcC*, Δ*manB*, and Δ*manC* mutants are hypercytotoxic and their LPS is lacking the O-antigen and has a defect in the lipid A core. The respective roles of the defect in the lipid A core and of the lack of O-antigen are unclear since *wbtA* mutants (which have a normal lipid A core but lack the O-antigen) have been reported as hypercytotoxic by Peng et al. (2011) but not hypercytotoxic by Lai et al. (2010). Conversely, a *htrB* mutant (deleted in a gene encoding a lipid A acylase) displays a LPS with a normal O-Antigen but is hypercytotoxic (Lai et al. 2010). Altogether, these data indicate that the lipid A structure is a key element to limit bacterial detection by the inflammasome pathway possibly through the maintenance of membrane integrity, resistance to cytosolic antimicrobial factors and spontaneous or host-induced bacteriolysis. MviN, a lipid II flippase, is an enzyme involved in peptidoglycan maturation. In contrast to *mviN* mutants in *E. coli*, an *mviN* mutant in LVS grows normally in axenic medium but presents an aberrant morphology suggesting it might be more susceptible to bacteriolysis than wild-type LVS (Ulland et al. 2010). While it is clear that envelope integrity and its regulation during infection are key factors to avoid recognition by TLR2 and the Aim2 inflammasome, most of the above-cited mutants have been shown by Peng et al. to display aberrant morphologies during growth in minimal medium and to lyse more in the host cell cytosol. This result suggests that under stressful conditions, numerous membrane proteins cooperate with peptidoglycan and LPS-biosynthetic proteins to ensure proper membrane integrity, a lack of PAMP exposure at the surface and a resistance to spontaneous and host-induced bacteriolysis inside the host cytosol (Peng et al. 2011). Altogether, the data from all these studies suggest that *Francisella* does not actively inhibit the inflammasome pathway but rather use a hiding strategy associated with tight envelope integrity and modification of the structure (e.g., LPS) or the level of expression of its surface exposed PAMP (e.g., TLR2-recognized bacterial lipoprotein). This stealth strategy is particularly striking for highly virulent *F. tularensis* subsp. *tularensis* strains.

7 Hypervirulent Strains Escape Inflammasome Detection

Considering the different *Francisella* species and the different *F. tularensis* subspecies, there is a good correlation between the ability of the various strains to escape recognition by the inflammasome and their ability to cause disease in humans (Table 1). As previously mentioned, *F. novicida* infection leads to a strong inflammasome activation although this activation occurs with a delay allowing significant replication in macrophages and in vivo (Mares et al. 2008). This delay in inflammasome activation is even stronger upon infection with *F. holarctica* LVS and is associated with a lower IL-1β release than upon *F. novicida* infection (Carlson et al. 2007; Ghonime et al. 2015). Finally, infection with *F. tularensis* type

A strains elicits no or a very weak inflammasome activation (Wickstrum et al. 2009; Bauler et al. 2011; Crane et al. 2014; Ghonime et al. 2015).

While numerous groups have observed caspase-1 activation, IL-1β release, and pyroptosis in macrophages infected with LVS (Mariathasan et al. 2005; Gavrilin et al. 2006; Li et al. 2006; Wickstrum et al. 2009; Rathinam et al. 2010), apoptosis is also observed in *F. holarctica* LVS-infected macrophages with an involvement of the intrinsic pathway proceeding through caspase-9 and caspase-3 activation and PARP cleavage (Lai and Sjostedt 2003). This apoptotic pathway has been observed in the J774 macrophage cell line, which is deficient for *F. novicida*-mediated inflammasome cell death (TH unpublished observation) but also in BMDMs (Doyle et al. 2014). These observations suggest that *F. holarctica* LVS may be less detected by the inflammasome and that another cell-intrinsic mechanism may trigger mitochondrial apoptosis to synergize with pyroptosis to fully remove the bacterial replicative niche. The ability of *F. holarctica* LVS to dampen inflammasome activation correlates with a general ability of this strain to dampen pro-inflammatory signaling (Telepnev et al. 2003). Indeed, infection with *F. holarctica* LVS inhibits LPS-induced TNF-α and IL-1β release. This inhibition is dependent on FPI-encoded genes, and it is still unclear whether it is a direct consequence of phagosomal escape or whether a specific T6SS-effector modulates this pro-inflammatory response (Telepnev et al. 2003; Broms et al. 2010, 2011).

As mentioned above, infection with *F. tularensis* subs. *tularensis* strains, leads to almost no IL-1β (undetectable to less than 20 ng/ml depending on the studies) and no IL-18 (undetectable to less than 50 ng/ml depending on the studies) release (Wickstrum et al. 2009; Crane et al. 2014; Ghonime et al. 2015). Death of SCHU S4-infected macrophages occurs very late, suggesting that pyroptosis does not play a major role in restricting SCHU S4 replication. The low inflammasome response is not due to an inability of the NLRP3 inflammasome to be primed upon infection. Indeed, when the NLRP3 priming event was evaluated using ATP as a second signal, *F. novicida*, LVS, and *F. tularensis* strains had a similar NLRP3 priming activity (Ghonime et al. 2015). This result suggests that infection with SCHU S4 does not provide the signal 2 required for inflammasome activation. Bosio and co-workers (Crane et al. 2014) demonstrated that *F. novicida* is more sensitive to H_2O_2-mediated killing than SCHU S4. As previously mentioned, ROS are required for *F. novicida*-mediated inflammasome activation. The resistance of SCHU S4 to ROS could thus limit ROS-mediated bacteriolysis of SCHU S4 in the cytosol, DNA release, and AIM2 inflammasome activation (Crane et al. 2014). While this model is appealing, further evidence remains to be provided to understand the role of mitochondrial ROS and the inability of *F. tularensis* strain to be detected by the inflammasome. In addition to the resistance to ROS-mediated killing, the ability of *F. tularensis* SCHU S4 to escape recognition by the inflammasome pathway might be related to its LPS and capsule structure. Indeed, mutants lacking genes in a locus required for O-antigen and capsule biosynthesis are hypercytotoxic upon macrophage infection (Lindemann et al. 2011).

The ability of SCHU S4 strain to modulate IL-1β response is also dependent on its entry pathway in human macrophages. Indeed, the efficient uptake of SCHU S4 strain by human macrophages requires C3 opsonization and the presence of the CR3 receptor at the macrophage surface. Despite an increase entry into host cells, infection with C3-opsonized *F. tularensis* leads to a decreased IL-1β release compared to an infection with non-opsonized *F. tularensis*. This CR3-mediated immune suppression is not specific for the inflammasome pathway but may act through inhibition of TLR2 signaling highlighting the connections between these pathways (Dai et al. 2013).

In vivo, in a mouse model of tularemia with a type A strain, an immunostaining approach to detect active caspases revealed a predominant activation of caspase-3 suggesting that the absence of inflammasome activation is also true in vivo. Although the role of caspase-1 has not been deeply investigated, the formation of granuloma and the frequency of TUNEL$^+$ cells were similar in the wild-type and *Casp1*$^{-/-}$ mice infected with type A strain (Wickstrum et al. 2009). Only a few cells displayed active caspase-1 staining both in the liver and in the spleen. Interestingly, active Caspase-1 was restricted to the splenic marginal zones (Wickstrum et al. 2009), which contain macrophage subsets with great ability to phagocyte pathogens (Borges da Silva et al. 2015). Altogether, while the results suggest that the inflammasome does not play a major role to fight highly virulent *F. tularensis* subsp. *tularensis* strains, inflammasome activation may be important in dedicated macrophage subsets localized in strategic places in the lymphoid organs to orchestrate downstream immune responses.

8 Concluding Remarks

Inflammasome-deficient mice have demonstrated the key role of the inflammasome to fight *F. novicida* infection. The mechanisms leading to inflammasome activation are now well established with the action of the interferon-inducible proteins Gbp2 and Gbp5 required to lyse *F. novicida* and to release its genomic DNA into the host cytosol. Aim2 is then the key sensor to initiate inflammasome complex formation and trigger the ensuing inflammatory cascade. Interestingly, the ability of *Francisella* strains to activate the inflammasome pathway is inversely correlated with their pathogenicity toward human. In addition to having a non-toxic lipid A escaping recognition by both TLR4 and caspase-11, *Francisella* has evolved a tight membrane associated with resistance to bacteriolysis and down-regulation of potential PAMPs. This strategy allows *F. novicida* and *F. holarctica* LVS to replicate in the cytosol before being eventually detected by the inflammasome sensors. This furtiveness is even higher in the highly virulent *F. tularensis* subsp. *tularensis* strains since macrophages infected with these strains do not demonstrate any inflammasome activation markers.

One striking feature of *F. novicida* and *F. holarctica* LVS inflammasome in murine macrophages is their exclusive dependence on the Aim2 receptor. This is unique since several inflammasome sensors are involved in the detection of most other intracellular bacteria, including *Listeria monocytogenes*, which is detected by

Nlrp3, Nlrc4, and Aim2 (Wu et al. 2010). This strict dependence is even more surprising considering that in human phagocytes, three different receptors (PYRIN, NLRP3, and AIM2) have been implicated in the detection of *Francisella* infection.

The inflammasome complex is highly interconnected with other immune pathways acting both upstream of Aim2 and downstream of caspase-1-dependent cytokines. Indeed, the TLR2 and the cGas/Ifi204-STING pathways potentiate or are required for Aim2 inflammasome activation in *Francisella*-infected cells. Surprisingly, for a cytosolic bacterium, the contribution of NOD1/2, the cytosolic sensors of peptidoglycan fragments, seems minimal. Downstream of caspase-1, there are clear evidences that IL-18 regulates IFN-γ level and that IL-1β regulates the IL-23/IL-17 cascade and a protective IgM response mediated by B1a B cells (del Barrio et al. 2015). IFN-γ (Anthony et al. 1989) and to a lesser extent IL-17 (Lin et al. 2009; Khader and Gopal 2010; Skyberg et al. 2013), are two very important cytokines to fight *Francisella* and other intracellular pathogens highlighting the key role of the inflammasome pathway in the orchestration of the immune response.

Of note, while this review focuses on mouse and human cells, *F. noatunensis*, a species responsible for the systemic granulomatous inflammatory disease, francisellosis in cod (Furevik et al. 2011) triggers IL-1β release in zebrafish leukocytes in a caspase-1-like dependent manner (Vojtech et al. 2012). The inflammasome pathway is thus likely to be very important in other *Vertebrates* species in response to other *Francisella* species.

Finally, while the understanding of the mechanisms controlling the inflammasome activation during *Francisella* infection has greatly progressed since the first report in 2005 (Mariathasan et al. 2005), there are still a number of questions to be solved, five of which are listed below.

First, while cytosolic DNA is detected early on during infection by cGAS and Ifi204, Aim2 activation by cytosolic DNA occurs much later during infection and requires the bacteriolytic action of Gbp2 and Gbp5. This observation suggests a hierarchy of sensing between the different DNA sensors, which is not understood.

Second, the mechanisms leading to the recruitment of Gbps onto cytosolic *Francisella* and to their bacteriolytic actions remain unknown. The presence of a CAAX-motif (prenylation motif) in the C-terminus of Gbp2 and Gbp5 suggests the involvement of an undefined membrane compartment. Furthermore, other interferon-inducible proteins might participate in Gbp2/5 action although such actors remain to be identified.

Third, the role of the autophagy pathway in counterbalancing inflammasome activation during *Francisella* remains to be assessed. Indeed, autophagy targets AIM2 inflammasome components during transfection of synthetic DNA (Shi et al. 2012). Furthermore, autophagy clears the cytosol from O-antigen mutants (Case et al. 2014) and from replication-deficient *Francisella* dying in the cytosol (Chong et al. 2012) possibly decreasing the source of cytosolic DNA to activate the Aim2 inflammasome.

Fourth, the specificity of the human inflammasomes remains to be understood. The activation of pyrin in monocytes remains mysterious in absence of any known modification of the actin cytoskeleton upon *Francisella* infections. Similarly, the danger signals or the PAMP leading to activation of the human NLRP3 receptor but not of the murine Nlrp3 remains to be identified.

Fifth, while in vivo experiments demonstrate that the inflammasome is an important mechanism to fight *F. novicida* and *F. holarctica* LVS infection, the repertoire of cells sustaining inflammasome activation in vivo during infection remains to be determined. Particularly, the role of the inflammasome pathway in neutrophils or in macrophages from the splenic marginal zones would be interesting to investigate to understand why despite efficient in vitro inflammasome activation, *F. novicida* kills wild-type mice at a very low infectious dose.

No doubt that the next decade will provide us with exiting findings regarding the activation of the inflammasome and how *Francisella* escapes this recognition to remain one of the most infectious pathogen.

References

Aachoui Y, Leaf IA, Hagar JA et al (2013) Caspase-11 protects against bacteria that escape the vacuole. Science 339:975–978. doi:10.1126/science.1230751

Akhter A, Caution K, Abu Khweek A et al (2012) Caspase-11 promotes the fusion of phagosomes harboring pathogenic bacteria with lysosomes by modulating actin polymerization. Immunity 37:35–47. doi:10.1016/j.immuni.2012.05.001

Anthony LS, Ghadirian E, Nestel FP, Kongshavn PA (1989) The requirement for gamma interferon in resistance of mice to experimental tularemia. Microb Pathog 7:421–428

Atianand MK, Duffy EB, Shah A et al (2011) *Francisella tularensis* reveals a disparity between human and mouse NLRP3 inflammasome activation. J Biol Chem 286:39033–39042. doi:10.1074/jbc.M111.244079

Auerbuch V, Brockstedt DG, Meyer-Morse N et al (2004) Mice lacking the type I interferon receptor are resistant to *Listeria monocytogenes*. J Exp Med 200:527–533

Barel M, Charbit A (2013) *Francisella tularensis* intracellular survival: to eat or to die. Microbes Infect Inst Pasteur. doi:10.1016/j.micinf.2013.09.009

Barker JR, Chong A, Wehrly TD et al (2009) The *Francisella tularensis* pathogenicity island encodes a secretion system that is required for phagosome escape and virulence. Mol Microbiol 74:1459–1470

Baron GS, Nano FE (1998) MglA and MglB are required for the intramacrophage growth of *Francisella novicida*. Mol Microbiol 29:247–259

Bauler TJ, Chase JC, Bosio CM (2011) IFN-beta mediates suppression of IL-12p40 in human dendritic cells following infection with virulent *Francisella tularensis*. J Immunol Baltim Md 1950 187:1845–1855. doi:10.4049/jimmunol.1100377

Bedoya F, Sandler LL, Harton JA (2007) Pyrin-only protein 2 modulates NF-kappaB and disrupts ASC:CLR interactions. J Immunol 178:3837–3845

Belhocine K, Monack DM (2012) Francisella infection triggers activation of the AIM2 inflammasome in murine dendritic cells. Cell Microbiol 14:71–80. doi:10.1111/j.1462-5822.2011.01700.x

Bokhari SM, Kim K-J, Pinson DM et al (2008) NK cells and gamma interferon coordinate the formation and function of hepatic granulomas in mice infected with the *Francisella tularensis* live vaccine strain. Infect Immun 76:1379–1389. doi:10.1128/IAI.00745-07

Borges da Silva H, Fonseca R, Pereira RM et al (2015) Splenic macrophage subsets and their function during blood-borne infections. Front Immunol 6:480. doi:10.3389/fimmu.2015.00480

Briken V, Ruffner H, Schultz U et al (1995) Interferon regulatory factor 1 is required for mouse Gbp gene activation by gamma interferon. Mol Cell Biol 15:975–982

Broms JE, Sjostedt A, Lavander M (2010) The role of the *Francisella tularensis* pathogenicity island in Type VI secretion, intracellular survival, and modulation of host cell signaling. Front Microbiol. doi:10.3389/fmicb.2010.00136

Broms JE, Lavander M, Meyer L, Sjostedt A (2011) IglG and IglI of the Francisella pathogenicity island are important virulence determinants of *Francisella tularensis* LVS. Infect Immun 79:3683–3696. doi:10.1128/IAI.01344-10

Burckstummer T, Baumann C, Bluml S et al (2009) An orthogonal proteomic-genomic screen identifies AIM2 as a cytoplasmic DNA sensor for the inflammasome. Nat Immunol 10: 266–272

Carlson PEJ, Carroll JA, O'Dee DM, Nau GJ (2007) Modulation of virulence factors in *Francisella tularensis* determines human macrophage responses. Microb Pathog 42:204–214. doi:10.1016/j.micpath.2007.02.001

Case EDR, Chong A, Wehrly TD et al (2014) The Francisella O-antigen mediates survival in the macrophage cytosol via autophagy avoidance. Cell Microbiol 16:862–877. doi:10.1111/cmi. 12246

Checroun C, Wehrly TD, Fischer ER et al (2006) Autophagy-mediated reentry of *Francisella tularensis* into the endocytic compartment after cytoplasmic replication. Proc Natl Acad Sci USA 103:14578–14583

Chong A, Wehrly TD, Child R et al (2012) Cytosolic clearance of replication-deficient mutants reveals *Francisella tularensis* interactions with the autophagic pathway. Autophagy 8:1342–1356. doi:10.4161/auto.20808

Clemens DL, Horwitz MA (2007) Uptake and intracellular fate of *Francisella tularensis* in human macrophages. Ann NY Acad Sci R 101196annals1409001 1105:160–186

Clemens DL, Lee BY, Horwitz MA (2004) Virulent and avirulent strains of *Francisella tularensis* prevent acidification and maturation of their phagosomes and escape into the cytoplasm in human macrophages. Infect Immun 72:3204–3217. doi:10.1128/IAI.72.6.3204-3217.2004

Clemens DL, Lee BY, Horwitz MA (2005) *Francisella tularensis* enters macrophages via a novel process involving pseudopod loops. Infect Immun 73:5892–5902. doi:10.1128/IAI.73.9.5892-5902.2005

Cole LE, Shirey KA, Barry E et al (2007) Toll-like receptor 2-mediated signaling requirements for *Francisella tularensis* live vaccine strain infection of murine macrophages. Infect Immun 75:4127–4137

Cole LE, Santiago A, Barry E et al (2008) Macrophage proinflammatory response to *Francisella tularensis* live vaccine strain requires coordination of multiple signaling pathways. J Immunol 180:6885–6891

Cole LE, Laird MHW, Seekatz A et al (2010) Phagosomal retention of *Francisella tularensis* results in TIRAP/Mal-independent TLR2 signaling. J Leukoc Biol 87:275–281. doi:10.1189/jlb.0909619

Crane DD, Bauler TJ, Wehrly TD, Bosio CM (2014) Mitochondrial ROS potentiates indirect activation of the AIM2 inflammasome. Front Microbiol 5:438. doi:10.3389/fmicb.2014.00438

Dai S, Rajaram MVS, Curry HM et al (2013) Fine tuning inflammation at the front door: macrophage complement receptor. PLoS Pathog 9:e1003114. doi:10.1371/journal.ppat.1003114

De Pascalis R, Taylor BC, Elkins KL (2008) Diverse myeloid and lymphoid cell subpopulations produce gamma interferon during early innate immune responses to *Francisella tularensis* live vaccine strain. Infect Immun 76:4311–4321

Degrandi D, Kravets E, Konermann C et al (2013) Murine guanylate binding protein 2 (mGBP2) controls *Toxoplasma gondii* replication. Proc Natl Acad Sci USA 110:294–299. doi:10.1073/pnas.1205635110

del Barrio L, Sahoo M, Lantier L et al (2015) Production of anti-LPS IgM by B1a B cells depends on IL-1beta and is protective against lung infection with *Francisella tularensis* LVS. PLoS Pathog 11:e1004706. doi:10.1371/journal.ppat.1004706

Dietrich N, Lienenklaus S, Weiss S, Gekara NO (2010) Murine toll-like receptor 2 activation induces type I interferon responses from endolysosomal compartments. PLoS ONE 5:e10250. doi:10.1371/journal.pone.0010250

Dotson RJ, Rabadi SM, Westcott EL et al (2013) Repression of inflammasome by *Francisella tularensis* during early stages of infection. J Biol Chem 288:23844–23857. doi:10.1074/jbc. M113.490086

Doyle CR, Pan J-A, Mena P et al (2014) TolC-dependent modulation of host cell death by the *Francisella tularensis* live vaccine strain. Infect Immun 82:2068–2078. doi:10.1128/IAI. 00044-14

Elkins KL, Rhinehart-Jones TR, Culkin SJ et al (1996) Minimal requirements for murine resistance to infection with *Francisella tularensis* LVS. Infect Immun 64:3288–3293

Fang FC (2004) Antimicrobial reactive oxygen and nitrogen species: concepts and controversies. Nat Rev Microbiol 2:820–832. doi:10.1038/nrmicro1004

Fernandes-Alnemri T, Yu JW, Datta P et al (2009) AIM2 activates the inflammasome and cell death in response to cytoplasmic DNA. Nature 458:509–513

Fernandes-Alnemri T, Yu JW, Juliana C et al (2010) The AIM2 inflammasome is critical for innate immunity to *Francisella tularensis*. Nat Immunol 11:385–393. doi:10.1038/ni.1859

Finethy R, Jorgensen I, Haldar AK et al (2015) Guanylate binding proteins enable rapid activation of canonical and noncanonical inflammasomes in chlamydia-infected macrophages. Infect Immun 83:4740–4749. doi:10.1128/IAI.00856-15

Furevik A, Pettersen EF, Colquhoun D, Wergeland HI (2011) The intracellular lifestyle of *Francisella noatunensis* in Atlantic cod (*Gadus morhua* L.) leucocytes. Fish Shellfish Immunol 30:488–494. doi:10.1016/j.fsi.2010.11.019

Gavrilin MA, Bouakl IJ, Knatz NL et al (2006) Internalization and phagosome escape required for Francisella to induce human monocyte IL-1beta processing and release. Proc Natl Acad Sci USA 103:141–146

Gavrilin MA, Mitra S, Seshadri S et al (2009) Pyrin critical to macrophage IL-1beta response to Francisella challenge. J Immunol Baltim Md 1950 182:7982–7989. doi:10.4049/jimmunol. 0803073

Ghonime MG, Mitra S, Eldomany RA et al (2015) Inflammasome priming is similar for Francisella species that differentially induce inflammasome activation. PLoS ONE 10: e0127278. doi:10.1371/journal.pone.0127278

Golovliov I, Baranov V, Krocova Z et al (2003) An attenuated strain of the facultative intracellular bacterium *Francisella tularensis* can escape the phagosome of monocytic cells. Infect Immun 71:5940–5950

Gunn J, Ernst R (2007) The structure and function of Francisella LPS

Gurcan S (2014) Epidemiology of tularemia. Balk Med J 31:3–10. doi:10.5152/balkanmedj.2014. 13117

Hagar JA, Powell DA, Aachoui Y et al (2013) Cytoplasmic LPS activates caspase-11: implications in TLR4-independent endotoxic shock. Science 341:1250–1253. doi:10.1126/science.1240988

Hajjar AM, Harvey MD, Shaffer SA et al (2006) Lack of in vitro and in vivo recognition of *Francisella tularensis* subspecies lipopolysaccharide by toll-like receptors. Infect Immun 74:6730–6738

Hall JD, Woolard MD, Gunn BM et al (2008) Infected-host-cell repertoire and cellular response in the lung following inhalation of *Francisella tularensis* Schu S4, LVS, or U112. Infect Immun 76:5843–5852

Henry T, Monack DM (2007) Activation of the inflammasome upon *Francisella tularensis* infection: interplay of innate immune pathways and virulence factors. Cell Microbiol 9: 2543–2551

Henry T, Brotcke A, Weiss DS et al (2007) Type I interferon signaling is required for activation of the inflammasome during Francisella infection. J Exp Med 204:987–994

Henry T, Kirimanjeswara GS, Ruby T et al (2010) Type I IFN signaling constrains IL-17A/F secretion by gammadelta T cells during bacterial infections. J Immunol 184:3755–3767. doi:10.4049/jimmunol.0902065

Hornung V, Ablasser A, Charrel-Dennis M et al (2009) AIM2 recognizes cytosolic dsDNA and forms a caspase-1-activating inflammasome with ASC. Nature 458:514–518

Huang MT-H, Mortensen BL, Taxman DJ et al (2010) Deletion of ripA alleviates suppression of the inflammasome and MAPK by *Francisella tularensis*. J Immunol Baltim Md 1950 (185):5476–5485. doi:10.4049/jimmunol.1002154

Jayakar HR, Parvathareddy J, Fitzpatrick EA et al (2011) A galU mutant of *Francisella tularensis* is attenuated for virulence in a murine pulmonary model of tularemia. BMC Microbiol 11: 1–16. doi:10.1186/1471-2180-11-179

Jin L, Hill KK, Filak H, et al (2011) MPYS is required for IFN response factor 3 activation and type I IFN production in the response of cultured phagocytes to bacterial second messengers cyclic-di-AMP and cyclic-di-GMP. J Immunol Baltim Md 1950 187:2595–2601. doi:10.4049/jimmunol.1100088

Jin T, Perry A, Jiang J et al (2012) Structures of the HIN domain: DNA complexes reveal ligand binding and activation mechanisms of the AIM2 inflammasome and IFI16 receptor. Immunity 36:561–571. doi:10.1016/j.immuni.2012.02.014

Jones J (2010) Molecular mechanisms of the innate immune response to *Francisella tularensis*. Stanford University

Jones CL, Weiss DS (2011) TLR2 signaling contributes to rapid inflammasome activation during *F. novicida* infection. PLoS ONE 6:e20609. doi:10.1371/journal.pone.0020609

Jones JW, Kayagaki N, Broz P et al (2010) Absent in melanoma 2 is required for innate immune recognition of *Francisella tularensis*. Proc Natl Acad Sci USA 107:9771–9776. doi:10.1073/pnas.1003738107

Jones CL, Sampson TR, Nakaya HI et al (2012) Repression of bacterial lipoprotein production by *F. novicida* facilitates evasion of innate immune recognition. Cell. doi:10.1111/j.1462-5822.2012.01816.x

Kanistanon D, Hajjar AM, Pelletier MR et al (2008) A Francisella mutant in lipid A carbohydrate modification elicits protective immunity. PLoS Pathog 4:e24. doi:10.1371/journal.ppat.0040024

Katz J, Zhang P, Martin M et al (2006) Toll-like receptor 2 is required for inflammatory responses to *Francisella tularensis* LVS. Infect Immun 74:2809–2816

Kayagaki N, Wong MT, Stowe IB et al (2013) Noncanonical inflammasome activation by intracellular LPS independent of TLR4. Science. doi:10.1126/science.1240248

Khader SA, Gopal R (2010) IL-17 in protective immunity to intracellular pathogens. Virulence 1:423–427. doi:10.4161/viru.1.5.12862

Kim BH, Shenoy AR, Kumar P et al (2011) A family of IFN-gamma-inducible 65-kD GTPases protects against bacterial infection. Science 332:717–721. doi:10.1126/science.1201711

Kingry LC, Petersen JM (2014) Comparative review of *Francisella tularensis* and *Francisella novicida*. Front Cell Infect Microbiol 4:35. doi:10.3389/fcimb.2014.00035

Lai XH, Sjostedt A (2003) Delineation of the molecular mechanisms of *Francisella tularensis*-induced apoptosis in murine macrophages. Infect Immun 71:4642–4646

Lai X-H, Shirley RL, Crosa L et al (2010) Mutations of *Francisella novicida* that alter the mechanism of its phagocytosis by murine macrophages. PLoS ONE 5:e11857. doi:10.1371/journal.pone.0011857

Larsson P, Elfsmark D, Svensson K et al (2009) Molecular evolutionary consequences of niche restriction in *Francisella tularensis*, a facultative intracellular pathogen. PLoS Pathog 5: e1000472. doi:10.1371/journal.ppat.1000472

Lemichez E, Aktories K (2013) Hijacking of Rho GTPases during bacterial infection. Exp Cell Res 319:2329–2336. doi:10.1016/j.yexcr.2013.04.021

Li H, Nookala S, Bina XR et al (2006) Innate immune response to *Francisella tularensis* is mediated by TLR2 and caspase-1 activation. J Leukoc Biol 80:766–773

Lin Y, Ritchea S, Logar A et al (2009) Interleukin-17 is required for T helper 1 cell immunity and host resistance to the intracellular pathogen *Francisella tularensis*. Immunity 31:799–810

Lindemann SR, Peng K, Long ME et al (2011) *Francisella tularensis* Schu S4 O-antigen and capsule biosynthesis gene mutants induce early cell death in human macrophages. Infect Immun 79:581–594. doi:10.1128/IAI.00863-10

Lloubes R, Cascales E, Walburger A et al (2001) The Tol-Pal proteins of the *Escherichia coli* cell envelope: an energized system required for outer membrane integrity? Res Microbiol 152:523–529

Luksch H, Romanowski MJ, Chara O et al (2012) Naturally occurring genetic variants of human caspase-1 differ considerably in structure and the ability to activate interleukin-1beta. Hum Mutat. doi:10.1002/humu.22169

Mahawar M, Atianand MK, Dotson RJ et al (2012) Identification of a novel *Francisella tularensis* factor required for intramacrophage survival and subversion of innate immune response. J Biol Chem 287:25216–25229. doi:10.1074/jbc.M112.367672

Man SM, Hopkins LJ, Nugent E et al (2014) Inflammasome activation causes dual recruitment of NLRC4 and NLRP3 to the same macromolecular complex. Proc Natl Acad Sci USA. doi:10.1073/pnas.1402911111

Man SM, Karki R, Malireddi RKS et al (2015) The transcription factor IRF1 and guanylate-binding proteins target activation of the AIM2 inflammasome by Francisella infection. Nat Immunol 16:467–475. doi:10.1038/ni.3118

Mares CA, Ojeda SS, Morris EG et al (2008) Initial delay in the immune response to *Francisella tularensis* is followed by hypercytokinemia characteristic of severe sepsis and correlating with upregulation and release of damage-associated molecular patterns. Infect Immun 76:3001–3010. doi:10.1128/IAI.00215-08

Mariathasan S, Weiss DS, Dixit VM, Monack DM (2005) Innate immunity against *Francisella tularensis* is dependent on the ASC/caspase-1 axis. J Exp Med 202:1043–1049

Mariathasan S, Weiss DS, Newton K et al (2006) Cryopyrin activates the inflammasome in response to toxins and ATP. Nature 440:228–232

McLendon MK, Apicella MA, Allen LA (2006) *Francisella tularensis*: taxonomy, genetics, and immunopathogenesis of a potential agent of biowarfare. Annu Rev Microbiol 60:167–185

Meunier E, Broz P (2016) Interferon-inducible GTPases in cell autonomous and innate immunity. Cell Microbiol 18:168–180. doi:10.1111/cmi.12546

Meunier E, Dick MS, Dreier RF et al (2014) Caspase-11 activation requires lysis of pathogen-containing vacuoles by IFN-induced GTPases. Nature 509:366–370. doi:10.1038/nature13157

Meunier E, Wallet P, Dreier RF et al (2015) Guanylate-binding proteins promote activation of the AIM2 inflammasome during infection with *Francisella novicida*. Nat Immunol 16:476–484. doi:10.1038/ni.3119

Miao EA, Leaf IA, Treuting PM et al (2010) Caspase-1-induced pyroptosis is an innate immune effector mechanism against intracellular bacteria. Nat Immunol 11:1136–1142. doi:10.1038/ni.1960

Miller CN, Steele SP, Brunton JC et al (2014) Extragenic suppressor mutations in DeltaripA disrupt stability and function of LpxA. BMC Microbiol 14:336. doi:10.1186/s12866-014-0336-x

Nano FE, Zhang N, Cowley SC et al (2004) A *Francisella tularensis* pathogenicity island required for intramacrophage growth. J Bacteriol 186:6430–6436

Okan NA, Kasper DL (2013) The atypical lipopolysaccharide of Francisella. Carbohydr Res 378:79–83. doi:10.1016/j.carres.2013.06.015

Pelegrin P, Barroso-Gutierrez C, Surprenant A (2008) P2X7 receptor differentially couples to distinct release pathways for IL-1beta in mouse macrophage. J Immunol 180:7147–7157

Peng K, Broz P, Jones J et al (2011) Elevated AIM2-mediated pyroptosis triggered by hypercytotoxic Francisella mutant strains is attributed to increased intracellular bacteriolysis. Cell Microbiol 13:1586–1600. doi:10.1111/j.1462-5822.2011.01643.x

Pierini R, Juruj C, Perret M et al (2012) AIM2/ASC triggers caspase-8-dependent apoptosis in Francisella-infected caspase-1-deficient macrophages. Cell Death Differ 19:1709–1721. doi:10.1038/cdd.2012.51

Pierini R, Perret M, Djebali S et al (2013) ASC controls IFN-gamma levels in an IL-18-Dependent Manner in Caspase-1-Deficient mice infected with *Francisella novicida*. J Immunol 191: 3847–3857. doi:10.4049/jimmunol.1203326
Pilla DM, Hagar JA, Haldar AK et al (2014) Guanylate binding proteins promote caspase-11-dependent pyroptosis in response to cytoplasmic LPS. Proc Natl Acad Sci USA 111:6046–6051. doi:10.1073/pnas.1321700111
Platz GJ, Bublitz DC, Mena P et al (2010) A tolC mutant of *Francisella tularensis* is hypercytotoxic compared to the wild type and elicits increased proinflammatory responses from host cells. Infect Immun 78:1022–1031. doi:10.1128/IAI.00992-09
Py BF, Kim M-S, Vakifahmetoglu-Norberg H, Yuan J (2013) Deubiquitination of NLRP3 by BRCC3 critically regulates inflammasome activity. Mol Cell 49:331–338. doi:10.1016/j.molcel.2012.11.009
Rajaram MVS, Butchar JP, Parsa KVL et al (2009) Akt and SHIP modulate Francisella escape from the phagosome and induction of the Fas-mediated death pathway. PLoS ONE 4:e7919. doi:10.1371/journal.pone.0007919
Rathinam VA, Jiang Z, Waggoner SN et al (2010) The AIM2 inflammasome is essential for host defense against cytosolic bacteria and DNA viruses. Nat Immunol 11:395–402. doi:10.1038/ni.1864
Roberts TL, Idris A, Dunn JA et al (2009) HIN-200 proteins regulate caspase activation in response to foreign cytoplasmic DNA. Science 323:1057–1060
Robertson GT, Case EDR, Dobbs N et al (2014) FTT0831c/FTL_0325 contributes to *Francisella tularensis* cell division, maintenance of cell shape, and structural integrity. Infect Immun 82:2935–2948. doi:10.1128/IAI.00102-14
Sagulenko V, Thygesen SJ, Sester DP et al (2013) AIM2 and NLRP3 inflammasomes activate both apoptotic and pyroptotic death pathways via ASC. Cell Death Differ (Epub May 2013) 3:1–12. doi:10.1038/cdd.2013.37
Sampson TR, Napier BA, Schroeder MR et al (2014) A CRISPR-Cas system enhances envelope integrity mediating antibiotic resistance and inflammasome evasion. Proc Natl Acad Sci USA 111:11163–11168. doi:10.1073/pnas.1323025111
Sandstrom G, Sjostedt A, Johansson T et al (1992) Immunogenicity and toxicity of lipopolysaccharide from *Francisella tularensis* LVS. FEMS Microbiol Immunol 5:201–210
Sandstrom A, Peigne C-M, Leger A et al (2014) The intracellular B30.2 domain of butyrophilin 3A1 binds phosphoantigens to mediate activation of human Vgamma9Vdelta2 T cells. Immunity 40:490–500. doi:10.1016/j.immuni.2014.03.003
Sha W, Mitoma H, Hanabuchi S et al (2014) Human NLRP3 inflammasome senses multiple types of bacterial RNAs. Proc Natl Acad Sci USA 111:16059–16064. doi:10.1073/pnas.1412487111
Shenoy AR, Wellington DA, Kumar P et al (2012) GBP5 promotes NLRP3 inflammasome assembly and immunity in mammals. Science 336:481–485. doi:10.1126/science.1217141
Shi CS, Shenderov K, Huang NN et al (2012) Activation of autophagy by inflammatory signals limits IL-1beta production by targeting ubiquitinated inflammasomes for destruction. Nat Immunol 13:255–263. doi:10.1038/ni.2215
Shi J, Zhao Y, Wang Y et al (2014) Inflammatory caspases are innate immune receptors for intracellular LPS. Nature 514:187–192. doi:10.1038/nature13683
Sjostedt A (2007) Tularemia: history, epidemiology, pathogen physiology, and clinical manifestations
Skyberg JA, Rollins MF, Samuel JW et al (2013) Interleukin-17 protects against the *Francisella tularensis* live vaccine strain but not against a virulent *F. tularensis* type A strain. Infect Immun 81:3099–3105. doi:10.1128/IAI.00203-13
Stanley SA, Johndrow JE, Manzanillo P, Cox JS (2007) The Type I IFN response to infection with *Mycobacterium tuberculosis* requires ESX-1-mediated secretion and contributes to pathogenesis. J Immunol 178:3143–3152
Storek KM, Gertsvolf NA, Ohlson MB, Monack DM (2015) cGAS and Ifi204 cooperate to produce Type I IFNs in response to Francisella infection. J Immunol Baltim Md 1950 194:3236–3245. doi:10.4049/jimmunol.1402764

Telepnev M, Golovliov I, Grundstrom T et al (2003) *Francisella tularensis* inhibits Toll-like receptor-mediated activation of intracellular signalling and secretion of TNF-alpha and IL-1 from murine macrophages. Cell Microbiol 5:41–51

Thakran S, Li H, Lavine CL et al (2008) Identification of *Francisella tularensis* lipoproteins that stimulate the toll-like receptor (TLR) 2/TLR1 heterodimer. J Biol Chem 283:3751–3760. doi:10.1074/jbc.M706854200

Ulland TK, Buchan BW, Ketterer MR et al (2010) Cutting edge: mutation of *Francisella tularensis* mviN leads to increased macrophage absent in melanoma 2 inflammasome activation and a loss of virulence. J Immunol 185:2670–2674. doi:10.4049/jimmunol.1001610

Vojtech LN, Scharping N, Woodson JC, Hansen JD (2012) Roles of inflammatory caspases during processing of zebrafish interleukin-1beta in Francisella noatunensis infection. Infect Immun 80:2878–2885. doi:10.1128/IAI.00543-12

Wang X, Ribeiro AA, Guan Z et al (2007) Attenuated virulence of a Francisella mutant lacking the lipid A 4′-phosphatase. Proc Natl Acad Sci USA 104:4136–4141. doi:10.1073/pnas.0611606104

Weiss DS, Brotcke A, Henry T et al (2007) In vivo negative selection screen identifies genes required for Francisella virulence. Proc Natl Acad Sci USA 104:6037–6042

Welch MD, Way M (2013) Arp2/3-mediated actin-based motility: a tail of pathogen abuse. Cell Host Microbe 14:242–255. doi:10.1016/j.chom.2013.08.011

West AP, Brodsky IE, Rahner C et al (2011) TLR signalling augments macrophage bactericidal activity through mitochondrial ROS. Nature 472:476–480. doi:10.1038/nature09973

Wickstrum JR, Bokhari SM, Fischer JL et al (2009) *Francisella tularensis* induces extensive caspase-3 activation and apoptotic cell death in the tissues of infected mice. Infect Immun 77:4827–4836. doi:10.1128/IAI.00246-09

Wu J, Fernandes-Alnemri T, Alnemri ES (2010) Involvement of the AIM2, NLRC4, and NLRP3 inflammasomes in caspase-1 activation by Listeria monocytogenes. J Clin Immunol 30: 693–702. doi:10.1007/s10875-010-9425-2

Xu H, Yang J, Gao W et al (2014) Innate immune sensing of bacterial modifications of Rho GTPases by the Pyrin inflammasome. Nature 513:237–241. doi:10.1038/nature13449

Yamamoto M, Okuyama M, Ma JS et al (2012) A cluster of interferon-gamma-inducible p65 GTPases plays a critical role in host defense against *Toxoplasma gondii*. Immunity 37: 302–313. doi:10.1016/j.immuni.2012.06.009

Yang H, Ji X, Zhao G et al (2012) Structural insight into HIV-1 capsid recognition by rhesus TRIM5alpha. Proc Natl Acad Sci USA 109:18372–18377. doi:10.1073/pnas.1210903109

Inflammasome Activation Can Mediate Tissue-Specific Pathogenesis or Protection in *Staphylococcus aureus* Infection

Jason H. Melehani and Joseph A. Duncan

Abstract *Staphylococcus aureus* is a Gram-positive coccus that interacts with human hosts on a spectrum from quiet commensal to deadly pathogen. *S. aureus* is capable of infecting nearly every tissue in the body resulting in cellulitis, pneumonia, osteomyelitis, endocarditis, brain abscesses, bacteremia, and more. *S. aureus* has a wide range of factors that promote infection, and each site of infection triggers a different response in the human host. In particular, the different patterns of inflammasome activation mediate tissue-specific pathogenesis or protection in *S. aureus* infection. Although still a nascent field, understanding the unique host–pathogen interactions in each infection and the role of inflammasomes in mediating pathogenesis may lead to novel strategies for treating *S. aureus* infections. Reviews addressing *S. aureus* virulence and pathogenesis (Thammavongsa et al. 2015), as well as epidemiology and pathophysiology (Tong et al. 2015), have recently been published. This review will focus on *S. aureus* factors that activate inflammasomes and their impact on innate immune signaling and bacterial survival.

J.H. Melehani · J.A. Duncan
Department of Pharmacology, School of Medicine,
University of North Carolina at Chapel Hill, Chapel Hill, NC, USA
e-mail: jason_melehani@med.unc.edu

J.A. Duncan (✉)
Department of Medicine, Division of Infectious Diseases,
School of Medicine, University of North Carolina at Chapel Hill,
Chapel Hill, NC, USA
e-mail: jaduncan@med.unc.edu

J.A. Duncan
Lineberger Comprehensive Cancer Center, School of Medicine,
University of North Carolina at Chapel Hill, Chapel Hill, NC, USA

S. Backert (ed.), *Inflammasome Signaling and Bacterial Infections*,
Current Topics in Microbiology and Immunology 397,
DOI 10.1007/978-3-319-41171-2_13

Contents

1 Introduction

Staphylococcus aureus has numerous mechanisms to evade and subvert the immune system allowing it to produce infection broadly in immune-competent hosts. Many virulence factors, such as pore-forming toxins (PFTs) and phenol-soluble modulins (PSMs), activate inflammasomes and will be discussed here (Fig. 1).

Most *S. aureus* PFTs (i.e., alpha-hemolysin [Hla], gamma-hemolysin [HlgAB, HlgCB], leukocidin AB [LukAB, also known as LukGH], leukocidin ED [LukED], and Panton-Valentine leukocidin [PVL, also known as LukFS]) are two-component toxins that initiate attack when one component binds a toxin-specific cell surface receptor and inserts itself into the host cell membrane. This is followed by recruitment of the second component then additional pairs to complete a hexameric beta-barrel pore (Badarau et al. 2015; Yamashita et al. 2011; Pedelacq et al. 1999;

Fig. 1 *S. aureus* PAMPs and virulence factors activate inflammasome signaling. A wide variety ▶ of *S. aureus* bacterial molecular patterns and virulence factors, including pore-forming toxins (PFTs), peptidoglycan (PGN), phenol-soluble modulins (PSMs and Hld), activate inflammasome signaling. The NLRP3 inflammasome is depicted as the primary example of inflammasome signaling in this figure. The PFTs—Hla, LukAB, LukED, PVL, HlgAB, and HlgCB—bind cellular receptors, as seen in the *center of the figure*, to promote pore formation. Pore formation causes potassium efflux and leads to NLRP3 inflammasome activation. When activated, NLRP3 binds ASC through PYD–PYD interactions and ASC binds pro-Caspase 1 through CARD–CARD interactions. Formation of the NLRP3 inflammasome complex causes activation of Caspase 1 that can process pro-IL-1β and pro-IL-18 for activation and secretion. Beta-hemolysin (Hlb), a sphingomyelinase, and the PSMs also activate inflammasome-like signaling. Purified Hlb and PSMs can trigger IL-1β and IL-18 secretion but direct activation of the NLRP3 inflammasome by these virulence factors has not yet been demonstrated

Guillet et al. 2004). The notable exception to this pattern is Hla which forms a pore by binding to its receptor then forming a heptameric beta-barrel pore made up of only a single component (Sugawara et al. 2015; Gouaux et al. 1997).

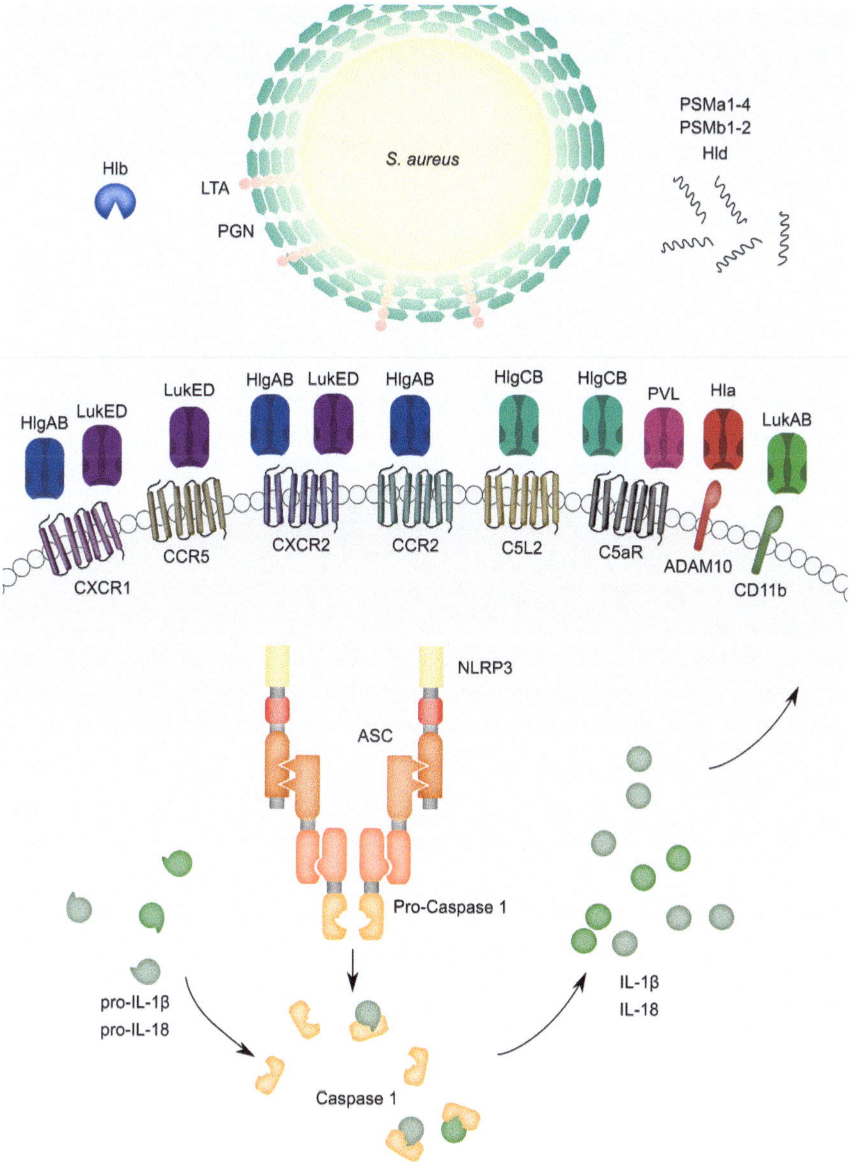

Each of these toxins is cytotoxic to a range of cell types based on expression of the toxins' cognate receptors: Hla and A Disintegrin And Metalloprotease 10 (ADAM10) (Wilke and Bubeck Wardenburg 2010); HlgAB and CXCR1, CXCR2, and CCR2; HlgCB and C5aR and C5L2 (Spaan et al. 2014); LukAB and CD11b (DuMont et al. 2013a), LukED and CCR5 (Alonzo et al. 2013), CXCR1 and CXCR2 (Reyes-Robles et al. 2013); and PVL and C5aR (Spaan et al. 2013). Many of these toxin–receptor pairs lend a level of species specificity as well. In primary human monocytes, the rank-order potency of the PFTs is LukAB and PVL (tied), Hla and HlgCB (tied), HlgAB, then LukED (Melehani et al. 2015). The *lukAB* operon promoter is preferentially activated when USA300 strain *S. aureus* is exposed to human polymorphonuclear leukocytes (PMNs) (DuMont et al. 2013b). Despite these hints at a prominent role for LukAB in *S. aureus* pathogenesis, the role of LukAB in infections has largely been overlooked because of its lack of effect in mice. LukAB binds to human CD11b with 1000-fold higher affinity than murine CD11b, rendering mouse leukocytes resistant to LukAB (DuMont et al. 2013a). Murine neutrophils are also completely resistant to HlgCB and are also partially resistant to HlgAB because CCR2 is the only compatible HlgAB murine receptor (Spaan et al. 2014). Both rabbit and human neutrophils are susceptible to PVL while murine and java monkey neutrophils are resistant (Loffler et al. 2010). As predicted by this susceptibility pattern, HEK cells exogenously expressing rabbit and human C5aR both bind the LukS component of PVL while cells expressing mouse or macaque C5aR do not bind LukS (Spaan et al. 2013). Thus, *S. aureus* PFTs have the potential to play varying roles depending on the site of *S. aureus* infection and species of the host.

Like alpha-hemolysin (Hla), beta-hemolysin (Hlb) and delta-hemolysin (Hld) are named for their ability to lyze red blood cells in vitro. However, they are functionally distinct toxins that do not form a beta-barrel pore like Hla. Hlb is a hemolytic and cytotoxic sphingomyelinase (Walev et al. 1996; Huseby et al. 2007). Delta-hemolysin is a 26-amino acid membrane-damaging peptide that is classified as a phenol-soluble modulin (Fitton et al. 1980).

Phenol-soluble modulins [PSMs—reviewed (Peschel and Otto 2013)] are cytotoxic and pro-inflammatory peptides produced by *S. aureus* (Wang et al. 2007). *S. aureus* produces four PSMα peptides (PSMα1-PSMα4), two PSMβ peptides (PSMβ1 and PSMβ2), and delta-hemolysin (Hld). All seven PSMs are sensed by formyl peptide receptor 2 (FPR2) in human neutrophils. Activation of FPR2 by PSMs promotes chemotaxis of neutrophils into a site of infection in response to *S. aureus* (Kretschmer et al. 2010). PSMs also have demonstrated roles in cytolysis, biofilm development, and immunomodulation and are thought to be partially responsible for the increased virulence of community-acquired *S. aureus* over hospital-acquired strains [reviewed (Otto 2010)].

Lipoteichoic acid (LTA) is a component of the bacterial wall in Gram-positive bacteria that is essential for bacterial growth, cell division, and survival (Grundling and Schneewind 2007). The innate immune system has evolved multiple strategies for detecting and responding to LTA, including Toll-like receptor 2 (TLR2), scavenger receptor A (Dunne et al. 1994; Greenberg et al. 1996), CD36 (Hoebe et al. 2005), CD14, and soluble CD14 (Hermann et al. 2002; Han et al. 2003), surfactant protein D (van de Wetering et al. 2001), paired Ig-like receptor B (Nakayama et al. 2012), and L-ficolin (Lynch et al. 2004; reviewed Weidenmaier and Peschel 2008). Having multiple methods for sensing LTA is probably important to fine-tune the immune response to infections. For example, LTA binding to TLR2 upregulates cytokine production through NF-κB signaling, while LTA binding to paired Ig-like receptor B suppresses IL-6 and IL-1β secretion (Nakayama et al. 2012). LTA is not unique to *S. aureus* and is also found in other Firmicutes, such as *Bacillus subtilis* and *Listeria monocytogenes* (Percy and Grundling 2014). Because these bacteria last shared a common ancestor approximately 1.5 billion years ago (Hedges et al. 2006; Kumar and Hedges 2011), the utilization of LTA likely evolved well before the establishment of multicellular life. This suggests that LTA sensing in humans is the result of selective pressures favoring LTA detection by the host rather than evolution of LTA to manipulate immune signaling.

Additionally, other integral bacterial components such as bacterial RNA (Sha et al. 2014; Eigenbrod et al. 2012), lipoproteins (Munoz-Planillo et al. 2009), and peptidoglycan (Muller et al. 2015; Shimada et al. 2010) are important molecular patterns detected by the immune system through inflammasomes. However, due to space limitations, these will not be discussed here.

2 Inflammasomes that are Activated by *S. aureus*

Multiple inflammasomes are activated during *S. aureus* infections. The first demonstration of inflammasome activation by *S. aureus* was the observation that NLRP3-deficient macrophages were dramatically impaired in secretion of IL-1β after exposure to live *S. aureus* (Mariathasan et al. 2006). Activation of the NLRP3 inflammasome involves two steps: (1) a priming step that involves NF-κB-mediated upregulation of pro-IL-1β transcription and post-translational modifications of inflammasome components and (2) NLRP3 inflammasome oligomerization leading to Caspase 1 activation and secretion of mature IL-1β and IL-18 and pyroptosis (Fig. 2).

In the case of an *S. aureus* infection, TLR2 detects LTA to provide an initial priming signal (Schwandner et al. 1999; Yoshimura et al. 1999). In experimental systems, cells are often primed by the addition of purified LTA (Melehani et al. 2015) or heat-killed *S. aureus* (HKSA) (Craven et al. 2009) or other TLR-ligand such as LPS.

Numerous studies have since delineated the contribution of various *S. aureus* virulence factors to NLRP3 inflammasome activation. All six PFTs from *S. aureus* activate the NLRP3 inflammasome in THP1 cells and primary human monocytes, leading to canonical cytokine processing and cell death (Melehani et al. 2015; Holzinger et al. 2012; Craven et al. 2009; Munoz-Planillo et al. 2009).

Caspase 1 is critical for processing IL-1β and IL-18 for secretion and is required for a type of necrotic cell death termed pyroptosis. However, Hla-, PVL-, or

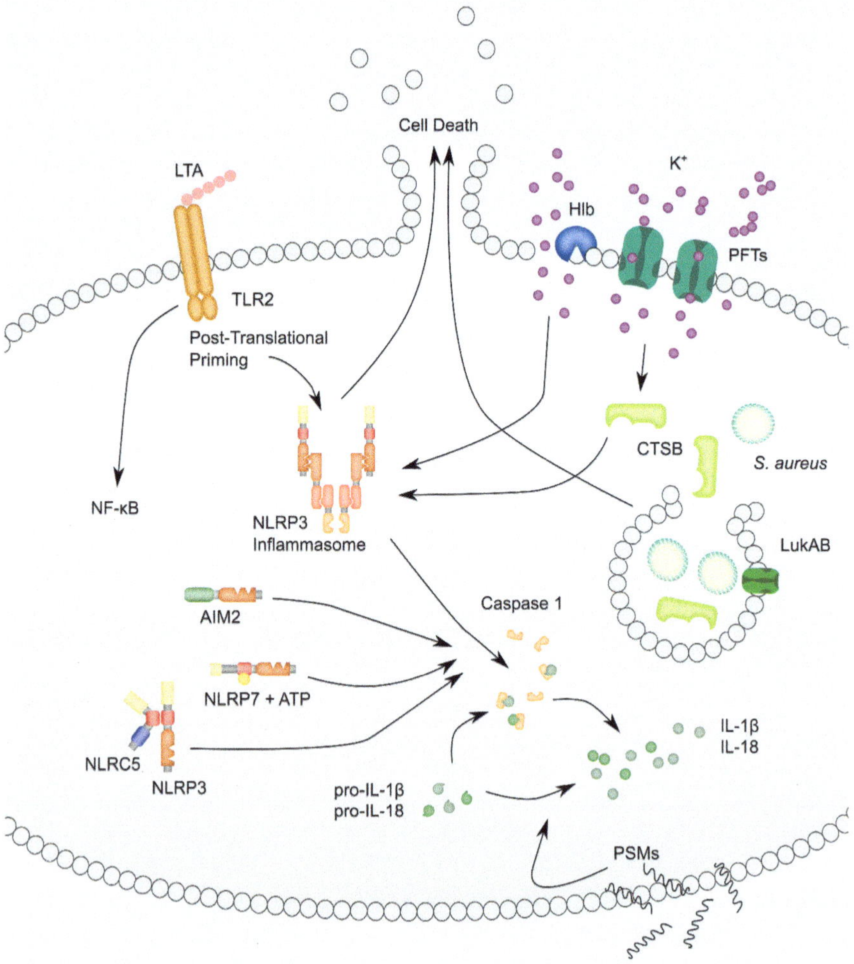

LukAB-triggered cell death is not blocked in human cells pretreated with Caspase 1 inhibitors (Craven et al. 2009; Melehani et al. 2015; Holzinger et al. 2012). Additionally, *Casp1/Casp11* knockout mouse macrophages are killed by treatment with Hla just as Caspase-1/Caspase 11 sufficient macrophages (Craven et al. 2009; Melehani et al. 2015; Holzinger et al. 2012). However, loss of NLRP3 protects cells against PFT-induced cell death. Therefore, *S. aureus* PFTs seem to trigger a host cell death pathway with an incompletely described molecular mechanism through NLRP3 inflammasome activation that abrogates a requirement for the classical Caspase 1-dependent pathway.

S. aureus activation of the NLRP3 inflammasome with PFTs, Hla (Craven et al. 2009), LukAB (Melehani et al. 2015), and PVL (Holzinger et al. 2012), is blocked

◀ **Fig. 2** NLRP3, NLRC5, NLRP7, AIM2, and other inflammasome-like signaling functions are activated in response to *S. aureus*. Activation of inflammasome signaling typically proceeds in two steps. First, TLRs engage PAMPS to prime inflammasome signaling. *S. aureus* lipoteichoic acid (LTA), a major constituent of the cell wall of Gram-positive bacteria, binds TLR2. TLR signaling activates NF-κB-mediated transcription of pro-IL-1β and also triggers post-translational modifications of the inflammasome signaling pathway. Activation of the NLRP3 inflammasome occurs through pore-mediated and phagocytosis-mediated processes. PFTs and Hlb damage the plasma membrane and cause potassium efflux necessary for inflammasome activation. PVL-induced NLRP3 inflammasome signaling requires CTSB activation. PVL, Hla, and LukAB also promote Caspase 1-independent cell death. PFTs may also destabilize lysophagosomes during phagocytosis of *S. aureus*. In particular, LukAB has been shown to bind CD11b on the phagosomal membrane to promote *S. aureus* escape. This triggers NLRP3-dependent cytokine secretion and NLRP3-independent cell death. Lysophagosomal rupture is thought to lead to CTSB leakage into the cytoplasm, though this has not been shown directly with *S. aureus*. A variety of other inflammasomes have also been implicated in sensing *S. aureus*. NLRC5 binds to NLRP3 and enhances cytokine secretion. NLRP7 and its ATP-binding activity are required for Caspase 1 activation in response to *S. aureus*. AIM2 is activated in response to *S. aureus* central nervous system abscess formation, though it is not clear whether AIM2 is sensing *S. aureus* or a danger-associated molecular pattern resulting from *S. aureus* infection. PSMs also trigger secretion of IL-1β and IL-18, but through a Caspase 1-independent mechanism

by supraphysiologic concentrations of extracellular potassium suggesting a requirement for potassium efflux. PVL also causes activation of Cathepsin B (CTSB) and inhibition of CTSB protease activity by CA-074, a CTSB inhibitor, blocks PVL-induced cell death (Holzinger et al. 2012). CTSB is a lysosomal cysteine protease required for NLRP3 activation by disruption of lysophagosomes after ingestion of inflammation inducing particulate matter like silica (Hornung et al. 2008). CTSB is thought to spill into to the cytoplasm after lysophagosomal disruption, but the mechanism by which CTSB activates NLRP3 is unknown. It is also not known whether the requirement for CTSB in PVL-induced NLRP3 activation is toxin specific or common to all *S. aureus* PFTs. One possible explanation for how CTSB is liberated from lysophagosomes is that *S. aureus* PFTs can bind to their cognate receptor on phagosomal membranes. This binding promotes *S. aureus* escape from phagocytosis to enable survival (DuMont et al. 2013b). Interestingly, the route of exposure (i.e., extracellular versus within a phagocytic vacuole) also seems to influence signaling in response to LukAB (Melehani et al. 2015). When LukAB interacts with CD11b at the plasma membrane, NLRP3 is activated, driving IL-1β and IL-18 secretion and cell death. Alternatively, when LukAB expressed from phagocytosed *S. aureus* interact with CD11b on the phagocytic vacuole membrane, cell death becomes independent of NLRP3 activation. Phagosomal delivery of LukAB does not generate an alternative Caspase 1 activating pathway as NLRP3 is still required for IL-1β and IL-18 secretion. It is not clear whether this separate cell death mechanism engaged by phagosome-localized LukAB is activated in addition to NLRP3-dependent pathways or if the NLRP3-dependent pathways activated by extracellular toxin are not activated by toxin in this setting (Melehani et al. 2015).

S. aureus can also trigger necroptosis, cell death dependent on RIPK1, RIPK3, and MLKL that can be blocked by necrostatin-1 and necrosulfonamide (NSA). Pretreatment of THP1 cells with NSA prior to exposure to *S. aureus* or purified Hla decreased cell death. NSA shifted the potency of Hla and at higher doses of Hla cells still died suggesting that Hla is able to engage additional cell death pathways at these doses. NSA diminished *S. aureus*-induced IL-1β secretion suggesting this agent or the necroptotic pathway may tie in with inflammasome activation (Kitur et al. 2015).

While the cytotoxicity of PFTs has been demonstrated widely in primary human monocytes, macrophages, neutrophils, and dendritic cells, most studies of PFTs and inflammasome activation have been carried out in THP1 cells, mouse macrophages, or human peripheral blood monocytes. The prevailing assumption is that the cytotoxicity of these toxins depends on host NLRP3 across all susceptible human cell types, though this has not been demonstrated directly (Dumont et al. 2011; Holzinger et al. 2012; Ventura et al. 2010).

Few other mechanistic details between PFT exposure and NLRP3 inflammasome activation have been determined. Many other host proteins have been implicated along the NLRP3 inflammasome signaling cascade with other triggers, but have yet to be confirmed for *S. aureus*.

Like the pore-forming Hla, purified Hlb also triggers human monocytes to secrete IL-1β (Walev et al. 1996). Though these studies preceded the discovery of the NLRP3 inflammasome, Hlb-induced IL-1β secretion was suppressed by extracellular potassium, which strongly suggests a role for the NLRP3 inflammasome. Despite purified toxin being sufficient to induce IL-1β secretion, *S. aureus* Hlb-deficient mutants had no observable defect in inducing IL-1β and IL-18 secretion in mouse macrophages. *S. aureus* lacking the combination of Hla, Hlb, and Hlg caused very little IL-1β and IL-18 secretion when compared to *S. aureus* lacking any pair of these hemolysins (Munoz-Planillo et al. 2009). These studies suggest that all three hemolysins in *S. aureus* have redundant capacity to activate the NLRP3 inflammasome. It is also possible that hemolysin production is context dependent and different hemolysins (or combinations of hemolysins) may be required in infection of separate tissues, which has not been recapitulated in in vitro cell culture of immune cells to date. Hlb also limits production of IL-8, a potent neutrophil chemoattractant, by human endothelial cells (Tajima et al. 2009) and as such may play a dual role in establishing and promoting infection.

Other NLR proteins may interact with NLRP3 to enhance inflammasome signaling. In one instance, NLRC5 was shown to co-immunoprecipitate with NLRP3 and knockdown of NLRC5 in THP1 cells by shRNA led to dramatic reduction in IL-1β secretion in response to live *S. aureus* (Davis et al. 2011). The *S. aureus* factors that might contribute to activation of NLRC5 in this context have not been identified.

In addition to NLRP3, NLRP7, a sensor of acylated lipopeptides, leads to signaling in response to *S. aureus*. Knockdown of NLRP7 in THP1 cells resulted in decreased *S. aureus*-induced IL-1β secretion and increased intracellular bacteria following infection (Khare et al. 2012). NLRP7 function requires ATP binding and hydrolysis. Overexpression of wild-type myc-tagged NLRP7, but not myc-tagged ATPase-deficient NLRP7, by lentiviral transduction led to an increase in *S. aureus*-induced IL-1β secretion as compared to transduction of myc alone (Radian et al. 2015). The role of NLRP7 in cell death is unclear. Depletion of NLRP7 by siRNA did not impact *S. aureus*-induced cell death as measured by LDH release (Khare et al. 2012) but viral transduction of THP1 cells to overexpress NLRP7 enhanced *S. aureus*-induced cell death as measured by PI uptake (Radian et al. 2015). More work is needed on endogenous NLRP7 to better understand its role in *S. aureus*-induced cell death.

AIM2, a sensor responsible for inflammasome formation in response to cytoplasmic DNA, may also detect *S. aureus*. Mice lacking AIM2 demonstrated reduced IL-1β recovered from central nervous system abscesses resulting from intracranial injection of *S. aureus*. *Asc*$^{-/-}$ mice, but not *Nlrp3*$^{-/-}$ mice, exhibited the same defect in IL-1β production during intracranial abscess formation. Interestingly, mice lacking AIM2 or ASC also demonstrated dramatically decreased production of non-inflammasome dependent cytokines including the following: IL-6, CXCL1, and CXCL10 production in this model. This suggests either production of these cytokines is influenced by production of IL-1β or that there are non-inflammasome-dependent signaling pathways activated by AIM2 and ASC. Because no studies of *S. aureus*-exposed AIM2-deficient cultured immune cells, including CNS microglia, have been published, it is not clear whether the defect in IL-1β production observed in infected brain tissues of AIM2-deficient mice is the result of a failure to sense a component of *S. aureus* or to sense a host factor produced as a result of *S. aureus* infection (Hanamsagar et al. 2014).

Inflammasome-like or non-canonical inflammasome responses may also initiate an immune response against *S. aureus*. When purified, each of the seven *S. aureus* PSMs is sufficient to induce IL-18 secretion from human keratinocytes. *S. aureus* with all seven PSMs deleted induced lower levels of IL-18 and IL-1β secretion from keratinocytes. Unexpectedly, secretion of IL-18 or IL-1β from keratinocytes in response to live *S. aureus* could not be blocked by a panel of caspase inhibitors (Syed et al. 2015). These data suggest that keratinocytes may rely upon a non-Caspase protease for processing and secretion of IL-1β and IL-18 in response to PSMs.

While many studies of inflammasome activation by *S. aureus* have been conducted in in vitro cell culture using primary human monocytes as well as other cell types (THP1 cells, mouse macrophages, etc.), in natural infections, *S. aureus* and its secreted toxins interact with multiple cell types each with a unique capacity to sense and respond to the bacteria. This leads to the possibility that a variety of inflammasomes may be activated in different cell types over the course of an infection with variable effects on infection outcomes. To understand the role of inflammasome activation in the natural course of infection, we turn our review to studies of *S. aureus* infection in animal models, primarily in mice. In most cases, mouse studies were carried out using only the C57BL/6 strain in which inflammasome components have been genetically deleted. Mice as a model for *S. aureus* infections have serious limitations as mice are relatively naturally resistant to *S. aureus* and non-responsive to many virulence factors that act specifically on human cells. These studies would benefit from incorporating more genetic diversity.

3 Role of Inflammasome Activation in Infection Models

3.1 *The NLRP3 Inflammasome Responds to Hemolysins to Control* S. aureus *Dermal Infections*

S. aureus is the number one cause of skin infections. These infections range in severity from pimples and boils to abscesses and cellulitis. A mouse model for *S. aureus* skin infections has been developed (Miller et al. 2006) and adapted to study the role of inflammasomes in this setting (Miller et al. 2007; Cho et al. 2012; Soong et al. 2015).

Tissue biopsies were collected from mice injected subcutaneously with a variety of isogenic hemolysin knockout *S. aureus* strains. When *hla*, *hlb*, and *hlg* genes were deleted together, secretion of IL-1β and IL-18 was almost completely eliminated, whereas only minor reductions were observed with any single or double hemolysin knockout as compared to the parental *S. aureus* strain. Subcutaneous injection of mice with individual deletions of *Nlrp3*, *Asc*, or *Casp1* did not elicit IL-18 secretion, demonstrating a role for the NLRP3 inflammasome in this setting (Munoz-Planillo et al. 2009).

Mice deficient in IL-1β (*Il1b*$^{-/-}$) develop larger lesions with higher bacterial burdens and reduced neutrophil recruitment following subcutaneous injection of *S. aureus* as compared to parental strain mice (Miller et al. 2007). Adoptive transfer of *Il1b*$^{-/-}$ bone marrow into irradiated parental strain mice mirrored the defect seen in wholly *Il1b*$^{-/-}$ mice, suggesting that immune cells are a critical source of IL-1β in this model skin infection. In *Il1b*$^{-/-}$ mice, recombinant active IL-1β helped to control infections and promote bacterial clearance (Miller et al. 2007). In the same subcutaneous injection model, multispectral noninvasive imaging during infection was used to localize different types of phagocytes and fluorescent staining of IL-1β revealed a stronger spatial correlation between neutrophils and IL-1β than between macrophages and IL-1β. Adoptive transfer of IL-1β-expressing neutrophils into an *Il1b*$^{-/-}$ host was sufficient to restore the impaired neutrophil abscess formation, suggesting neutrophils are the critical source of immune cell-derived IL-1β in this model. *S. aureus* induced IL-1β production during subcutaneous injection in mice in a Hla-dependent manner that required intact genes encoding *Tlr2*, *Nod2*, *Fpr1*, *Asc*, *Casp1*, and *Nlrp3* (Cho et al. 2012). In humans, neutrophils are an important component of controlling cutaneous *S. aureus* infections (Borregaard 2010) as neutrophil functional defects, such as chronic granulomatous disease, result in susceptibility to invasive *S. aureus* cutaneous infections (Miller and Cho 2011).

The importance of neutrophils during subcutaneous infection of mice suggests that the mouse model is appropriately recapitulating this feature of the human infection.

Although these studies of *S. aureus* skin infection in mice have been done with direct inoculation of bacteria through the epidermal barrier, *S. aureus* can cause infection by invading through a human keratinocyte barrier. In an organotypic culture of human keratinocytes grown at an air–liquid interface on a dermal substitute matrix, *S. aureus* can be seen by microscopy to disrupt the apical keratinocyte layer. Hla triggers Caspase 1 and Calpain (Ca(2+)-dependent intracellular protease) activation leading to IL-1β secretion and cell death. If Caspase 1 or Calpain is inhibited during this process, invasion efficiency is partially decreased (Soong et al. 2012).

Patients with atopic dermatitis and psoriasis are commonly colonized with *S. aureus*; however, only those with atopic dermatitis suffer from increased risk of bacterial superinfections. Expression of NLRP3 and Caspase 1 by immunohistochemistry was reduced in skin in a cohort of patients with lesional atopic dermatitis as compared to patients with psoriatic or healthy skin. Primary human keratinocytes from healthy skin and primary human monocytes have been found to downregulate NLRP3 and ASC expression in response to IL-4, IL-5, and IL-13 (cytokines abundant in atopic dermatitis). Accordingly, Hla-induced IL-1β secretion was diminished in monocytes from patients with atopic dermatitis compared to patients with psoriasis and healthy controls (Niebuhr et al. 2014). These data suggest that the Th2-dominant immunologic milieu of patients with atopic dermatitis leads to defects in inflammasome signaling promoting susceptibility to *S. aureus* infection. As epidemiologic data also indicates patients with these conditions have increased risk of *S. aureus* infection (Juhn 2014).

PSMs also play a role in pathogenesis of *S. aureus*-induced skin inflammation. Primary human keratinocytes exposed to *S. aureus* with genes encoding all seven PSMs knocked out by allelic replacement or start codon disruption secreted significantly less IL-1β, IL-18, and LDH, a marker of cell death, than keratinocytes exposed to wild-type *S. aureus*. Keratinocytes exposed to any purified PSM secreted IL-18 but only the four PSMαs and delta-toxin were cytotoxic. Interestingly, caspase inhibitors and extracellular potassium were unable to block cytokine secretion or cytotoxicity, suggesting that this combination of cell death and cytokine secretion that is normally associated with inflammasome activation of Caspase 1 is triggered by a Caspase 1-independent process in keratinocytes that remains unidentified. In mice challenged epicutaneously with *S. aureus* and an isogenic mutant lacking PSM, the parental *S. aureus* strain triggered a predominantly neutrophil infiltrate and the PSM-deficient *S. aureus* did not. This difference in host neutrophil recruitment to infection site persisted in mice lacking IL-18, suggesting that PSM-mediated killing of keratinocytes led to the release of chemoattractive factors or that PSM themselves are chemoattractive (Syed et al. 2015). Previous research has shown that blocking formyl peptide receptor 2, the receptor for *S. aureus* PSMs, blocked PSM-mediated leukocyte infiltration in a mouse air pouch and mouse peritoneal infection model, suggesting that PSMs themselves are chemoattractive or trigger a chemoattractive cascade (Kretschmer et al. 2010).

Currently though, the translatability of these models to the human condition remains an open question. There is limited published evidence for the role of IL-1β or inflammasome activation in cellulitis and other skin infections in humans. In clinical trials of IL-1β antagonists, a trend toward increased risk of serious infections was observed in patients with inflammatory conditions treated with these agents (Galloway et al. 2011). There are no reports regarding the use of these agents and specific risk of *S. aureus* infection of the skin. Experimental models in mice offer an opportunity to study this but face major limitations in their relevance. Mouse skin differs significantly from human skin. Mouse epidermis is made up of only three cell layers and is less than 25 μm thick, whereas human epidermis is usually 6–10 cell layers and is greater than 50 μm thick. The epithelial turnover in mouse skin is faster and mouse skin regenerates without significant scarring. These differences and more are reviewed in Gudjonsson et al. (2007).

Considerable interest exists in humanized mouse models of skin infections, and these have been deployed to study the human-specific pathogen *Neisseria meningitides* (Melican et al. 2013). While human skin grafting on mice may eventually be a useful model for studying *S. aureus* skin infections, this methodology has yet to be deployed in this arena. Humanized NSG mice with humanized immune systems have been used in *S. aureus* skin infection models. A *S. aureus* inoculum one to two log lower induced consistent skin lesions as compared with non-humanized mice. This model aided in studying the role of PVL in dermonecrosis. Blocking the PVL receptor, human C5aR, with PMX53 or an anti-C5aR antibody eliminated the enhanced cytotoxicity of PVL-positive *S. aureus* but also reduced recruitment of neutrophils and exacerbated the infection (Tseng et al. 2015). Caution must be taken in interpreting results, however, given changes observed in skin grafts over time (Kappes et al. 2004). Additionally, mice with human skin grafts still carry myeloid-derived cells from mice, which may be a significant cause of differences seen in *S. aureus* susceptibility between human and mice.

With those caveats in mind, current evidence suggests that *S. aureus* activates inflammasomes in keratinocytes to promote invasion and establish infection. PSMs that induce NLRP3-independent cell death in keratinocytes may provide a redundant mechanism to support invasion. Once infection is established, neutrophils are recruited and secrete IL-1β, probably through NLRP3 inflammasome activation, to promote clearance of *S. aureus* infections (Fig. 3, top panels). If this holds true, therapeutic intervention to suppress inflammasome signaling may have a desirable effect in preventing invasion through keratinocytes but may unduly hinder clearance of already invaded *S. aureus*.

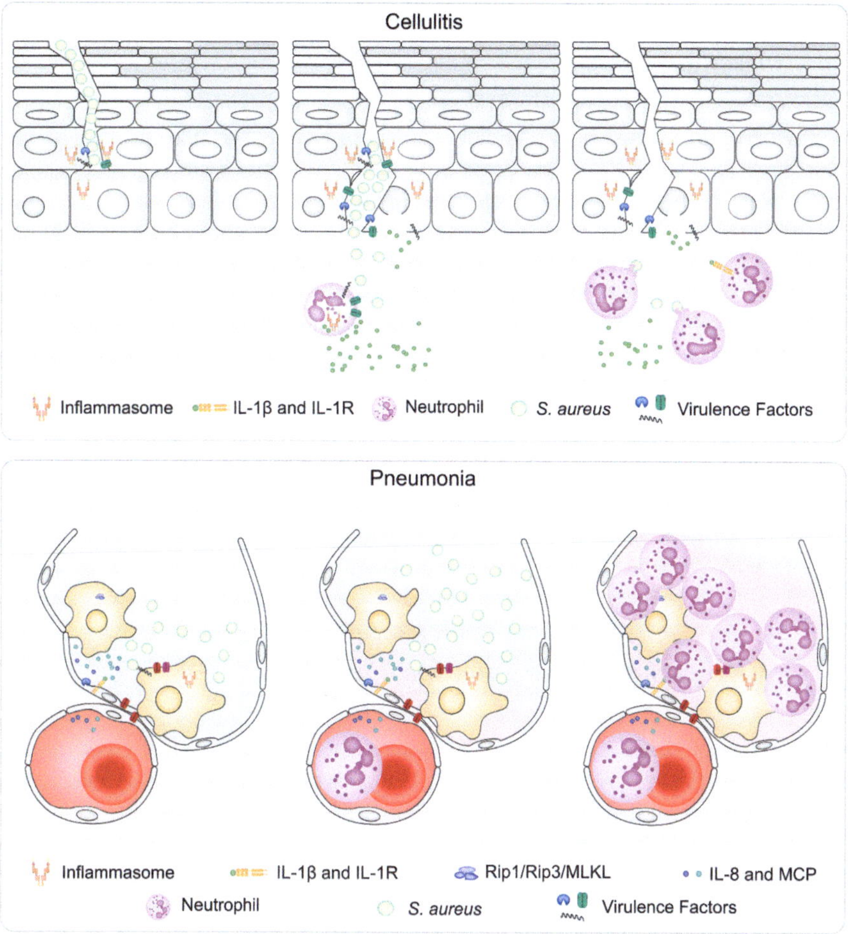

Fig. 3 NLRP3 inflammasome signaling plays unique rolls in skin and lung infections. Skin (*top row*) and lung (*bottom row*) infections are modeled here as an example of the unique effects of inflammasome signaling during *S. aureus* infections. In model skin infection (*top row*), *S. aureus* is cocultured with an organotypic culture of human keratinocytes grown at an air–liquid interface on a dermal substitute matrix, *S. aureus* utilizes Hla to activate inflammasome signaling to enhance invasion through the keratinocyte barrier (*left*). Once invasion has occurred (experimentally induced by subcutaneous injection), neutrophils are recruited to the site of infection and activate the NLRP3 inflammasome to promote IL-1β secretion (*center*). Neutrophil-derived IL-1β is required for abscess formation and bacterial clearance (*right*). In the lung (*bottom row*), *S. aureus* triggers alveolar macrophages to secrete IL-1β that promotes secretion of the neutrophil attracting chemokines, IL-8 and MCP-1, from human lung epithelial cells (*left*). Neutrophil recruitment leads to further NLRP3 inflammasome activation by Hla (*middle*) that ultimately destroys the lung tissue (*right*), as loss of NLRP3 in mice improves pathology with no change in bacterial burden. Loss of the IL-1 receptor does not impact the infection suggesting that IL-1 signaling is not required. Hla, LukAB, and PSMs also induce necroptosis, a RIPK1-, RIPK3-, MLKL-dependent cell death process that hinders clearance of *S. aureus*, as deletion of RIPK3 improves bacterial control. MLKL also plays a role in IL-1β secretion as loss of MLKL blocks Caspase 1 activation suggesting an intersection between the necroptosis signaling pathway and the inflammasome signaling pathway

3.2 *IL-1β Signaling Is Critical for Combating Soft Tissue Infections*

Staphylococcus aureus is a common culprit in soft tissue and medical device infections. Patients with ulcers, commonly resulting from advanced complications of diabetes, deep injuries, recent surgery, or indwelling medical devices, are particularly at risk. Recent work has demonstrated a differential effect of *S. aureus* strains in inducing inflammasome activation and IL-1β secretion in surgical site infections, with strain PS80 leading to substantially more IL-1β than SH1000. In this study, loss of NLRP3 did not completely eliminate IL-1β secretion, suggesting a coordinated role with other inflammasomes. However, deletion of *Nlrp3* or IL-1 receptor (*Il1r*) compromised control of bacterial burden in infected surgical wounds (Maher et al. 2013). Little else has been done to characterize the role of inflammasome signaling in soft tissue infections, though it is tempting to assume these would be similar to dermal infections once *S. aureus* crosses the keratinocyte barrier.

3.3 S. aureus *Hijacks the NLRP3 Inflammasome to Exacerbate Lung Infection Pathology*

S. aureus pneumonia is a difficult disease to effectively treat. Even with proper antimicrobial therapy, *S. aureus* triggers a massive immune response that causes significant tissue destruction with substantial lethality.

Hla is an essential virulence factor in mouse models of severe pneumonia. *S. aureus* lacking Hla has reduced bacterial burden and causes less lung pathology and host death when delivered intranasally or intratracheally to the mouse lung when compared to *S. aureus* with an intact Hla gene (Bubeck Wardenburg and Schneewind 2008). Blocking Hla with an antibody also promotes bacterial clearance from the lung, indicating that expression of Hla is important in the ability of the bacteria to cause lung infection in this model (Bubeck Wardenburg and Schneewind 2008; Ragle and Bubeck Wardenburg 2009). Highly purified recombinant Hla, alone or in conjunction with HKSA, delivered intratracheally causes a clinical syndrome resembling *S. aureus* pulmonary infection in mice. Pathologic changes in the lung including destruction of normal alveolar architecture, tissue necrosis, hemorrhage, and inflammation are consistent with pathologic findings in human lungs with severe *S. aureus* pneumonia. This syndrome is largely abrogated when Hla is administered to mice lacking NLRP3 (Kebaier et al. 2012). There was no difference in bacterial burdens between mice with and without NLRP3 inoculated with intratracheal *S. aureus*. Thus, unlike the skin where signaling from the

NLRP3 inflammasome contributes to host elimination of the bacteria, NLRP3 inflammasome signaling is not essential for clearing *S. aureus* from the mouse lung. There was no difference in bacterial burden or lung inflammation between wild-type and NLRP3 null mice that were infected with *S. aureus* lacking Hla, suggesting Hla-induced activation of NLRP3 is a major cause of pneumonia pathology in this model of *S. aureus* infection (Fig. 3, bottom panels). IL-1 receptor-deficient mice exhibited severe inflammatory responses to pulmonary delivery of Hla; thus, tissue destruction mediated by this toxin in the lung did not depend on IL-1β signaling (Kebaier et al. 2012). These findings suggest that cell death and/or other Caspase 1-dependent cytokines downstream of NLRP3 activation were sufficient to cause severe lung pathology in the setting of *S. aureus* infection.

There is significant debate as to whether PVL is a critical virulence factor in pneumonia. Mice infected intranasally with PVL-expressing *S. aureus* experienced substantial neutrophil recruitment, inflammation in the lung parenchyma, bronchial epithelial damage, tissue necrosis, and hemorrhage. These same effects were seen upon administration of both components of the purified PVL toxin but were absent in isogenic PVL-negative *S. aureus* or upon administration of a single component of the PVL toxin (Labandeira-Rey et al. 2007). A second report replicated these findings, demonstrating that deletion of PVL from the LAC *S. aureus* strain improves Balb/c mouse survival following intranasal inoculation. Additionally, immunization against PVL showed a trend toward protection in subsequent *S. aureus* challenge by intranasal infection (Brown et al. 2009). These results are in contrast to separate findings that deletion of PVL from the LAC *S. aureus* strain actually enhances virulence and decreases survival of Balb/c mice infected using the same protocol (Bubeck Wardenburg et al. 2008; Villaruz et al. 2009). Given the general resistance of mouse neutrophils to *S. aureus* PVL (Loffler et al. 2010), the differences in these studies may result from differences in susceptibility of non-immune cells to PVL. However, the use of Balb/c mice in both groups of studies suggests the difference is not strain-intrinsic. In general, because of the species specificity of *S. aureus* PFTs, humanized mice may be required to more fully appreciate the impact of PVL in *S. aureus* pneumonia as was seen when applied to the humanized mouse cellulitis model (Tseng et al. 2015).

In primary human monocyte-derived macrophages, HKSA followed by administration of recombinant PVL induced secretion of high levels of IL-1β. IL-1β secretion was enhanced by co-administration of PVL and the PSMs Hld and PSMalpha3, the PFTs Hlg and LukED, and Hlb, greater than the sum of their parts (Perret et al. 2012). PVL also triggered IL-1β secretion in human alveolar macrophages that goes on to stimulate the secretion of neutrophil attracting chemokines IL-8 and MCP-1 by A549 human lung epithelial cells in a mixed culture model. Blocking the IL-1 receptor with IL-1R antagonist abolishes PVL-induced secretion of IL-8 and MCP-1 (Perret et al. 2012). These data suggest that blockade of IL-1β signaling may diminish pathology in *S. aureus* pulmonary infection. However, a subsequent study in rabbits demonstrated that anakinra, an IL-1 receptor antagonist, could block pathology induced by treatment with recombinant PVL and HKSA but not in live *S. aureus* infection. Furthermore, treatment of infected rabbits with anakinra led to increased bacterial burden in the lungs (Labrousse et al. 2014). These data demonstrate that IL-1β secretion triggered by PVL plays a protective role in *S. aureus* pneumonia in rabbits, where the toxin is active. While these findings seem at odds with the apparent pathologic role of NLRP3 inflammasome activation in mice with *S. aureus* pneumonia, the studies in IL-1 receptor-deficient mice demonstrate that toxin-mediated activation of NLRP3 drove pathogenic findings in the lung and clinical severity of disease independently of IL-1β production. Additionally, because mice are deficient in NLRP3 through the entire course of the model pneumonia, there is no way to separate effects of NLRP3 on the ability of *S. aureus* to initiate infection from the effects of NLRP3 on the response to ongoing infection.

Recently, *S. aureus* toxins were shown to induce necroptosis in the lung through RIPK1/RIPK3/MLKL signaling. In primary human macrophage co-culture with *S. aureus*, inhibitors of necroptosis, including necrostatin-1 and necrosulfonamide (NSA), or siRNA-mediated knockdown of RIPK3 or MLKL blocked *S. aureus*-induced cell death. NSA also decreased purified Hla-induced THP1 cell death. In murine *S. aureus* pneumonia, blocking necroptosis with NSA or knockout of RIPK3 improved bacterial clearance. RIPK3-deficient mice had improved lung architecture and less disruption of the pulmonary barrier, resulting in less protein in bronchoalveolar lavage fluid. Interestingly, NSA pretreatment of THP1 cells exposed to *S. aureus* supernatants in culture or RIPK3 knockout in *S. aureus* pneumonia led to decreased IL-1β secretion, suggesting a possible link between necroptosis and inflammasome activation (Kitur et al. 2015).

Currently, there is strong evidence supporting the use of systemic corticosteroids in hospitalized patients with severe community-acquired pneumonia (Siemieniuk et al. 2015). Interestingly, a glucocorticosteroid-responsive negative regulatory element has been identified just upstream of the IL-1β transcription start site (Zhang et al. 1997) and, as such, suppression of IL-1β production might be contributing to improved outcomes in these patients. However, conflicting results as to whether IL-1 signaling enhances or diminishes lung pathology will be important to clarify before targeted therapeutic immunosuppression is used clinically in the setting of severe *S. aureus* pneumonia. Clinical trials of anakinra suggest patients with asthma or other pulmonary comorbidities might be at increased risk of infectious complications, though results were not statistically significant (Schiff et al. 2004; Fleischmann et al. 2003). Suppression of necroptosis through RIPK3 inhibition may also be a viable strategy but understanding the contributions of necroptosis and NLRP3 inflammasome-mediated cell death to pathology and bacterial clearance is still in its infancy. Finally, activation of IL-1 independent effects of NLRP3, particularly inflammatory cell death, by *S. aureus* virulence factors may be sufficient to drive severe lung pathology even in the setting of neutralizing IL-1. Better animal models of *S. aureus* pneumonia that are responsive to the major PFT and other *S. aureus* virulence factors are needed to propel this field forward.

3.4 *Microglia Activate NLRP3 In Vitro but Depend on AIM2 to Clear* S. aureus *Central Nervous System Infections*

S. aureus can also cause brain abscesses usually as a complication of surgery, trauma, or bacteremia. Microglia are immunologically competent cells in the brain activated early in the process of *S. aureus* abscess formation (Kielian 2004). Exposure of primary microglia isolated from C57BL/6 mice to live *S. aureus* strain USA300 induced both IL-1β and IL-18 secretion. Microglia from mice lacking NLRP3 or ASC were deficient in IL-1β secretion but responded to live *S. aureus* with similar levels of IL-18 as wild-type mice. Deletion of Hla or Hlg, but not LukAB or LukED, reduced Newman strain *S. aureus*-induced IL-1β secretion from mouse-derived microglia (Hanamsagar et al. 2011). However, this is not surprising given the general resistance of mouse to LukAB (DuMont et al. 2013a) and the low level of expression of LukED (DuMont et al. 2013b) under the conditions the bacteria were cultured. Caspase 1 and CTSB inhibitors blocked *S. aureus*-induced microglial IL-1β secretion in these in vitro cell culture experiments. Consistent with a second pathway controlling IL-18 secretion in this setting, the CTSB inhibitor had no effect on IL-18 secretion (Hanamsagar et al. 2011).

In a model of brain abscess, *Asc* and *Casp1/Casp11* knockout mice had earlier mortality and diminished detectable IL-1β recovered from abscesses as compared to the parental mouse strain. Unexpectedly, the survival of *Nlrp3* knockout mice in this brain abscess model was identical to the parental strain. Upon further investigation, loss of AIM2 mimicked the sensitivity to *S. aureus* infection and diminished IL-1β production seen in ASC-deficient mice, suggesting that AIM2 is the primary upstream sensor for the inflammasome activating ASC and Caspase 1 in this model. Besides IL-1β, other key inflammatory mediators, including IL-6, CXCL1, CXCL10, and CCL2, were significantly reduced in the CNS of *Aim2* and *Asc* knockout mice (Hanamsagar et al. 2014). Immune cell infiltrates, including neutrophils and macrophages, and other cytokines, such as IL-10, TNF-α and IFN-γ, were not changed between *Asc* knockout and the parental strain mice. Also, the bacterial burden of *Aim2* knockout mice and the parental strain were equivalent in the first 18 h of infection. Most *Aim2* knockout mice that died were recorded as having died at approximately 20 h post-infection, leaving us without a satisfying explanation for why these mice suddenly succumbed to disease given the similarities in bacterial burden and immune cell infiltrates (Hanamsagar et al. 2014). Previous studies of *S. aureus* brain abscesses have demonstrated damage to the blood–brain barrier (Kielian et al. 2001) and impaired capillary perfusion and parenchymal cell death has been noted in soft tissue infections (Harding et al. 2014), both explanation that might be worth further investigation in this setting. The discrepancy between IL-1β production being NLRP3 dependent in cultured microglia exposed to *S. aureus* and AIM2-dependent after intracranial inoculation of *S. aureus* points to the need to push through additional mechanistic studies to understand the role of inflammasome activation in brain abscess. Host-derived DAMPs, like extracellular chromatin, arising from cellular injury during inoculation or cytotoxic factors from the bacteria may be responsible for activating AIM2 in the in vivo brain abscess model. Further investigation of AIM2-deficient microglia in vitro would be beneficial for characterizing the response seen in vivo. Also, study of *Aim2* knockout mice in the context of other *S. aureus* infections would help determine whether the role of AIM2 is brain/microglia-specific or is a general feature of the immune response to *S. aureus*.

4 How the Host Inflammasome Can Affect Other Inflammatory Processes

The Th17 response has emerged as a major focus of studies looking downstream of inflammasome activation. The most direct hint that IL-17-driven responses are critical for defense against *S. aureus* is seen in patients with hyper IgE syndrome (HIES). In HIES, mutation in *STAT3* causes impaired Th17 cell function and recurrent and severe *S. aureus* infections (Milner et al. 2008). Similar impairments have been recapitulated in mouse models of *S. aureus* infection.

NLRP3-mediated secretion of IL-1β by bone marrow dendritic cells promotes secretion of IL-17 by γδ-T cells, as loss of the IL-1 receptor limits this response. Importantly, the ability to control surgical site infections is enhanced by IL-1β stimulated γδ-T cells as bacterial burden is increased in δTCR-deficient and IL-17R-deficient mice (Maher et al. 2013). In *S. aureus* cutaneous infections, loss of γδ-T cells led to larger skin lesions with higher bacterial counts as a result of impaired neutrophil recruitment. IL-17R-deficient mice had a similar phenotype. Treatment of γδ-T cell-deficient mice with a single dose of recombinant IL-17 rescued the impaired immune response to *S. aureus* (Cho et al. 2010). A population of CD44+ CD27− memory γδ-T cells that is expanded upon peritoneal infection of C57BL/6 mice with *S. aureus* produces high levels of IL-17 and promotes bacterial clearance during reinfections. IL-1 signaling was not required for activation or expansion of memory γδ-T cells during reinfection (Murphy et al. 2014). γδ-T cells also play a prominent role in defense against *S. aureus* pneumonia. γδ-T cell-deficient mice had impaired neutrophil recruitment and an increased bacterial burden in the lung. However, the absence of γδ-T cells decreases lung pathology and improves survival (Cheng et al. 2012), similar to studies with loss of NLRP3 (Kebaier et al. 2012).

5 Integrating Inflammasome Studies to Improve Patient Care

Early evidence suggests that vaccines against PFTs may be beneficial in reducing virulence and severity of infections. In hospitalized patients with *S. aureus* infections, the risk of sepsis was significantly lower in those patients with higher levels of IgG against Hla, Hld, PVL, staphylococcal enterotoxin C-1, and PSMα3 (Adhikari et al. 2012). Some children with culture-proven *S. aureus* infection developed anti-LukAB antibodies that potently neutralize cytotoxicity in vitro, suggesting that the toxin is produced in vivo and that it elicits a humoral response (Thomsen et al. 2014). Immunization of mice with a mutant form of Hla that cannot form pores, Hla H35L, generates antigen-specific immunoglobulin G responses and affords protection against staphylococcal pneumonia. Additionally, transfer of Hla-specific antibodies or anti-Hla monoclonal antibodies also protects naïve animals against *S. aureus* challenge and prevents injury of human lung epithelial cells during infection (Bubeck Wardenburg and Schneewind 2008; Ragle and Bubeck Wardenburg 2009). This is presumably the result of decreased NLRP3 inflammasome activation in this setting, however, that has not been measured in vivo directly.

Therapies targeting specific virulence factors may provide a useful adjuvant to antibiotic therapy. As recently declared in the Annals of Internal Medicine, many believe it is "time to change clinical practice" and recommend systemic administration of corticosteroids in patients with severe community-acquired pneumonia. Theoretically, tailoring therapy more narrowly to reduce virulence and hyperactive immune responses without compromising bacterial clearance would provide further benefit over this broadly immunosuppressive therapy.

For understanding tissue-specific pathogenesis, we will need improved infection models that better recapitulate human tissues and immune cells to overcome the natural resistance of mice to *S. aureus*. Tissue-specific knockouts and bone marrow chimeras will also help differentiate tissue-driven immune responses from those of bone marrow origin. Ultimately, investigations into inflammasome-mediated responses to *S. aureus* infections may provide us with the therapeutic tools we need to push back against this growing threat.

6 Concluding Remarks

The rapid pace of discovery of new *S. aureus* toxins and other virulence factors has led to a dramatically improved understanding of the interactions between *S. aureus* and the innate immune system during infection. The NLRP3 inflammasome is now recognized as a critical driver of pro-inflammatory signaling in *S. aureus* infections. Additional studies provide evidence that other inflammasomes can also contribute to the pro-inflammatory milieu during *S. aureus* infection. However, significant shortcomings in the frequently used mouse models raise doubts about relative importance of many potential *S. aureus* virulence factors as well as the role of host inflammasome activation during infection. Novel approaches in studying *S. aureus* in clinical infections would provide much needed insight as to the relevance of preclinical studies of S. aureus pathogenesis. Further clarification of the protective and pathogenic roles of virulence factor-induced inflammasome signaling in *S.*

aureus infections is needed to guide the development of novel therapeutics and vaccines to diminish the virulence of and treat *S. aureus* infections.

Acknowledgments The authors' research has been supported by the National Institutes of Health: JHM received support from UNC MSTP T32GM008719 and UNC Predoctoral Training in Pharmacologic Sciences T32GM007040, and JAD was supported by research grant R01AI088255. JAD is an awardee of the Burroughs Wellcome Fund Career Award for Medical Scientists.

References

Adhikari RP, Ajao AO, Aman MJ, Karauzum H, Sarwar J, Lydecker AD, Johnson JK, Nguyen C, Chen WH, Roghmann MC (2012) Lower antibody levels to *Staphylococcus aureus* exotoxins are associated with sepsis in hospitalized adults with invasive *S. aureus* infections. J Infect Dis 206(6):915–923. doi:10.1093/infdis/jis462

Alonzo F III, Kozhaya L, Rawlings SA, Reyes-Robles T, DuMont AL, Myszka DG, Landau NR, Unutmaz D, Torres VJ (2013) CCR5 is a receptor for *Staphylococcus aureus* leukotoxin ED. Nature 493(7430):51–55. doi:10.1038/nature11724

Badarau A, Rouha H, Malafa S, Logan DT, Hakansson M, Stulik L, Dolezilkova I, Teubenbacher A, Gross K, Maierhofer B, Weber S, Jagerhofer M, Hoffman D, Nagy E (2015) Structure-function analysis of heterodimer formation, oligomerization, and receptor binding of the *Staphylococcus aureus* bi-component toxin LukGH. J Biol Chem 290(1):142–156. doi:10.1074/jbc.M114.598110

Borregaard N (2010) Neutrophils, from marrow to microbes. Immunity 33(5):657–670. doi:10.1016/j.immuni.2010.11.011

Brown EL, Dumitrescu O, Thomas D, Badiou C, Koers EM, Choudhury P, Vazquez V, Etienne J, Lina G, Vandenesch F, Bowden MG (2009) The Panton-Valentine leukocidin vaccine protects mice against lung and skin infections caused by Staphylococcus aureus USA300. Clin Microbiol Infect: Official Publ Eur Soc Clin Microbiol Infect Dis 15(2):156–164. doi:10.1111/j.1469-0691.2008.02648.x

Bubeck Wardenburg J, Schneewind O (2008) Vaccine protection against *Staphylococcus aureus* pneumonia. J Exp Med 205(2):287–294. doi:10.1084/jem.20072208

Bubeck Wardenburg J, Palazzolo-Ballance AM, Otto M, Schneewind O, DeLeo FR (2008) Panton-Valentine leukocidin is not a virulence determinant in murine models of community-associated methicillin-resistant *Staphylococcus aureus* disease. J Infect Dis 198 (8):1166–1170. doi:10.1086/592053

Cheng P, Liu T, Zhou WY, Zhuang Y, Peng LS, Zhang JY, Yin ZN, Mao XH, Guo G, Shi Y, Zou QM (2012) Role of gamma-delta T cells in host response against *Staphylococcus aureus*-induced pneumonia. BMC immunology 13:38. doi:10.1186/1471-2172-13-38

Cho JS, Pietras EM, Garcia NC, Ramos RI, Farzam DM, Monroe HR, Magorien JE, Blauvelt A, Kolls JK, Cheung AL, Cheng G, Modlin RL, Miller LS (2010) IL-17 is essential for host defense against cutaneous *Staphylococcus aureus* infection in mice. J Clin Investig 120 (5):1762–1773. doi:10.1172/JCI40891

Cho JS, Guo Y, Ramos RI, Hebroni F, Plaisier SB, Xuan C, Granick JL, Matsushima H, Takashima A, Iwakura Y, Cheung AL, Cheng G, Lee DJ, Simon SI, Miller LS (2012) Neutrophil-derived IL-1beta is sufficient for abscess formation in immunity against *Staphylococcus aureus* in mice. PLoS Pathog 8(11):e1003047. doi:10.1371/journal.ppat.1003047

Craven RR, Gao X, Allen IC, Gris D, Bubeck Wardenburg J, McElvania-Tekippe E, Ting JP, Duncan JA (2009) *Staphylococcus aureus* alpha-hemolysin activates the

NLRP3-inflammasome in human and mouse monocytic cells. PLoS ONE 4(10):e7446. doi:10.1371/journal.pone.0007446
Davis BK, Roberts RA, Huang MT, Willingham SB, Conti BJ, Brickey WJ, Barker BR, Kwan M, Taxman DJ, Accavitti-Loper MA, Duncan JA, Ting JP (2011) Cutting edge: NLRC5-dependent activation of the inflammasome. J Immunol 186(3):1333–1337. doi:10.4049/jimmunol.1003111
Dumont AL, Nygaard TK, Watkins RL, Smith A, Kozhaya L, Kreiswirth BN, Shopsin B, Unutmaz D, Voyich JM, Torres VJ (2011) Characterization of a new cytotoxin that contributes to *Staphylococcus aureus* pathogenesis. Mol Microbiol 79(3):814–825. doi:10.1111/j.1365-2958.2010.07490.x
DuMont AL, Yoong P, Day CJ, Alonzo F III, McDonald WH, Jennings MP, Torres VJ (2013a) *Staphylococcus aureus* LukAB cytotoxin kills human neutrophils by targeting the CD11b subunit of the integrin Mac-1. Proc Natl Acad Sci USA 110(26):10794–10799. doi:10.1073/pnas.1305121110
DuMont AL, Yoong P, Surewaard BG, Benson MA, Nijland R, van Strijp JA, Torres VJ (2013b) *Staphylococcus aureus* elaborates leukocidin AB to mediate escape from within human neutrophils. Infect Immun 81(5):1830–1841. doi:10.1128/IAI.00095-13
Dunne DW, Resnick D, Greenberg J, Krieger M, Joiner KA (1994) The type I macrophage scavenger receptor binds to gram-positive bacteria and recognizes lipoteichoic acid. Proc Natl Acad Sci USA 91(5):1863–1867
Eigenbrod T, Franchi L, Munoz-Planillo R, Kirschning CJ, Freudenberg MA, Nunez G, Dalpke A (2012) Bacterial RNA mediates activation of caspase-1 and IL-1beta release independently of TLRs 3, 7, 9 and TRIF but is dependent on UNC93B. J Immunol 189(1):328–336. doi:10.4049/jimmunol.1103258
Fitton JE, Dell A, Shaw WV (1980) The amino acid sequence of the delta haemolysin of *Staphylococcus aureus*. FEBS Lett 115(2):209–212
Fleischmann RM, Schechtman J, Bennett R, Handel ML, Burmester GR, Tesser J, Modafferi D, Poulakos J, Sun G (2003) Anakinra, a recombinant human interleukin-1 receptor antagonist (r-metHuIL-1ra), in patients with rheumatoid arthritis: a large, international, multicenter, placebo-controlled trial. Arthritis Rheum 48(4):927–934. doi:10.1002/art.10870
Galloway JB, Hyrich KL, Mercer LK, Dixon WG, Watson KD, Lunt M, Consortium BCC, Symmons DP, British Society for Rheumatology Biologics R (2011) The risk of serious infections in patients receiving anakinra for rheumatoid arthritis: results from the British Society for Rheumatology Biologics Register. Rheumatology 50(7):1341–1342. doi:10.1093/rheumatology/ker146
Gouaux E, Hobaugh M, Song L (1997) alpha-Hemolysin, gamma-hemolysin, and leukocidin from Staphylococcus aureus: distant in sequence but similar in structure. Protein Sci: Publ Protein Soc 6(12):2631–2635. doi:10.1002/pro.5560061216
Greenberg JW, Fischer W, Joiner KA (1996) Influence of lipoteichoic acid structure on recognition by the macrophage scavenger receptor. Infect Immun 64(8):3318–3325
Grundling A, Schneewind O (2007) Synthesis of glycerol phosphate lipoteichoic acid in *Staphylococcus aureus*. Proc Natl Acad Sci USA 104(20):8478–8483. doi:10.1073/pnas.0701821104
Gudjonsson JE, Johnston A, Dyson M, Valdimarsson H, Elder JT (2007) Mouse models of psoriasis. J Invest Dermatol 127(6):1292–1308. doi:10.1038/sj.jid.5700807
Guillet V, Roblin P, Werner S, Coraiola M, Menestrina G, Monteil H, Prevost G, Mourey L (2004) Crystal structure of leucotoxin S component: new insight into the Staphylococcal beta-barrel pore-forming toxins. J Biol Chem 279(39):41028–41037. doi:10.1074/jbc.M406904200
Han SH, Kim JH, Martin M, Michalek SM, Nahm MH (2003) Pneumococcal lipoteichoic acid (LTA) is not as potent as staphylococcal LTA in stimulating Toll-like receptor 2. Infect Immun 71(10):5541–5548
Hanamsagar R, Torres V, Kielian T (2011) Inflammasome activation and IL-1beta/IL-18 processing are influenced by distinct pathways in microglia. J Neurochem 119(4):736–748. doi:10.1111/j.1471-4159.2011.07481.x

Hanamsagar R, Aldrich A, Kielian T (2014) Critical role for the AIM2 inflammasome during acute CNS bacterial infection. J Neurochem 129(4):704–711. doi:10.1111/jnc.12669

Harding MG, Zhang K, Conly J, Kubes P (2014) Neutrophil crawling in capillaries; a novel immune response to *Staphylococcus aureus*. PLoS Pathog 10(10):e1004379. doi:10.1371/journal.ppat.1004379

Hedges SB, Dudley J, Kumar S (2006) TimeTree: a public knowledge-base of divergence times among organisms. Bioinformatics 22(23):2971–2972. doi:10.1093/bioinformatics/btl505

Hermann C, Spreitzer I, Schroder NW, Morath S, Lehner MD, Fischer W, Schutt C, Schumann RR, Hartung T (2002) Cytokine induction by purified lipoteichoic acids from various bacterial species–role of LBP, sCD14, CD14 and failure to induce IL-12 and subsequent IFN-gamma release. Eur J Immunol 32(2):541–551. doi:10.1002/1521-4141(200202)32:2<541:AID-IMMU541>3.0.CO;2-P

Hoebe K, Georgel P, Rutschmann S, Du X, Mudd S, Crozat K, Sovath S, Shamel L, Hartung T, Zahringer U, Beutler B (2005) CD36 is a sensor of diacylglycerides. Nature 433(7025):523–527. doi:10.1038/nature03253

Holzinger D, Gieldon L, Mysore V, Nippe N, Taxman DJ, Duncan JA, Broglie PM, Marketon K, Austermann J, Vogl T, Foell D, Niemann S, Peters G, Roth J, Loffler B (2012) *Staphylococcus aureus* Panton-Valentine leukocidin induces an inflammatory response in human phagocytes via the NLRP3 inflammasome. J Leukoc Biol 92(5):1069–1081. doi:10.1189/jlb.0112014

Hornung V, Bauernfeind F, Halle A, Samstad EO, Kono H, Rock KL, Fitzgerald KA, Latz E (2008) Silica crystals and aluminum salts activate the NALP3 inflammasome through phagosomal destabilization. Nat Immunol 9(8):847–856. doi:10.1038/ni.1631

Huseby M, Shi K, Brown CK, Digre J, Mengistu F, Seo KS, Bohach GA, Schlievert PM, Ohlendorf DH, Earhart CA (2007) Structure and biological activities of beta toxin from *Staphylococcus aureus*. J Bacteriol 189(23):8719–8726. doi:10.1128/JB.00741-07

Juhn YJ (2014) Risks for infection in patients with asthma (or other atopic conditions): is asthma more than a chronic airway disease? J Allergy Clin Immunol 134(2):247–257; quiz 258–249. doi:10.1016/j.jaci.2014.04.024

Kappes U, Schliemann-Willers S, Bankova L, Heinemann C, Fischer TW, Ziemer M, Schubert H, Norgauer J, Fluhr JW, Elsner P (2004) The quality of human skin xenografts on SCID mice: a noninvasive bioengineering approach. Br J Dermatol 151(5):971–976. doi:10.1111/j.1365-2133.2004.06191.x

Kebaier C, Chamberland RR, Allen IC, Gao X, Broglie PM, Hall JD, Jania C, Doerschuk CM, Tilley SL, Duncan JA (2012) *Staphylococcus aureus* alpha-hemolysin mediates virulence in a murine model of severe pneumonia through activation of the NLRP3 inflammasome. J Infect Dis 205(5):807–817. doi:10.1093/infdis/jir846

Khare S, Dorfleutner A, Bryan NB, Yun C, Radian AD, de Almeida L, Rojanasakul Y, Stehlik C (2012) An NLRP7-containing inflammasome mediates recognition of microbial lipopeptides in human macrophages. Immunity 36(3):464–476. doi:10.1016/j.immuni.2012.02.001

Kielian T (2004) Immunopathogenesis of brain abscess. J Neuroinflammation 1(1):16. doi:10.1186/1742-2094-1-16

Kielian T, Barry B, Hickey WF (2001) CXC chemokine receptor-2 ligands are required for neutrophil-mediated host defense in experimental brain abscesses. J Immunol 166(7):4634–4643

Kitur K, Parker D, Nieto P, Ahn DS, Cohen TS, Chung S, Wachtel S, Bueno S, Prince A (2015) Toxin-induced necroptosis is a major mechanism of *Staphylococcus aureus* lung damage. PLoS Pathog 11(4):e1004820. doi:10.1371/journal.ppat.1004820

Kretschmer D, Gleske AK, Rautenberg M, Wang R, Koberle M, Bohn E, Schoneberg T, Rabiet MJ, Boulay F, Klebanoff SJ, van Kessel KA, van Strijp JA, Otto M, Peschel A (2010) Human formyl peptide receptor 2 senses highly pathogenic *Staphylococcus aureus*. Cell Host Microbe 7(6):463–473. doi:10.1016/j.chom.2010.05.012

Kumar S, Hedges SB (2011) TimeTree2: species divergence times on the iPhone. Bioinformatics 27(14):2023–2024. doi:10.1093/bioinformatics/btr315

Labandeira-Rey M, Couzon F, Boisset S, Brown EL, Bes M, Benito Y, Barbu EM, Vazquez V, Hook M, Etienne J, Vandenesch F, Bowden MG (2007) *Staphylococcus aureus* Panton-Valentine leukocidin causes necrotizing pneumonia. Science 315(5815):1130–1133. doi:10.1126/science.1137165

Labrousse D, Perret M, Hayez D, Da Silva S, Badiou C, Couzon F, Bes M, Chavanet P, Lina G, Vandenesch F, Croisier-Bertin D, Henry T (2014) Kineret(R)/IL-1ra blocks the IL-1/IL-8 inflammatory cascade during recombinant Panton Valentine Leukocidin-triggered pneumonia but not during *S. aureus* infection. PLoS ONE 9(6):e97546. doi:10.1371/journal.pone.0097546

Loffler B, Hussain M, Grundmeier M, Bruck M, Holzinger D, Varga G, Roth J, Kahl BC, Proctor RA, Peters G (2010) *Staphylococcus aureus* panton-valentine leukocidin is a very potent cytotoxic factor for human neutrophils. PLoS Pathog 6(1):e1000715. doi:10.1371/journal.ppat.1000715

Lynch NJ, Roscher S, Hartung T, Morath S, Matsushita M, Maennel DN, Kuraya M, Fujita T, Schwaeble WJ (2004) L-ficolin specifically binds to lipoteichoic acid, a cell wall constituent of Gram-positive bacteria, and activates the lectin pathway of complement. J Immunol 172(2):1198–1202

Maher BM, Mulcahy ME, Murphy AG, Wilk M, O'Keeffe KM, Geoghegan JA, Lavelle EC, McLoughlin RM (2013) Nlrp-3-driven interleukin 17 production by gammadeltaT cells controls infection outcomes during *Staphylococcus aureus* surgical site infection. Infect Immun 81(12):4478–4489. doi:10.1128/IAI.01026-13

Mariathasan S, Weiss DS, Newton K, McBride J, O'Rourke K, Roose-Girma M, Lee WP, Weinrauch Y, Monack DM, Dixit VM (2006) Cryopyrin activates the inflammasome in response to toxins and ATP. Nature 440(7081):228–232. doi:10.1038/nature04515

Melehani JH, James DB, DuMont AL, Torres VJ, Duncan JA (2015) *Staphylococcus aureus* leukocidin A/B (LukAB) kills human monocytes via host NLRP3 and ASC when extracellular, but not intracellular. PLoS Pathog 11(6):e1004970. doi:10.1371/journal.ppat.1004970

Melican K, Michea Veloso P, Martin T, Bruneval P, Dumenil G (2013) Adhesion of Neisseria meningitidis to dermal vessels leads to local vascular damage and purpura in a humanized mouse model. PLoS Pathog 9(1):e1003139. doi:10.1371/journal.ppat.1003139

Miller LS, Cho JS (2011) Immunity against *Staphylococcus aureus* cutaneous infections. Nat Rev Immunol 11(8):505–518. doi:10.1038/nri3010

Miller LS, O'Connell RM, Gutierrez MA, Pietras EM, Shahangian A, Gross CE, Thirumala A, Cheung AL, Cheng G, Modlin RL (2006) MyD88 mediates neutrophil recruitment initiated by IL-1R but not TLR2 activation in immunity against *Staphylococcus aureus*. Immunity 24(1):79–91. doi:10.1016/j.immuni.2005.11.011

Miller LS, Pietras EM, Uricchio LH, Hirano K, Rao S, Lin H, O'Connell RM, Iwakura Y, Cheung AL, Cheng G, Modlin RL (2007) Inflammasome-mediated production of IL-1beta is required for neutrophil recruitment against *Staphylococcus aureus* in vivo. J Immunol 179(10):6933–6942

Milner JD, Brenchley JM, Laurence A, Freeman AF, Hill BJ, Elias KM, Kanno Y, Spalding C, Elloumi HZ, Paulson ML, Davis J, Hsu A, Asher AI, O'Shea J, Holland SM, Paul WE, Douek DC (2008) Impaired T(H)17 cell differentiation in subjects with autosomal dominant hyper-IgE syndrome. Nature 452(7188):773–776. doi:10.1038/nature06764

Muller S, Wolf AJ, Iliev ID, Berg BL, Underhill DM, Liu GY (2015) Poorly cross-linked peptidoglycan in MRSA due to mecA induction activates the inflammasome and exacerbates immunopathology. Cell Host Microbe 18(5):604–612. doi:10.1016/j.chom.2015.10.011

Munoz-Planillo R, Franchi L, Miller LS, Nunez G (2009) A critical role for hemolysins and bacterial lipoproteins in *Staphylococcus aureus*-induced activation of the Nlrp3 inflammasome. J Immunol 183(6):3942–3948. doi:10.4049/jimmunol.0900729

Murphy AG, O'Keeffe KM, Lalor SJ, Maher BM, Mills KH, McLoughlin RM (2014) *Staphylococcus aureus* infection of mice expands a population of memory gammadelta T cells that are protective against subsequent infection. J Immunol 192(8):3697–3708. doi:10.4049/jimmunol.1303420

Nakayama M, Kurokawa K, Nakamura K, Lee BL, Sekimizu K, Kubagawa H, Hiramatsu K, Yagita H, Okumura K, Takai T, Underhill DM, Aderem A, Ogasawara K (2012) Inhibitory receptor paired Ig-like receptor B is exploited by *Staphylococcus aureus* for virulence. J Immunol 189(12):5903–5911. doi:10.4049/jimmunol.1201940

Niebuhr M, Baumert K, Heratizadeh A, Satzger I, Werfel T (2014) Impaired NLRP3 inflammasome expression and function in atopic dermatitis due to Th2 milieu. Allergy 69 (8):1058–1067. doi:10.1111/all.12428

Otto M (2010) Basis of virulence in community-associated methicillin-resistant *Staphylococcus aureus*. Annu Rev Microbiol 64:143–162. doi:10.1146/annurev.micro.112408.134309

Pedelacq JD, Maveyraud L, Prevost G, Baba-Moussa L, Gonzalez A, Courcelle E, Shepard W, Monteil H, Samama JP, Mourey L (1999) The structure of a *Staphylococcus aureus* leucocidin component (LukF-PV) reveals the fold of the water-soluble species of a family of transmembrane pore-forming toxins. Structure 7(3):277–287

Percy MG, Grundling A (2014) Lipoteichoic acid synthesis and function in gram-positive bacteria. Annu Rev Microbiol 68:81–100. doi:10.1146/annurev-micro-091213-112949

Perret M, Badiou C, Lina G, Burbaud S, Benito Y, Bes M, Cottin V, Couzon F, Juruj C, Dauwalder O, Goutagny N, Diep BA, Vandenesch F, Henry T (2012) Cross-talk between Staphylococcus aureus leukocidins-intoxicated macrophages and lung epithelial cells triggers chemokine secretion in an inflammasome-dependent manner. Cell Microbiol 14(7):1019–1036. doi:10.1111/j.1462-5822.2012.01772.x

Peschel A, Otto M (2013) Phenol-soluble modulins and staphylococcal infection. Nat Rev Microbiol 11(10):667–673. doi:10.1038/nrmicro3110

Radian AD, Khare S, Chu LH, Dorfleutner A, Stehlik C (2015) ATP binding by NLRP7 is required for inflammasome activation in response to bacterial lipopeptides. Mol Immunol 67(2 Pt B):294–302. doi:10.1016/j.molimm.2015.06.013

Ragle BE, Bubeck Wardenburg J (2009) Anti-alpha-hemolysin monoclonal antibodies mediate protection against *Staphylococcus aureus* pneumonia. Infect Immun 77(7):2712–2718. doi:10.1128/IAI.00115-09

Reyes-Robles T, Alonzo F III, Kozhaya L, Lacy DB, Unutmaz D, Torres VJ (2013) *Staphylococcus aureus* leukotoxin ED targets the chemokine receptors CXCR1 and CXCR2 to kill leukocytes and promote infection. Cell Host Microbe 14(4):453–459. doi:10.1016/j.chom.2013.09.005

Schiff MH, DiVittorio G, Tesser J, Fleischmann R, Schechtman J, Hartman S, Liu T, Solinger AM (2004) The safety of anakinra in high-risk patients with active rheumatoid arthritis: six-month observations of patients with comorbid conditions. Arthritis Rheum 50(6):1752–1760. doi:10.1002/art.20277

Schwandner R, Dziarski R, Wesche H, Rothe M, Kirschning CJ (1999) Peptidoglycan- and lipoteichoic acid-induced cell activation is mediated by toll-like receptor 2. J Biol Chem 274 (25):17406–17409

Sha W, Mitoma H, Hanabuchi S, Bao M, Weng L, Sugimoto N, Liu Y, Zhang Z, Zhong J, Sun B, Liu YJ (2014) Human NLRP3 inflammasome senses multiple types of bacterial RNAs. Proc Natl Acad Sci USA 111(45):16059–16064. doi:10.1073/pnas.1412487111

Shimada T, Park BG, Wolf AJ, Brikos C, Goodridge HS, Becker CA, Reyes CN, Miao EA, Aderem A, Gotz F, Liu GY, Underhill DM (2010) *Staphylococcus aureus* evades lysozyme-based peptidoglycan digestion that links phagocytosis, inflammasome activation, and IL-1beta secretion. Cell Host Microbe 7(1):38–49. doi:10.1016/j.chom.2009.12.008

Siemieniuk RA, Meade MO, Alonso-Coello P, Briel M, Evaniew N, Prasad M, Alexander PE, Fei Y, Vandvik PO, Loeb M, Guyatt GH (2015) Corticosteroid therapy for patients hospitalized with community-acquired pneumonia: a systematic review and meta-analysis. Ann Int Med 163(7):519–528. doi:10.7326/M15-0715

Soong G, Chun J, Parker D, Prince A (2012) *Staphylococcus aureus* activation of caspase 1/calpain signaling mediates invasion through human keratinocytes. J Infect Dis 205 (10):1571–1579. doi:10.1093/infdis/jis244

Soong G, Paulino F, Wachtel S, Parker D, Wickersham M, Zhang D, Brown A, Lauren C, Dowd M, West E, Horst B, Planet P, Prince A (2015) Methicillin-resistant *Staphylococcus aureus* adaptation to human keratinocytes. mBio 6(2). doi:10.1128/mBio.00289-15

Spaan AN, Henry T, van Rooijen WJ, Perret M, Badiou C, Aerts PC, Kemmink J, de Haas CJ, van Kessel KP, Vandenesch F, Lina G, van Strijp JA (2013) The staphylococcal toxin Panton-Valentine Leukocidin targets human C5a receptors. Cell Host Microbe 13(5):584–594. doi:10.1016/j.chom.2013.04.006

Spaan AN, Vrieling M, Wallet P, Badiou C, Reyes-Robles T, Ohneck EA, Benito Y, de Haas CJ, Day CJ, Jennings MP, Lina G, Vandenesch F, van Kessel KP, Torres VJ, van Strijp JA, Henry T (2014) The staphylococcal toxins gamma-haemolysin AB and CB differentially target phagocytes by employing specific chemokine receptors. Nat Commun 5:5438. doi:10.1038/ncomms6438

Sugawara T, Yamashita D, Kato K, Peng Z, Ueda J, Kaneko J, Kamio Y, Tanaka Y, Yao M (2015) Structural basis for pore-forming mechanism of staphylococcal alpha-hemolysin. Toxicon: Official J Int Soc Toxinol 108:226–231. doi:10.1016/j.toxicon.2015.09.033

Syed AK, Reed TJ, Clark KL, Boles BR, Kahlenberg JM (2015) *Staphlyococcus aureus* phenol-soluble modulins stimulate the release of proinflammatory cytokines from keratinocytes and are required for induction of skin inflammation. Infect Immun 83(9):3428–3437. doi:10.1128/IAI.00401-15

Tajima A, Iwase T, Shinji H, Seki K, Mizunoe Y (2009) Inhibition of endothelial interleukin-8 production and neutrophil transmigration by *Staphylococcus aureus* beta-hemolysin. Infect Immun 77(1):327–334. doi:10.1128/IAI.00748-08

Thammavongsa V, Kim HK, Missiakas D, Schneewind O (2015) Staphylococcal manipulation of host immune responses. Nat Rev Microbiol 13(9):529–543. doi:10.1038/nrmicro3521

Thomsen IP, Dumont AL, James DB, Yoong P, Saville BR, Soper N, Torres VJ, Creech CB (2014) Children with invasive *Staphylococcus aureus* disease exhibit a potently neutralizing antibody response to the cytotoxin LukAB. Infect Immun 82(3):1234–1242. doi:10.1128/IAI.01558-13

Tong SY, Davis JS, Eichenberger E, Holland TL, Fowler VG Jr (2015) *Staphylococcus aureus* infections: epidemiology, pathophysiology, clinical manifestations, and management. Clin Microbiol Rev 28(3):603–661. doi:10.1128/CMR.00134-14

Tseng CW, Biancotti JC, Berg BL, Gate D, Kolar SL, Muller S, Rodriguez MD, Rezai-Zadeh K, Fan X, Beenhouwer DO, Town T, Liu GY (2015) Increased susceptibility of humanized NSG mice to panton-valentine leukocidin and *Staphylococcus aureus* skin infection. PLoS Pathog 11(11):e1005292. doi:10.1371/journal.ppat.1005292

van de Wetering JK, van Eijk M, van Golde LM, Hartung T, van Strijp JA, Batenburg JJ (2001) Characteristics of surfactant protein A and D binding to lipoteichoic acid and peptidoglycan, 2 major cell wall components of gram-positive bacteria. J Infect Dis 184(9):1143–1151. doi:10.1086/323746

Ventura CL, Malachowa N, Hammer CH, Nardone GA, Robinson MA, Kobayashi SD, DeLeo FR (2010) Identification of a novel *Staphylococcus aureus* two-component leukotoxin using cell surface proteomics. PLoS ONE 5(7):e11634. doi:10.1371/journal.pone.0011634

Villaruz AE, Bubeck Wardenburg J, Khan BA, Whitney AR, Sturdevant DE, Gardner DJ, DeLeo FR, Otto M (2009) A point mutation in the agr locus rather than expression of the Panton-Valentine leukocidin caused previously reported phenotypes in *Staphylococcus aureus* pneumonia and gene regulation. J Infect Dis 200(5):724–734. doi:10.1086/604728

Walev I, Weller U, Strauch S, Foster T, Bhakdi S (1996) Selective killing of human monocytes and cytokine release provoked by sphingomyelinase (beta-toxin) of *Staphylococcus aureus*. Infect Immun 64(8):2974–2979

Wang R, Braughton KR, Kretschmer D, Bach TH, Queck SY, Li M, Kennedy AD, Dorward DW, Klebanoff SJ, Peschel A, DeLeo FR, Otto M (2007) Identification of novel cytolytic peptides as key virulence determinants for community-associated MRSA. Nat Med 13(12):1510–1514. doi:10.1038/nm1656

Weidenmaier C, Peschel A (2008) Teichoic acids and related cell-wall glycopolymers in Gram-positive physiology and host interactions. Nat Rev Microbiol 6(4):276–287. doi:10.1038/nrmicro1861

Wilke GA, Bubeck Wardenburg J (2010) Role of a disintegrin and metalloprotease 10 in *Staphylococcus aureus* alpha-hemolysin-mediated cellular injury. Proc Natl Acad Sci USA 107 (30):13473–13478. doi:10.1073/pnas.1001815107

Yamashita K, Kawai Y, Tanaka Y, Hirano N, Kaneko J, Tomita N, Ohta M, Kamio Y, Yao M, Tanaka I (2011) Crystal structure of the octameric pore of staphylococcal gamma-hemolysin reveals the beta-barrel pore formation mechanism by two components. Proc Natl Acad Sci USA 108(42):17314–17319. doi:10.1073/pnas.1110402108

Yoshimura A, Lien E, Ingalls RR, Tuomanen E, Dziarski R, Golenbock D (1999) Cutting edge: recognition of Gram-positive bacterial cell wall components by the innate immune system occurs via Toll-like receptor 2. J Immunol 163(1):1–5

Zhang G, Zhang L, Duff GW (1997) A negative regulatory region containing a glucocorticosteroid response element (nGRE) in the human interleukin-1beta gene. DNA Cell Biol 16(2):145–152

MIX
Papier aus verantwortungsvollen Quellen
Paper from responsible sources
FSC® C105338

If you have any concerns about our products,
you can contact us on
ProductSafety@springernature.com

In case Publisher is established outside the EU,
the EU authorized representative is:
Springer Nature Customer Service Center GmbH
Europaplatz 3, 69115 Heidelberg, Germany

Printed by Libri Plureos GmbH
in Hamburg, Germany